Cellular Interactions in Symbiosis and Parasitism

The editors regret that
the illustrations for
figures 4 and 7 on
pages 283 and 292 have
been transposed.

Genetics and Biogenesis of
Mitochondria and Chloroplasts
5–7 September 1974
C. W. Birky, Jr., P. S. Perlman, and T. J. Byers

Regulatory Biology
4–6 September 1975
J. C. Copeland and G. A. Marzluf

Analysis of Ecological Systems
29 April–May 1976
D. J. Horn, G. R. Stairs, and R. D. Mitchell

Plant Cell and Tissue Culture:
Principles and Applications
6–9 September 1977
W. R. Sharp, P. O. Larsen, E. F. Paddock,
and V. Raghaven

Cellular Interactions in Symbiosis and Parasitism
7–9 September 1978
C. B. Cook, P. W. Pappas, and E. D. Rudolph

EDITED BY CLAYTON B. COOK,
PETER W. PAPPAS, AND
EMANUEL D. RUDOLPH

Cellular Interactions in Symbiosis and Parasitism

OHIO STATE UNIVERSITY PRESS: COLUMBUS

Library of Congress Cataloguing in Publication Data

Biosciences Colloquium, 5th, Ohio State University, 1978.
  Cellular interactions in symbiosis and parasitism.

  (Ohio State University biosciences colloquia)
  Includes index.
  1. Cell interaction—Congresses. 2. Symbiosis—Congresses.
3. Parasitism—Congresses. I. Cook, Clayton B. II. Pappas,
Peter W. III. Rudolph, Emanuel David, 1927– IV. Title.
V. Series: Ohio. State University, Columbus. Ohio State Uni-
versity biosciences colloquia.
QH604.2.B56   1978   574.5'24   79-23304
ISBN 0-8142-0315-9

Contents

Introduction

During the past fifteen years, the study of symbiotic associations has increasingly emphasized cellular aspects; it is now clear that cellular phenomena provide the basis for understanding many symbiotic and parasitic interactions. Symbiologists and parasitologists routinely apply the techniques of cell biology to their particular disciplines, and as a result new questions develop both about symbiotic relationships and about relevant cellular events. The present volume is the product of a colloquium entitled "Cellular Interactions in Symbiotic and Parasitic Associations," which was held at the Ohio State University 7–9 September 1978. The colloquium, one of the continuing series sponsored by the College of Biological Sciences, consisted of sessions of major invited papers and shorter contributed papers. The contributed papers are listed at the end of this introduction; the invited papers make up this volume. We are indebted to our invited speakers for their contributions of time and effort in making this publication possible, and we are grateful to each of them for the care with which their manuscripts were prepared.

The colloquium examined cellular processes in several well-studied symbiotic systems: lichens, host-parasite relationships, algae-invertebrate symbioses, endomycorrhizae, and microbial associations. We considered three general kinds of interactions: (1) the establishment of relationships; (2) surface interactions in intact associations, including nutrient uptake and translocation between partners; and (3) aspects of genetic and metabolic integration. By focusing attention on the interactions between symbionts, we believe that the colloquium contrasts with previous symposia and collections of reviews on symbiosis, which either have emphasized particular kinds of associations (e.g., host-parasite, lichens, endomycorrhizae) or have dealt in a general way with a variety of systems. We hope that this approach will stimulate additional interest in the cellular aspects of symbiosis that are readily amenable to experimental analysis.

Several cellular themes are evident in the papers contained in this volume; these are themes that we feel point the way to future research. The first of these is the recognition of symbionts and parasites at the cellular level. During the formation of intracellular relationships, immunological surface interactions appear to function as stimuli for the endocytosis of appropriate symbionts by host cells. M. Aikawa documents the morphological events that occur during the penetration of erythrocytes by *Plasmodium*, and C. B. Cook discusses the events that accompany the endocytosis of potential algal symbionts by invertebrate cells. Both of these systems involve specific receptor-receptor interactions; evidence for lectin-sensitive sites on the surface of infective algae was presented during the contributed paper sessions by R. R. Pool, R. H. Meints and D. S. Weis (see contributed papers). In an established system the immunological responses of trypanosomes to host-produced antibodies represent another aspect of immunological recognition in symbiotic systems (see paper by J. R. Seed et al.). Specificity is also a feature of the resynthesis of lichens. V. Ahmadjian summarizes studies of isolated and recombined lichen symbionts; though surface recognition is involved, the exact mechanisms remain obscure.

The second theme that emerges from these papers is the interaction of symbionts and parasites with cellular defense mechanisms of hosts. Many intracellular symbionts subvert hydrolytic activities of host cells; K. W. Jeon suggests how this may occur in the *Amoeba*-bacteria symbiosis, and Cook reviews how hydrolase activity may be counteracted by algal symbionts in invertebrate cells. T. C. Hohman (contributed paper) discussed the interaction of lysosomes in green hydra cells with symbiotic *Chlorella*. The well-known abilities of parasites to foil defensive strategies of hosts are summarized in two papers: J. R. Seed and his colleagues discuss the immunological defenses of trypanosomes, and P. W. Pappas reviews the mechanisms by which helminth parasites cope with digestive enzymes produced by the vertebrate gut.

Cellular mechanisms for nutrient uptake from the environment and for nutrient translocation between partners provided the third major theme of the colloquium. The elegant morphological and biochemical modifications of the cestode surface that enable these worms to capitalize on the nutrients available in the intestinal lumen are discussed by R. Lumsden. Specialized uptake sites that provide symbiotic partners with dissolved nutrients are found in vesicular arbuscular mycorrhizae (L. Rhodes and J. W. Gerdemann), and in corals and sea anemones that contain symbiotic zooxanthellae (L. Muscatine). Lichens possess well-known abilities to absorb and accumulate heavy metals and other ions; D. H. S. Richardson and E. Nieboer consider the chemical properties of the lichen surface associated with these phenomena and attempt to distinguish general patterns. The transport of nutrients between symbiotic partners may involve a number of transport mechanisms; aspects of such transport are discussed by Rhodes and Gerdemann with respect to mycorrhizal associations, and by Muscatine in the context of nutrient cycling within invertebrate-zooxanthellae systems. The cellular mechanisms that may operate in the translocation of carbohydrate by lichen algae are reviewed by D. C. Smith, who proposes a new model to explain this phenomenon.

Genetic interactions between partners provide the fourth theme of these papers. Changes due either to differences in phenotypic expression or to selection of symbiont genotypes may occur following the establishment of a symbiosis. The *Amoeba*-bacteria symbiosis provides a historical record of how such changes may occur (K. W. Jeon). Genetic interactions between hosts and symbionts are considered by R. K. Trench and L. Margulis. Trench summarizes work on *Cyanophora* and its symbiotic cyanellae which suggests that the interactions between nuclear and cytoplasmic genomes that occur in eukaryotic cells may have parallels in endosymbiotic systems, and Margulis speculates on the broad implications of symbioses as multigenomic systems.

It was apparent to all who attended the colloquium that detailed information about cellular interactions in various symbiotic and parasitic associations is a fruitful approach to understanding fundamental problems connected to these modes of life. Excitement comes when phenomena at the cellular level are translated to higher levels of integration, and when this knowledge increases our appreciation of fundamental cellular processes. We hope that this volume will transmit some of this excitement.

ACKNOWLEDGMENTS

We are grateful to all those who assisted in the planning and operation of the colloquium. In particular, S. B. Cook contributed much administrative and editorial assistance, and M. J. Blevins and the staff of Dean R. O. Moore's office handled many of the secretarial details. J. Kreier and L. Rhodes assisted with the planning. We also thank the staff of the Fawcett Center for Tomorrow for their assistance with physical facilities. Financial assistance was provided in part by grants from the National Science Foundation (DEB 78-12028) and the College of Biological Sciences. Additional support was provided by the Merck Company, Fisher Scientific Company, Preiser Scientific Company, and The Columbus Distributing Company.

Clayton B. Cook
Department of Zoology

Peter W. Pappas
Department of Zoology

Emanuel D. Rudolph
Department of Botany

Ohio State University
March 1979

Contributed Papers

Robert R. Pool, University of California, Los Angeles. The role of the cell surface in the phagocytic recognition of algal symbionts by cells of *Chlorohydra*.

Paul McNeil, University of California, Los Angeles. Digestive cell surface transformations during symbiont phagocytosis by *Hydra*: a scanning electron microscope study.

Russel H. Meints, University of Nebraska. Recognition and uptake of symbiotic algae by *Hydra*: an examination of host-symbiont cell surfaces.

Stephen P. Fracek, Boston University. Effects of the microtubule inhibitors colchicine, nococazole and trifluralin on the uptake and migration of endosymbiotic *Chlorella* by *Hydra viridis*.

Alfred Ayala and Dale S. Weis, Cleveland State University. Stability of the host-symbiont association of *Paramecium bursaria* under super-infective conditions.

Dale S. Weis, Cleveland State University. Correlation of infectivity, concanavalin A agglutinability and sugar excretion of algae exsymbiotic from *Paramecium bursaria*.

J. Malachowski and K. Baker, Michigan State University. Scanning and transmission electron microscopy of algal-fungal interactions in the lichen genus *Usnea*.

S. C. Rabatin, University of Pittsburgh. Field observations relating to the ecology of the vesicular-arbuscular mycorrhizal fine endophyte *Glomis tenuis*.

J. H. Melhuish, Jr., Northeast Forest Experiment Station, Beltsville, Md. Effect of temperature on lipid production in mycorrhizal fungi.

M. S. Bogucki, Texas A & M University. Indirect immunoradiometric assay for trypanosome-bound, noncytotoxic host immunoglobins.

R. D. Herd and S. Weisbrode, Ohio State University. Changes in the tegument of adult *Echinococcus granulosus* during complement-mediated lysis *in vitro*.

R. D. Steele, University of the West Indies. Variations in the symbiotic associations between sea anemones and their zooxanthellae.

D. A. Schoenberg, Lehigh University. Surface ultrastructure of zooxanthellae.

J. J. Lee, City College of CUNY. Algal symbionts in various species of giant Foraminifera.

T. C. Hohman, University of California, Los Angeles. Intracellular digestion and symbiosis in *Hydra viridis*.

R. L. Pardy, University of Nebraska, Lincoln. Factors affecting cell size distribution of symbionts from *Hydra*.

R. W. Zei, University of Pittsburgh. Bleaching *Hydra viridis* of its endosymbionts

Fifth Annual Biosciences Colloquium
College of Biological Sciences
Ohio State University
7–9 September 1978

CELLULAR INTERACTIONS IN SYMBIOSIS AND PARASITISM

Organizers

Clayton B. Cook, Department of Zoology, Ohio State University
Peter W. Pappas, Department of Zoology, Ohio State University
Emanuel D. Rudolph, Department of Botany, Ohio State University

Speakers

Vernon Ahmadjian, Department of Biology, Clark University
Masamichi Aikawa, Institute of Pathology, Case Western Reserve University
Clayton B. Cook, Department of Zoology, Ohio State University
David H. S. Richardson, Department of Biology, Laurentian University
Richard D. Lumsden, Department of Biology, Tulane University
John R. Seed, Laboratory of Microbiology and Parasitology, Texas A & M University
Peter W. Pappas, Department of Zoology, Ohio State University
L. H. Rhodes, Department of Plant Pathology, Ohio State University
D. C. Smith, Department of Botany, University of Bristol
L. Muscatine, Department of Biology, University of California, Los Angeles
Kwang W. Jeon, Department of Zoology, University of Tennessee
Lynn Margulis, Department of Biology, Boston University
Robert K. Trench, Department of Biological Sciences and The Marine Science Institute, University of California, Santa Barbara

OHIO STATE UNIVERSITY BIOSCIENCES COLLOQUIA

Cellular Interactions in Symbiosis and Parasitism

VERNON AHMADJIAN

Separation and Artificial Synthesis of Lichens

1

INTRODUCTION

Progress in lichenology in the last five years has occurred rapidly and over a broad front. Three major books on lichens have been published (Ahmadjian and Hale, 1973; Brown et al., 1976; Seaward, 1977), and new concepts and theories have been proposed. Our understanding of these unique symbiotic systems has been significantly advanced by discoveries on the nutrient transfer patterns between autotroph and heterotroph (Smith, 1975), by findings in the area of physiological ecology (Kershaw and Dzikowski, 1977; Lange et al., 1977), and from taxonomists intrigued by the relationships of chemical substances to the evolution of lichens (Culberson and Culberson, 1976). A new controversy has replaced the one that surrounded chemotaxonomy: can one fungus form different lichens when it combines with different algae? James and Henssen (1976) revealed in the lichen families Pannariaceae and Peltigerineae the existence of related pairs of different thalli, each pair, according to their anatomical observations, presumably formed by one mycobiont. That is, one fungus can form two different, independent thalli, the type of thallus depending on whether the fungus associates with a green or a blue-green algal symbiont. According to James and Henssen, the algal symbiont plays a significant role in the morphogenesis of the fungus, a view contrary to the traditional one that the phycobiont does not influence thallus development.

Many of the theories proposed in different areas of lichenology cannot be resolved because we are unable to routinely culture lichen symbionts and to resynthesize them into symbiotic associations. The need to accomplish this is recognized by many lichenologists. For example, Brodo

(1978) stated that methods to prove the existence of genetic heterogeneity and hybridization in lichens can come most conclusively from culturing, and James and Henssen recognized that culturing experiments were needed to determine the processes that underlie the morphogenesis of their related morphotypes of lichens. Hawksworth (1976) recognized the need for culture studies on lichen algae to determine what role the phycobiont has in different chemical races of lichens. There is currently a logjam of hypotheses and theories in lichenology that can be cleared only by the successful culture and resynthesis of lichens. Unfortunately, such a general solution is not as yet available, although progress is being made.

Lichen thalli cannot be cultured for long periods of time under laboratory-controlled conditions. The reasons for this are that the symbiotic balance is delicate, the thallus harbors foreign microorganisms, and lichens appear to be individualistic in their growth requirements. Pearson and Benson (1977) tried to maintain lichens under controlled conditions with varying degrees of success. Attempts to resynthesize the separate symbionts have been even more difficult than culturing lichens. The history of such attempts has been summarized earlier (Ahmadjian, 1973a).

This paper describes recent studies in my laboratory on lichen resynthesis, and examines the characteristics of fungal and algal symbionts. We have examined the cultural behavior, development, ultrastructure, and chemistry of these organisms.

CHARACTERISTICS OF THE SYMBIONTS IN CULTURE

The techniques we use to isolate and culture lichen symbionts have been described earlier (Ahmadjian, 1973b). I will not repeat such information, but rather present new information on this subject both from my laboratory and from other investigators. I present also some general ideas and features that stand out from my years of experience in trying to culture the symbionts of lichens.

Mycobiont

Cultural morphology. There are several general points that I have noted from isolating the fungal symbionts of lichens. First, the mycobionts appear to be individualistic in their culture requirements, much as Pearson and Benson (1977) noted for some natural thalli that they tried to maintain in the laboratory. The symbionts of many lichens, especially the foliose types, have not been isolated into culture because of specific requirements that are not yet understood. Some lichen spores will not germinate unless they are near cells of the phycobiont, which suggests that some diffusible stimulatory substance is secreted by the algal cells (Lallemant and Bernard, 1977). Second, symbionts that can be cultured assume a growth form that is unlike that of the natural thallus. On an organic agar medium, for example, the fungal symbionts assume a compact, elevated form. The external shape of the colony appears unique for each particular mycobiont, but internally I have not seen a differentiation of layers comparable to those of the lichen thallus. Some investigators (Lallemant, 1977), however, disagree with my observations and have reported specific layers in the colonies of some lichen fungi. The issue appears to be one of interpretation.

Most of our cultural studies with lichen fungi were done on media that contained organic supplements, either undefined extracts of malt and yeast or different sugars and nitrogen sources. The mycobionts grew well on these media, and invariably they all had the same compact colonial growth. Even in liquid organic media the fungi formed tight, round colonies. On natural substrates, however, the growth form of the mycobionts was strikingly different. In our synthesis studies with *Cladonia cristatella*, the isolated mycobiont formed loose, cottony mycelia on rotted wood fragments and on garden soil (fig. 1; figures follow page 25). There was no indication of colonial growth on natural substrates. The growth form was reversible in that pieces of the loose mycelium on natural substrates that developed originally from a colony taken from organic media would develop back into the compact form when returned to an organic growth medium. From our ultrastructural observations, I feel that the compact form of the mycobiont is due to the fungus parasitizing itself. We found intrahyphal hyphae to be common in the mycobiont when it was cultured in organic medium (fig. 13). In natural situations lichen fungi bind the algal cells initially by means of a thigmatropic response, i.e., a response to a physical surface (Ahmadjian, 1959). Mycobionts have been shown to encircle inanimate objects, and one fungus, as seen with the scanning electron microscope, formed appressoria on its own hyphae (fig. 8) much as it would on the surface of an algal cell. Perhaps the organic compounds in the culture medium simulate the compounds the fungus receives from the algae in a thallus and trigger a particular growth response from the fungus—it begins to form appressoria and haustoria that then penetrate neighboring hyphae. Such a generalized condition would result in the compact form of the mycobionts and would account for their outwardly slow growth.

The lichen fungi isolated into culture have not fruited. Some have formed conidia (Ahmadjian, 1964; Lallemant, 1977), and I was able to induce the *C. cristatella* mycobiont to form juvenile podetia and apothecia. The young apothecia showed the developing hymenial layer and ascogonial filaments—spermatia were seen attached to the tips of the trichogynes (Ahmadjian, 1966), but no ascospores developed. Indeed, after the initial appearance of these young apothecia, their growth stopped. Anderson and Ahmadjian (1962) observed that podetia appeared in cultured discs of *Cladonia coniocraea* within a few days, but there was no further growth after this rapid development. Weise (1936) described the development of podetia from squamules of *Cladonia* in culture. The podetia reached sizes of 3–9 mm after one year and then stopped growing. Letrouit-Galinou (1964) described the development of podetia in *Cladonia floerkeana*. She observed that a podetium attained a height of 2–5 mm, after which an ascogonial complex formed at the tip of the podetium. Such a complex consisted of ascogonial filaments, each with an ascogonium and trichogyne. It may be possible that the young structures in our cultures of *C. cristatella* mycobiont were protoapothecia (Burnett, 1968, p. 415), i.e., structures where the ascogonial filaments form and further development of the apothecium depends on successful fertilization. The role of the pycniospores in lichens has not been fully clarified. The limited evidence (Ahmadjian, 1966; Jahns, 1970) plus comparison with other ascomycetous fungi suggest that these spores are spermatia, although in some crustose lichens they have germinated (Vobis, 1977); thus, they may serve also as a means of reproduction. We have not been able to induce germination of the pycniospores of *C. cristatella* despite several attempts on soil extract and on malt-yeast extract agar.

Ultrastructure. One of the most prominent organelles seen in ultrastructural observations of lichens are the concentric bodies (figs. 9–10). These unusual structures occur in the hyphae of the mycobiont, and they have been reported from about fifty lichens. They have also been found in twenty free-living ascomycetes, all plant pathogens (Griffiths and Greenwood, 1972; Bellemère, 1973; Tu and Colotelo, 1973; Garnett, 1974; Slocum, pers. comm.). They have been reported as absent in *Hydrothyria venosa*, a permanently submerged lichen that grows on rocks in streams at high elevations in the United States (Jacobs and Ahmadjian, 1973) and absent in two species of *Gonohymenia* from the Middle East (Paran et al., 1971)–all three lichens have blue-green phycobionts. Concentric bodies have been seen in mycobionts of other blue-green lichens (Peveling, 1973; Spector and Jensen, 1977). Each concentric body measures about 300 nm in diameter, and has a central electron-transparent core that is surrounded by several layers of different electron densities. An additional clear zone or "halo" may surround the body. The bodies are associated with membranes that may be connected to the internal membrane system of the fungus (Jacobs and Ahmadjian, 1971). The concentric bodies have a proteinaceous nature (Galun et al., 1974), and they occur singly or in clusters of up to a dozen or more. They have been found in hyphae from all parts of the thallus, and one report indicates their presence within ascospores (Ascaso and Galvan, 1975). There appears to be a close physical relationship between the concentric bodies and the cell nucleus (Griffiths and Greenwood, 1972).

We have not seen concentric bodies in the hyphae of cultured mycobionts. When the mycobiont is cultured, these bodies disappear, and we have not seen their reappearance except in one instance. The only appearance of these bodies in a cultured mycobiont was in a colony of *C. cristatella* that formed pycnidia during a period of slow drying (Jacobs and Ahmadjian, 1971). Spores of the mycobiont were allowed to discharge onto circular, sterile filter pads (4.5 cm diameter) that were then placed onto the surface of a malt-yeast extract agar. The spores were incubated at 19° C for 30 days. The pads with the developing fungal colonies were then soaked in liquid citrate-phosphate buffer (pH = 2.8) for 5 days. The pads were removed from the buffer solution, transferred to a 1.5% agar medium prepared with the same buffer solution, and incubated at 19° C, 500 lux, 16/8 light-dark diurnal cycle. About one month later, numerous pycnidia appeared on the fungal colonies, and in many instances the pycnidia had extruded strings of spores in a gelatinous matrix. Ultrastructural observations of the hyphae of one of these pycnidia-producing colonies showed that they contained concentric bodies.

Apart from the absence of concentric bodies, ultrastructural details of the free-living fungus cultured in malt-yeast extract medium (fig. 14), and of the fungus in the synthesized squamules (figs. 11-12), are not significantly different from that of the natural lichenized fungus. Details of the cultured *C. cristatella* mycobiont are given in Jacobs and Ahmadjian (1971).

Chemistry. One of the most enigmatic problems in experimental lichenology has been the failure of the separate mycobionts to synthesize in culture the same chemical compounds they do when they are in symbiosis with algae. For about the last ten years, I have sent to Chicita Culberson (Duke University) cultures of different lichen fungi grown under different conditions for chemical analyses. The results have been consistently negative; i.e., no symbiotic substances were produced by the lichen fungi alone. Dr. Culberson summarizes her findings as follows. (a) There is no direct relationship between the compounds that the lichen fungi produce alone in culture and the substances that they produce in symbiosis with the algae. Indeed, the fungi produce different compounds in culture. For example, the mycobionts *Stereocaulon vulcani* and *Graphis tenella* produced pigments in culture that were not detected in the natural thalli

(Ahmadjian, 1964). Mycobionts of *Acarospora fuscata* (gyrophoric acid), *Baeomyces roseus* (atranorin, baeomycesic acid), and *Cladonia arbuscula* (fumarprotocetaric acid, usnic acid) did not produce their natural compounds in culture but rather formed different substances. The mycobiont *Cladonia coniocraea* had a chemical complexity in culture that was greater than that of the mycobionts *C. rangiferina* and *C. arbuscula*, and yet, under natural conditions, the situation is the opposite, i.e., the lichen *C. coniocraea* has the least complex chemistry. (b) Lichen fungi that produce different characteristic lichen products in symbiosis (tested forms were *A. fuscata, B. roseus,* and *C. arbuscula*) were remarkably similar in compounds that they (the mycobionts) produced in culture. Some of the compounds produced were identical to those formed by species of marine *Verrucaria*, which not only have a unique ecology but also do not produce any of the characteristic lichen substances. Thus, there seems to be some underlying common pattern among lichen fungi in culture. The compounds produced in culture were polycyclic aliphatics, e.g., steroids, triterpenes. According to Dr. Culberson, "This would mean that they would be derived by the mevalonic acid pathway rather than the acetate-polymalonate pathway that leads to all of the characteristic aromatic lichen products—like depsides, depsidones, and dibenzofurans. Although both of these pathways start from acetate, the mevalonic acid pathway requires phosphorylation and the acetate-polymalonate pathway requires the formation of malonate from acetate and CO_2. Biotin and Mn^{++} are also required for the formation of malonate from acetate." The growth medium that we have been using for our studies has been Lilly and Barnett medium, which contains biotin and Mn^{++}. However, we have not increased the concentration of CO_2 in the culture flasks, and this may be a factor in the acid production. In a lichen thallus the most typical lichen products are produced by the medullary hyphae, which lie below the cortex and the algal layer. It is possible that the medullary hyphae are growing in a higher CO_2/O_2 atmosphere than the cortical layer. The cortex produces compounds that are similar to those reported from mycobionts in culture.

Although results of our studies on this subject have been negative, results of other investigators have been positive (table 1)—but only in isolated instances; that is, there has not been a general breakthrough that would apply to all lichen fungi. We cannot, at present, induce any mycobiont to form in culture its symbiotic compounds. Shibata and his colleagues have reported most of the success, in this area. They induced the cultured mycobiont *Cladonia crispata* to produce the depside squamatic acid by irradiating it with ultraviolet light (Ejiri and Shibata, 1975). Earlier, Komiya and Shibata (1969) had reported that (+) usnic acid and the

TABLE 1

REPORTS OF LICHEN SUBSTANCES SYNTHESIZED
BY MYCOBIONTS IN CULTURE

Lichen Fungus	Substances Produced in Culture	Investigators
Candelariella vitellina	pulvic acid, pulvic dilactone, calcycin, vulpinic acid	Thomas (1939); Mosbach (1969)
Xanthoria spp., *Caloplaca* spp.	parietin, erythroglaucin, erythroglaucin carboxylic acid, emodin, fallacinal, fallacinol, fragilin, 2-chloroemodin	Thomas (1939); Tomaselli (1957); Piatelli & deNicola (1968); Nakana et al. (1972)
Cladonia macilenta, Cladonia pyxidata, Cladonia chlorophaea	orsellinic acid, haematomic acid	Hess (1959)
Baeomyces roseus, Collema tenax, Lecidea coarctata, Lecidea enteromorpha, Cladonia cristatella	carotenoids	Henriksson & Pearson (1968)
Lecanora rupicola	roccellic acid, eugenitol, eugenetin, rupicolon	Fox & Huneck (1969)
Lecanora chlarotera	gangaleodin[1]	Santesson (1969)
Cladonia cristatella	usnic acid,[3] didymic acid[3] rhodocladonic acid	Castle & Kubsch (1949)
Ramalina crassa, Ramalina yasudae	usnic acid,[3] salazinic acid[1]	Komiya & Shibata (1969)
SW265 (unidentified)	steroids	Culberson & Ahmadjian (unpublished)
Bombyliospora japonica, Anaptychia hypoleuca	zeorin	Ejiri & Shibata (1974); Shibata (1974)
Cladonia crispata	squamatic acid[2]	Ejiri & Shibata (1975)
Acroscyphus sphaerophoroides	calycin, skyrin	Shibata (1974)
Cladonia bellidiflora	bellidiflorin, skyrin	Shibata (1974)

[1] Depsidone

[2] Depside

[3] Dibenzofuran

depsidone salazinic acid were formed by the mycobiont *Ramalina crassa* (with no prior treatment). Another report (Kurokawa et al., 1969) contradicted the above results—salazinic acid was not found in cultures of the mycobiont *R. crassa* (the presence of usnic acid was still noted). Santesson (1969) reported the presence of the depsidone gangaleoidin in cultures of the mycobiont *Lecanora chlarotera*, and Castle and Kubsch (1939) reported that usnic acid and didymic acid were formed by the mycobiont *Cladonia cristatella* in culture. Chicita Culberson and I could not duplicate the results of Castle and Kubsch with the same lichen (Ahmadjian, 1964). Bloomer and colleagues (1970) reported negative results in their search for lichen substances produced by isolated fungal symbionts. They did not detect protolichesterinic acid in submerged cultures of the mycobionts *Cetraria islandica* and *Cladonia papillaria*. Fox and Huneck (1969) found that the mycobiont *Lecanora rupicola* in separate culture produced roccellic acid but not atranorin and chloroatranorin, its other natural products. Ejiri and Shibata (1974) detected zeorin in cultures of the mycobiont *Anaptychia hypoleuca* but not the other natural compounds, i.e., atranorin, norstictic acid, and salazinic acid. Recent analyses of our mycobionts were made by Chicita Culberson and Siegfried Huneck. No lichen substances were detected in cultures of *Cladonia cristatella* mycobiont grown in different media. Cultures of the same mycobiont grown together with autoclaved, ethanol-killed, and living cells of its phycobiont (*Trebouxia erici*) also yielded negative findings, as did a culture of the phycobiont alone. Tests were made with thin-layer chromatography high-speed liquid chromotography, and, with one mycobiont culture, mass spectroscopy. A culture of *Lecidea albocaerulescens* mycobiont tested with TLC also gave negative results. Thus, the conflicting results and the limited studies on lichen fungi producing identifiable substances in culture do not rule out a role of the algal symbiont.

It is possible that isolated mycobionts do produce their chemical substances in culture but that the compounds readily hydrolyze and thus cannot be detected. This possibility was suggested by Umezawa et al. (1974), who found that lecanoric acid produced by a free-living fungus (*Pyricularia* sp.) readily hydrolyzed in aqueous solution to orsellinic acid and orcinol. However, since these compounds are not detected in mycobiont cultures, it may be possible that they are converted to other compounds such as the various pigments produced by mycobionts in culture.

Phycobiont

Trebouxia and *Pseudotrebouxia*. Twenty-eight genera of algae have been reported as phycobionts of lichens. The two most common genera of symbionts are *Trebouxia* and *Pseudotrebouxia*, and it is on these algae that most of my research has focused. The distinction between these genera is based primarily on the presence or absence of vegetative cell division (figs. 15-16). This distinction, recognized by earlier workers on lichen algae (Warén, 1918-19), was formalized recently by Archibald (1975). She assigned all previous species of *Trebouxia* that could reproduce by means of vegetative cell division to the new genus *Pseudotrebouxia*. Those species without vegetative cell division were retained in the genus *Trebouxia*. The distinction is a fundamental one since it places the two genera in separate orders; i.e., *Trebouxia* remains in the Chlorococcales and *Pseudotrebouxia* is placed in the Chlorosarcinales. One of our long-term studies is to identify the algal symbionts of different lichens, especially as to whether they belong to *Trebouxia* or *Pseudotrebouxia*. Table 2 summarizes the present status of this study. Seven genera of lichens are represented under both genera of phycobionts, as are two specific lichens (i.e., *Parmelia sulcata* and *Xanthoria parietina*). Species of *Lecidea* are also represented under seven other genera of green algae, and recently I found that the phycobionts of *Stereocaulon ramulosum* and *Stereocaulon strictum* collected from Costa Rica, although they looked trebouxoid in crushed thallus preparations, belonged to the genus *Pseudochlorella*, a genus heretofore found only in several species of *Lecidea*. It is of interest to note that *Pseudotrebouxia* has not yet been reported from *Cladonia* (17 species examined). This observation ties in with our synthesis results which showed that a *Cladonia cristatella* mycobiont entered into symbiosis only with isolates of *Trebouxia* and not with species of *Pseudotrebouxia* (table 3).

TABLE 2

Lichens from which the Algal Symbionts Have Been Cultured and Examined for the Presence or Absence of Vegetative Cell Division

Trebouxia (Vegetative cell division absent)	*Pseudotrebouxia* (Vegetative cell division present)
Cladonia bacillaris	*Acarospora fuscata*
Cladonia balfourii f. *chlorophaeoides*	*Alectoria implexa*[4]
Cladonia boryi	*Alectoria jubata*[4]
Cladonia coccifera[4]	*Aspicilia calcarea*
Cladonia convoluta	*Buellia pernigra*
Cladonia cornuta[4]	*Buellia punctata*[1]
Cladonia cristatella[1]	*Buellia russa*
Cladonia deformis[4]	*Buellia straminea*
Cladonia fimbriata[4]	*Buellia subdisciformis*
Cladonia furcata[4]	*Caloplaca cerina*
Cladonia gracilis v. *chordalis*[4]	*Caloplaca holocarpa*
Cladonia leporina	*Candelaria concolor* v. *antarctica*
Cladonia macilenta[4]	*Cetraria pinastri*[4]
Cladonia pyxidata[4]	*Diploschistes scruposus*[7]
Cladonia rangiferina[4]	*Dirina* sp.
Cladonia squamosa[1]	*Gyrophora flocculosa*[4]
Cladonia tenuis	*Gyrophora polyphylla*[4]
Lecidea albocaerulescens	*Lecania* sp.
Lecidea crustulata	*Lecanora actophila*

TABLE 2—*Continued*

Trebouxia (Vegetative cell division absent)	*Pseudotrebouxia* (Vegetative cell division present)
Lecidea erratica	*Lecanora dispersa*[1]
Lecidea lithophila	*Lecanora hageni*
Lecidea tumida	*Lecanora rubina*[1]
Lepraria membranacea	*Lecanora subfusca*[4]
Lepraria zonata	*Lecidea fuscoatra*
Parmelia acetabulum[3]	*Lecidea insularis*
Parmelia caperata[1,3]	*Lecidea sarcogynoides*
Parmelia rudecta[1]	*Lecidea tenebrosa*
Parmelia saxatilis[3]	*Lecidea tessellata*
Parmelia scortea[3]	*Parmelia conspersa*[5]
Parmelia sulcata[3]	*Parmelia furfuracea*[4]
Physcia millegrana	*Parmelia sulcata*
Physcia pulverulenta[1]	*Parmelia tinctina*[6]
Pilophoron acicularis[1]	*Parmelia tinctorum*
Protoblastenia metzleria	*Pertusaria* sp.
Ramalina usnea[8]	*Physcia caesia*
Stereocaulon dactylophyllum	*Physcia ciliaris*[4]
var. *occidentale*[1]	*Physcia obscura*[4]
Stereocaulon pileatum[1]	*Physcia pulverulenta*[4]
Stereocaulon saxile[1]	*Physcia stellaris*[1,4]
Umbilicaria mammulata	*Protoblastenia testacea*[9]
Umbilicaria papulosa	*Ramalina complanata*
Xanthoria auroela[1]	*Ramalina fraxinea*[4,5]
Xanthoria parietina[1]	*Ramalina subgeniculata*[10]
	Rhizocarpon geographicum
	Rinodina frigida
	Toninia caerulonigricans
	Umbilicaria antarctica
	Usnea (Neuropogon) antarctica
	Xanthoria mawsoni
	Xanthoria parietina [1,2,4,5]

[1] Archibald, 1975
[2] Hedlund, 1949
[3] Jaag, 1929
[4] Warén, 1918–1919
[5] Werner, 1927
[6] Werner, 1958
[7] Werner, 1961
[8] Werner, 1965
[9] Werner, 1967
[10] Werner, 1968

SYNTHESIS OF LICHENS

Introduction

The literature on previous attempts to synthesize lichens has been summarized recently by Ahmadjian (1973a). In this section I will

TABLE 3

ARTIFICIAL RESYNTHESES BETWEEN A CLADONIA CRISTATELLA MYCOBIONT AND DIFFERENT SPECIES OF TREBOUXIA AND PSEUDOTREBOUXIA PHYCOBIONTS[1]

Mycobiont	Phycobiont	Lichen from Which Phycobiont Was Isolated	Degree of Synthesis[2]
Cladonia cristatella	*Trebouxia* sp.	*Cladonia bacillaris*	+++
	Trebouxia sp.	*Cladonia boryi*	+++
	Trebouxia erici	*Cladonia cristatella*	+++
	Trebouxia sp.	*Cladonia leporina*	+++
	Trebouxia sp.	*Cladonia tenuis*	+++
	Trebouxia sp.	*Lecidea albocaerulescens*	+++
	Trebouxia sp.	*Lecidea tumida*	+++
	Trebouxia sp.	*Lepraria membranacea*	++
	Trebouxia sp.	*Lepraria zonata*	+++
	Trebouxia sp.	*Protoblastenia metzleri*	+++
	Trebouxia sp.	*Stereocaulon* sp.	
	Trebouxia sp.	*Stereocaulon evolutoides*	++
	Trebouxia sp.	*Stereocaulon pileatum*	+++
Cladonia cristatella	*Pseudotrebouxia* sp.	*Buellia straminaea*	–
	Pseudotrebouxia sp.	*Caloplaca cerina*	–
	Pseudotrebouxia sp.	*Lecania* sp.	–
	Pseudotrebouxia sp.	*Lecanora hageni*	–
	Pseudotrebouxia sp.	*Lecidea fuscoatra*	–
	Pseudotrebouxia sp.	*Lecidea tenebrosa*	–
	Pseudotrebouxia sp.	*Pertusaria* sp.	–
	Pseudotrebouxia sp.	*Physcia stellaris*	–
	Pseudotrebouxia sp.	*Ramalina complanata*	–
	Pseudotrebouxia sp.	*Usnea* sp.	–
	Pleurastrum terrestre	Free-living alga	–

[1] Observations after seven months culture.

[2] +++ Squamules present; barbatic and didymic acids detected; pycnidia and young podetia.

++ Squamules present but no pycnidia or podetia.

– Initial contacts between symbionts; no soredial formation.

concentrate on our latest and as yet unpublished synthesis studies. The substrates we used to grow our mixed cultures were purified agar and soil contained within small clay flowerpots. These substrates gave satisfactory results since they provided the conditions necessary for successful synthesis, that is, nutrient deficiency and alternating cycles of slow drying and wetting. We have not been able as yet to achieve a complete synthesis on a defined medium or even on an inert substrate such as silica gel. The most successful substrate is soil that is contained within small clay pots—such a combination was used first by Stahl (1877) in his culture of *Endocarpon pusillum*. In our recent studies, our primary goals were as follows: (1) to determine the range of suitable phycobionts for one mycobiont (how restrictive is a lichen fungus to its algal symbiont?); (2) to see if, in a successful resynthesis, a lichen fungus regained the characteristics that it had in symbiosis and that it lost when isolated into separate culture (i.e., can the mycobiont once again synthesize its symbiotic chemical substances, and will its concentric bodies reappear?); and (3) to ascertain if the algal symbiont influences the chemistry and morphology of the fungal symbiont. The syntheses also revealed aspects of how appressoria are formed by hyphae and a possible basis for the specificity between the symbionts.

Culture

Our first synthesis attempt was between a *Cladonia cristatella* mycobiont and the *Trebouxia* phycobiont of *Lecidea albocaerulescens.* The two lichens differ in morphology (one is fruticose and the other crustose) and in habitat (the former grows on soil and rotted wood, and the latter grows on rocks). Both lichens are common throughout the northeastern United States. The symbionts were growing separately in axenic cultures before they were mixed together on a purified agar surface. No supplements were added to the agar, since earlier studies have shown that nutrient media that support independent growth of the symbionts will not induce symbiotic interactions. The mycobiont had been growing on an organic liquid medium, but the mycelium was washed thoroughly with distilled water to minimize nutrient carry-over into the mixed culture. Algal cells were taken directly from the surface of an inorganic agar medium.

The first contacts between the symbionts were observed about four weeks after they were inoculated onto the purified agar. Fungal hyphae formed appressoria on the algal cells, and many of the algae were penetrated by haustoria. Small soredia were observed that contained numerous crystals. Although small pycnidia, short podetia, and immature apothecia appeared on the soredia after twelve weeks in culture, the development did not proceed beyond this point; i.e., even after sixteen weeks on the agar surface, the next stage of lichen development—thallus formation—was not achieved. The fungus formed loose, aerial hyphae over the surface of its original colony, which was a tight, compacted mass of hyphae. The fungus also grew within the purified agar, its hyphae radiating from the original colony. Thus, there was an advancing edge of fungal mycelium behind which the soredia were formed.

Pieces of mixed symbionts which did not form soredia on agar were placed in clay pots containing soil. These cultures subsequently formed soredia within one week, and within eight weeks squamules had formed (fig. 4). Development of the squamules was dependent on the alternating cycles of drying and wetting that we allowed the mixed cultures to experience on the soil surface. The necessity of drying and wetting cycles on lichen development and maintenance has now been well established by different investigators (Dibben, 1971; Galun, et al., 1972; Pearson and Benson, 1977). On the agar surface the condition was one of sustained moisture, and development did not proceed beyond the soredial stage. Ultrastructural comparisons between fungus from the agar cultures and fungus exposed to the wetting and drying cycles did not reveal any noticeable differences that could be correlated with different cultural conditions. The drying period triggers a response from the fungus to differentiate and form squamules. Such structures contain an upper cortex and would serve as a better means of protection for the algal population under difficult environmental conditions such as drying.

Our second series of synthesis experiments again used a *C. cristatella* mycobiont, but this time it was tested with a wide range of *Trebouxia* and *Pseudotrebouxia* phycobionts (table 3). Fragmented pieces of fungal mycelia were mixed together with algal cells of each phycobiont tested, and the mixture was then pipetted over the surface of soil contained within clay flowerpots. After three weeks of incubation, the initial fungal-algal contacts were seen in all cultures and soredia were seen in a few. After four weeks soredia had formed on almost all the cultures that contained the fungus and isolates of *Trebouxia*. After eight weeks all the *Trebouxia* cultures with the fungus had formed squamules, except the combination with the phycobiont of *Stereocaulon evolutoides*, which formed squamules after ten weeks. Pycnidia appeared on squamules of all the *Trebouxia* combinations after five months (fig. 3), and some squamules showed the formation of small podetia with the beginnings of apothecia (fig. 2). The squamules produced by the different *Trebouxia* cultures appeared identical—they began as rounded structures and then became flattened and upraised at one end. Sections of the squamules showed the presence of an upper cortex, algal layer, and medulla.

In all the synthesis pots, the growth of the fungus on the soil was loosely mycelial (fig. 1). Hyphae grew into the soil to a depth of about 5 mm, and the open network of hyphae was an effective soil binder. In some of the synthesis combinations that were about six months old, there was an outer zone of mycelial growth similar to that of a hypothallus; behind this there were well-developed squamules, and farther behind were squamules with pycnidia and young podetia. Galløe (1954) observed essentially the same stages in the natural development of *Cladonia pyxidata*.

Observations from these studies reveal that a lichen fungus (*C. cristatella*) is not necessarily specific to one species of *Trebouxia* symbiont. Table 3 shows that this mycobiont formed lichenized associations with *Trebouxia* phycobionts isolated from thirteen widely different lichens. Our tentative species determinations of these phycobionts indicate that at least three different species of *Trebouxia* are present among the isolates listed in table 3. The fungus did not form lichenized unions with any of the ten different isolates of *Pseudotrebouxia* tested. The initial contacts of appressoria and haustoria were prevalent, but development did not proceed further. There was no well-developed envelopment of algal cells by fungal hyphae. Growth of both symbionts in these mixed cultures was poor in terms of color of the algal cells (yellow-green) and growth of the fungal mycelium. The fungus also did not associate with *Pleurastrum terrestre*, an alga that has been proposed as a progenitor or ancestor of *Trebouxia* (Molnar et al., 1975).

Chemistry of the Synthesized Squamules

Chemical analyses of our synthesis cultures were made by Chicita Culberson, whose findings are summarized as follows. All cultures of the *C. cristatella* mycobiont with its own phycobiont (*Trebouxia erici*) and twelve other strains of *Trebouxia* isolated from different lichens (table 3) showed the presence of barbatic acid. In nine of the cultures, didymic acid was also detected. Several other depsides, closely related to barbatic acid, were tentatively identified from these cultures. Usnic acid, which is also present in the natural lichen, was not detected in the cultures. The lack of usnic acid, or its presence in quantities too small to be detected by thin-layer chromatography, could be a function of culture conditions. In its natural habitats the lichen is exposed to high light intensities (about 100,000 lux), whereas under laboratory incubation the synthesis cultures received about 4,500 lux. The lichen substances were detected in the earliest stages of lichen formation, i.e., in soredia.

Cultures of the *C. cristatella* mycobiont with seven strains of *Pseudotrebouxia* did not reveal the usual chemistry of the natural lichen with thin-layer chromatography. The mycobiont did not develop lichenized structures with these algae. The initial contacts between the symbionts were present but no further development occurred. The lack of the natural lichen compounds in these cultures might indicate that a physiological relationship between the symbionts was not established.

These findings, together with those on the chemistry of the isolated mycobionts, suggest that the algal symbiont has a general role in the synthesis of lichen substances; that is, the phycobiont is necessary for the synthesis of these particular lichen substances—they are products of symbiotic interaction. In terms of a direct relationship between algal species and type of chemical substances produced, however, there was little correlation. The mycobiont in our synthesis cultures with different species of *Trebouxia* formed the same morphology and produced some of the same chemical substances that it does in natural conditions with its own phycobiont.

Ultrastructure of the Synthesized Squamules

One of the questions that we posed in examining the ultrastructure of the squamules was to see if there were any differences in cellular detail between the symbionts in (a) resynthesized squamules, (b) naturally occurring squamules, or (c) separate culture. The most obvious organelle missing in the fungal hyphae of our resynthesized squamules as well as in the hyphae of the fungus in culture was the concentric body (figs. 11-14). Since this organelle is such an integral part of the lichenized fungus, being observed commonly in virtually all lichens examined with the electron microscope (figs. 9-10), we expected it to reappear in the synthesized squamules. Its absence suggests that these bodies are not involved in the morphological development of the thallus nor in the synthesis of lichen substances, since several of the naturally occurring lichen acids were found in the synthesized squamules. The concentric bodies may be associated with fruiting of the fungus. Jacobs and Ahmadjian (1971) found the bodies in hyphae of a cultured mycobiont that had formed pycnidia, and several workers (Tu and Colotelo, 1973; Granett, 1974) associated the concentric bodies of some free-living ascomycetes with the sexual or fruiting stages of these fungi. We found concentric bodies in newly synthesized squamules of *Endocarpon pusillum* (Ahmadjian and Jacobs, 1970), but, unfortunately, we did not study the ultrastructure of the separate fungus in culture to determine if this organelle was present. The squamules we observed had mature perithecia. Other workers (Peveling, 1969; Brown and Wilson, 1968; Ellis and Brown, 1972) have suggested a metabolic role for these bodies; i.e., they may transport materials within the cell or function as membrane synthesizers.

The fungal hyphae of the synthesized squamules contained unusually large mitochondria (fig. 12), larger than those in hyphae of the cultured mycobiont and in the natural lichen. Such giant mitochondria suggest that the mycobiont is in a state of high metabolic activity, which could be caused either by the artificial conditions of culture or more likely by the active development of the fungus as it forms the squamules.

Specificity

The question of how specific a particular fungal symbiont is to its algal partner is one in which we have maintained a long-term interest. We can examine this question from two sides: (1) from isolating and identifying the algal symbionts within lichen thalli, and (2) by attempting to synthesize one lichen fungus with a variety of algal symbionts. The first approach is the one from which we have the most evidence, and the second is one that has become newly feasible and from which we have no data. To date we have found specificity mostly at the generic level but not at the species level. That is, only two lichen species have been found that can associate with two separate genera of phycobionts (excluding cephalodia) (table 2). All the lichens we tested and from which we identified algal symbionts had either *Trebouxia* or *Pseudotrebouxia* as phycobionts. At the algal species level, however, there was variation—this has been observed in the past, and is confirmed by our recent studies. Single species of *Trebouxia* or *Pseudotrebouxia* have been isolated from different lichens and, conversely, individual lichen species have been found to contain different species of *Trebouxia* or *Pseudotrebouxia*.

The reason for the specificity of a lichen fungus to a particular type or group of algae is not known. In our scanning electron microscope studies with the symbionts of *Lecidea albocaerulescens* (Ahmadjian et al., 1978), we observed a gelatinous sheath around the cells of the phycobiont. The sheath bound hyphae to the algal cells of the fungus (figs. 5-7). Whether this type of adhesive substance is required to permanently fix the algal-fungus relationship and allow the fungus to differentiate is not known. It is intriguing to speculate on the possibility of a substance that allows the fungus and alga to recognize and accept each other. Such recognition compounds have been found in other lichens (Lockhart et al., 1978), and in symbiotic systems (Dazzo et al., 1978).

Our synthesis attempts with *Cladonia cristatella* revealed that the fungus did not form lichenized structures (soredia and squamules) with species of *Pseudotrebouxia* phycobionts isolated from other lichens, although the initial contacts were made, but did so readily with species of *Trebouxia* phycobionts. It will be important to determine the differences between these two genera to see if there are any clues that might explain the suitability of one algal genus over another. From the point of view of carbohydrate transfer, the list of lichens examined to date (Hill, 1976) indicates that the mobile carbohydrate from alga to fungus is ribitol in both genera.[1] Also, there were no significant differences in nutritional preferences between species of these two genera (Ahmadjian, 1977; Archibald, 1977).

Synthesis in Natural Situations

It is not known if synthesis occurs in natural situations and, if it does, how frequently and how widely. Although a trend toward the sterilization of lichen thalli appears to be evident (Poelt, 1970; Bowler and Rundel, 1975), large numbers of lichens form fruiting bodies and discharge spores. It would seem illogical to think that these spores cannot reestablish the symbiosis with free-living algae. The loosely mycelial growth of the fungus in our syntheses has been observed in lichen fungi by other investigators. A drawing on the cover of the *International Lichenological Newsletter* (vol. 10, no. 2, Oct., 1977) illustrates the lichen *Cladonia macrophylla* with an extensive underground mycelial network. According to the artist Per-Jan Thøgersen, most species of *Cladonia* have such a mycelial base. Galløe (1954) described a group of *Cladonia* lichens as "hypothallus-wanderers." He stated that an ascospore from such a lichen gives rise to a mycelium that spreads rapidly through the substratum, and where the hyphae encounter suitable algae, lichenization begins. Such a mycelial growth, he felt, could wander over and through the soil for years, always producing new squamules. In one of our synthesis pots where the symbionts of *Cladonia cristatella* were recombined, I observed an outer green zone of algal growth around the mixed growth of the symbionts. This zone consisted only of phycobiont cells which had formed from zoospores from the original algal inoculum. Thus, the zoospores and the mycelium represent a continuous source of the free-living symbionts. The type of long-lasting and wandering mycelium which Galløe (1954) described is bound to encounter cells of algal symbionts from other lichens that have been dispersed as part of asexual propagules or through thallus fragmentation. Since the evidence shows that one lichen may contain different species of *Trebouxia* or *Pseudotrebouxia*, it would seem that such a mycelium could lichenize different species of a particular algal genus.

If we can establish the presence of free-living mycelia of lichen fungi, then what about free-living algal symbionts? Most of the phycobionts I listed in my algal guide (Ahmadjian, 1967) are found in the free-living state. With regard to *Trebouxia* and *Pseudotrebouxia*, however, the evidence is not clear-cut. For years I have maintained that these two common genera of lichen phycobionts do not exist in the free-living state, but rather that they are genera that have been derived as a result of their fungal associations (Ahmadjian, 1970). Indeed, such a view seems to be supported by recent findings of Molnar and colleagues (1975), who used ultrastructural similarities to establish a close affinity between the green, filamentous alga *Pleurastrum* and *Trebouxia*. These authors felt that *Trebouxia* must be a reduced form of *Pleurastrum*.

Some recent papers have reported the existence of free-living colonies of *Trebouxia* (Nakano, 1971a,b; Tschermak-Woess, 1978). Tschermak-Woess (1978) observed that free-living colonies of algal symbionts such as *Myrmecia* and *Trebouxia* were often found near lichen thalli. She indicated that hyphae from neighboring thalli or spores of the mycobiont could reestablish the symbiosis with free-living algae. I feel that these free-living phycobionts could be microcolonies derived from zoospores from algae within existing lichenized associations. Such colonies are free-living, but only in a secondary sense. I have observed the late stages of zoosporogenesis in crushed preparations of lichens (Ahmadjian, 1970), and more recently we have observed various stages of zoosporogenesis in *Parmelia caperata* collected from its natural habitat at regular intervals throughout one year. No free-swimming zoospores were observed when we examined crushed preparations of the thallus with a light microscope, even though what appeared to be mature zoosporangia were observed. However, when fragments of the thallus and soredia were placed onto a mineral medium and illuminated at 2,400 lux (16/8 light-dark diurnal cycle), the algal cells produced numerous zoosporangia and free zoospores within 5–7 days. In natural situations lichens undergo frequent drying and wetting periods so that the sustained moist periods necessary for the complete process of zoosporogenesis are rarely present. Slocum (pers. comm.) made ultrastructural observations of *Parmelia caperata* and noted that flagella were present even during the early stages of internal cell cleavage in the *Trebouxia* phycobiont; such flagella are undoubtedly reabsorbed when the division products of the algal cell become transformed into the more drought resistant aplanospores. It would appear likely, however, that sometimes the environmental conditions of a particular microhabitat a thallus occupies will allow a small number of algal cells to complete zoosporogenesis and release zoospores. Such free-swimming spores could find their way out of the thallus through various cracks or holes and establish new colonies of the phycobiont away from the parent thallus. Spores from lichen fungi or hyphae of the mycobiont could then form new associations with these algae. Such free-living phycobiont colonies might explain the genetic diversity found in some lichen populations; i.e., fungal ascospores that are products of genetic recombination could introduce new traits into existing lichen populations by combining with free-living microcolonies of phycobionts.

Our SEM study with *Lecidea albocaerulescens* gave us a clue as to how a lichen fungus in its early stages of growth may obtain the autotrophic population it needs to form the lichen thallus. In our mixed cultures on agar, the hyphae grew over the algal cells and on each cell formed an appressorium. The appressorium did not terminate the hyphal filament, but rather the fungus kept growing and secured other algal cells in a similar manner (fig. 7).

SUMMARY

The slow growth and compact, colonial form of lichen fungi on organic nutrient media may be due to the fungus parasitizing itself. Intrahyphal hyphae were common in mycobiont cultures. Isolated mycobionts did not produce mature apothecia nor were their symbiotic chemical substances detected in culture. Concentric bodies were not found in the hyphae of mycobionts in culture or in hyphae of the resynthesized squamules. A list of lichens, grouped according to whether they have *Trebouxia* or *Pseudotrebouxia* as phycobionts, revealed some specificity among lichen genera; i.e., seventeen tested species of *Cladonia* had only *Trebouxia* as phycobiont, and five species of *Lecanora* had only *Pseudotrebouxia* phycobionts. Seven other genera of lichens were represented under both genera of phycobionts. Successful resyntheses were established between a *Cladonia cristatella* mycobiont and thirteen strains of *Trebouxia* isolated from different lichens. The same mycobiont would not recombine with any of the ten tested strains of *Pseudotrebouxia* phycobionts or with the free-living alga *Pleurastrum terrestre*. In all of the successful synthesis cultures the thallus morphology was identical to that of the lichen *C. cristatella*. The synthesized squamules contained barbatic and didymic acids, but not usnic acid. It is concluded that the algal symbiont, at the species level, does not influence the type of morphology of a mycobiont or the kinds of chemical substances it synthesizes.

ACKNOWLEDGMENTS

This research was supported by grant PCM 77–11715 from the National Science Foundation. Technical assistance was provided by Amelia H. Janeczek. Some of the findings reported here are based on studies by two of my graduate students: Katherine C. Hildreth (algal symbionts; zoosporogenesis) and Lorraine A. Russell (resynthesis). The SEM micrographs were taken by Jerome B. Jacobs. I thank Chicita F. Culberson for her continuing cooperation in analyzing mycobionts and resynthesized associations for chemical substances and for reading portions of this manuscript. My thanks also to Siegfried Huneck for doing chemical analyses of mycobiont cultures.

---

1. The lichens listed by Hill were not specifically examined to determine the identity of the phycobiont. From my own experience in algal identification, it appears that both *Trebouxia* and *Pseudotrebouxia* are represented among his short list of lichens with green algae. See also table 2.

LITERATURE CITED

Ahmadjian, V. 1959. A contribution toward lichen synthesis. Mycologia 51: 56–60.

Ahmadjian, V. 1964. Further studies on lichenized fungi. Bryologist 67: 87–98.

Ahmadjian, V. 1966. Artificial reestablishment of the lichen *Cladonia cristatella*. Science 151: 199–201.

Ahmadjian, V. 1967. A guide to the algae occurring as lichen symbionts: isolation, culture, cultural physiology, and identification. Phycologia 6: 127–60.

Ahmadjian, V. 1970. The lichen symbiosis: its origin and evolution. *In* Th. Dobzhansky, M. K. Hecht, and W. C. Steere (eds), Evolutionary biology, 4: 163–84. Appleton-Century-Crofts Publishers, New York.

Ahmadjian, V. 1973a. Resynthesis of lichens. *In* V. Ahmadjian and M. E. Hale (eds.), The lichens, pp. 565–79. Academic Press, New York.

Ahmadjian, V. 1973b. Methods of isolating and culturing lichen symbionts and thalli. *In* V. Ahmadjian and M. E. Hale (eds.), The lichens, pp. 653–59. Academic Press, New York.

Ahmadjian, V. 1977. Qualitative requirements and utilization of nutrients: lichens. *In* M. Rechcigl, Jr. (ed.), CRC Handbook Series in Nutrition and Food, Section D: Nutritional requirements, 1:203–15. CRC Press, Cleveland, Ohio.

Ahmadjian, V., and M. E. Hale (eds.). 1973. The lichens. Academic Press, New York. 697 pp.

Ahmadjian, V., and J. B. Jacobs. 1970. The ultrastructure of lichens. III. *Endocarpon pusillum*. Lichenologist 4:268–70.

Ahmadjian, V., J. B. Jacobs, and L. A. Russell. 1978. Scanning electron microscope study of early lichen synthesis. Science 200:1062–64.

Anderson, K. A., and V. Ahmadjian. 1962. Investigation on the development of lichen structures in laboratory controlled cultures. Sven. Bot. Tidskr. 56:501–6.

Archibald, P. A. 1975. *Trebouxia* de Puymaly (Chlorophyceae, Chlorococcales) and *Pseudotrebouxia* gen. nov. (Chlorophyceae, Chlorosarcinales). Phycologia 14:125–37.

Archibald, P. A. 1977. Physiological characteristics of *Trebouxia* (Chlorophyceae, Chlorococcales) and *Pseudotrebouxia* (Chlorophyceae, Chlorosarcinales). Phycologia 16:295–300.

Ascaso, C., and J. Galvan. 1975. Concentric bodies in three lichen species. Arch. Mikrobiol. 105:129–30.

Bellemère, A. 1973. Observation de "corps concentriques," semblables à ceux des lichens, dans certaines cellules de plusieurs ascomycètes non lichénisants. C. R. Hebd. Séances Acad. Sci. Ser. D. Sci. Nat. 276D:949–52.

Bloomer, J. L., W. R. Eder, and W. F. Hoffman. 1970. Some problems in lichen metabolism: studies with the mycobionts *Cetraria islandica* and *Cladonia papillaria*. Bryologist 73:586–91.

Bowler, P. A., and P. W. Rundel. 1975. Reproductive strategies in lichens. Bot. J. Linn. Soc. 70:325–40.

Brodo, I. M. 1978. Changing concepts regarding chemical diversity in lichens. Lichenologist 10:1–11.

Brown, D. H., D. L. Hawksworth, and R. H. Bailey (eds.). 1976. Lichenology: progress and problems. Academic Press, London. 551 pp.

Brown, R. M., and R. Wilson. 1968. Electron microscopy of the lichen *Physcia aipolia* (Ehrh.) Nyl. J. Phycol. 4:230–40.

Burnett, J. H. 1968. Fundamentals of mycology. St. Martin's Press, New York. 546 pp.

Castle, H., and F. Kubsch. 1949. The production of usnic, didymic, and rhodocladonic acids by the fungal component of the lichen *Cladonia cristatella*. Arch. Biochem. 23:158–59.

Culberson, C. F., and W. L. Culberson. 1976. Chemosyndromic variation in lichens. Syst. Bot. 1:324–39.

Dazzo, F. B., W. E. Yanke, and W. J. Brill. 1978. Trifoliin: *Rhizobium* recognition protein from white clover. Biochim. Biophys. Acta, 539:276–86.

Dibben, M. J. 1971. Whole-lichen culture in a phytotron. Lichenologist 5:1–10.

Ejiri, H., and S. Shibata. 1976. Zeorin from the mycobiont of *Anaptychia hypoleuca*. Phytochemistry 13:2871.

Ejiri, H., and S. Shibata. 1975. Squamatic acid from the mycobiont of *Cladonia crispata*. Phytochemistry 14:2505.

Ellis, E. A., and R. M. Brown, Jr. 1972. Freeze-etch ultrastructure of *Parmelia caperata* (L.) Ach. Trans. Amer. Microsc. Soc. 91:411–21.

Fox, C. H., and S. Huneck. 1969. The formation of roccellic acid, eugenitol, eugenetin, and rupicolon by the mycobiont *Lecanora rupicola*. Phytochemistry 8:1301–4.

Galløe, O. 1954. Natural history of the Danish lichens. Part 9. *Cladonia*. Ejnar Munksgaard, Copenhagen. 74 pp.; 194 pls.

Galun, M., L. Behr, and Y. Ben-Shaul. 1974. Evidence for protein content in concentric bodies of lichenized fungi. J. Microsc. (Paris) 19:193–96.

Galun, M., K. Marton, and L. Behr. 1972. A method for the culture of lichen thalli under controlled conditions. Arch. Mikrobiol. 83:189–92.

Granett, A. L. 1974. Ultrastructural studies of concentric bodies in the ascomycetous fungus *Venturi inaequalis*. Can. J. Bot. 52:2137–39.

Griffiths, H. B., and A. D. Greenwood. 1972. The concentric bodies of lichenized fungi. Arch. Mikrobiol. 87:285–302.

Hawksworth, D. L. 1976. Lichen chemotaxonomy. *In* D. H. Brown, D. L. Hawksworth, and R. H. Bailey (eds.), Lichenology: progress and problems, pp. 139–84. The Systematics Association Special Volume No. 8. Academic Press, New York.

Hedlund, T. 1949. A contribution to the knowledge of the development of the aerophile Chlorophyceae. Bot. Not. 3:173–200.

Henricksson, E., and L. C. Pearson. 1968. Carotenoids extracted from mycobionts of *Collema tenax, Baeomyces roseus*, and some other lichens. Sven. Bot. Tidskr. 62:441–47.

Hess, D. 1959. Untersuchungen über die Bildung von Phenolkörpen durch isolierte Flechtenpilze. Z. Naturforsch. 14b:345–47.

Hill, D. J. 1976. The physiology of lichen symbiosis. *In* D. H. Brown, D. L. Hawksworth, and R. H. Bailey (eds.), Lichenology: progress and problems, pp. 457–96. The Systematics Association Special Volume No. 8. Academic Press, New York.

Jaag, O. 1929. Recherches expérimentales sur les gonidies des lichens appartenant aux genres *Parmelia* et *Cladonia*. Bull. Soc. Botan. Genève, 21:1–119.

Jacobs, J. B., and V. Ahmadjian. 1971. The ultrastructure of lichens. II. *Cladonia cristatella*: The lichen and its isolated symbionts. J. Phycol. 7:71–82.

Jacobs, J. B., and V. Ahmadjian. 1973. The ultrastructure of lichens. V. *Hydrothyria venosa*, a freshwater lichen. New Phytol. 72:155–60.

Jahns, H. M. 1970. Untersuchungen zur Entwicklungeschichte der Cladoniaceen unter besonderer Berücksichtigung des Podetien-Problems. Nova Hedwigia 20:1–177.

James, P. W., and A. Henssen. 1976. The morphological and taxonomic significance of cephalodia. *In* D. H. Brown, D. L. Hawksworth and R. H. Bailey (eds.), Lichenology: progress and problems, pp. 27–77. The Systematics Association Special Volume No. 8. Academic Press, New York.

Kershaw, K. A., and P. A. Dzikowski. 1977. Physiological-environmental interactions in lichens. VI. Nitrogenase activity in *Peltigera polydactyla* after a period of desiccation. New Phytol. 79:417–21.

Komiya, T., and S. Shibata. 1969. Formation of lichen substances by mycobionts of lichens. Isolation of (+) usnic acid and salazinic acid from mycobionts of *Ramalina* spp. Chem. Pharm. Bull. 17(6):1305–6.

Kurokawa, S., S. Shibata, and T. Komiya. 1969. Isolation of algal and fungal components of lichens and their chemical products [translated from the Japanese]. Misc. Bryol. Lichenol. 5:8–9.

Lallemant, R. 1977. Recherches sur le développement en cultures pures *in vitro* du mycobionte du discolichen *Pertusaria pertusa* (L.) Tuck. Rev. Bryol. Lichenol. 42:255–82 (81–108).

Lallemant, R., and T. Bernard. 1977. Obtention de cultures pures des mycosymbiotes du *Lobaria laetevirens* (Lightf.) Zahlbr. et du *Lobaria pulmonaria* (L.) Hoffm.: le role des gonidies. Rev. Bryol. Lichenol. 43:303–8 (129–34).

Lange, O. L., I. L. Geiger, and E.-D. Schulze. 1977. Ecophysiological investigations on lichens of the Negev Desert. V. A model to simulate net photosynthesis and respiration of *Ramalina maciformis*. Oecologia (Berl.) 28:247–59.

Letrouit-Galinou, M.-A. 1964. Sur le développement des podétions et des apothécies du Lichen *Cladonia floerkeana* (Fr.) Sommf. Bull. Soc. Bot. Fr. 111:248–54.

Lockhart, C. M., P. Rowell, and W. D. P. Stewart 1978. Phytohaemagglutinins from the nitrogen-fixing lichens *Peltigera canina* and *P. polydactyla*. FEMS Microbiology Letters 3:127–30.

Molnar, K. E., K. D. Stewart, and K. R. Mattox. 1975. Cell division in the filamentous *Pleurastrum* and its comparison with the unicellular *Platymonas* (Chlorophyceae). J. Phycol. 11:287–96.

Mosbach, K. 1969. Zur Biosynthese von Flechtenstoffen, Produkten einer symbiotischen Lebensgemeinschaft. Angew. Chem. 81:233–44.

Nakano, T. 1971a. Subaerial algae of Patagonia, South America I. Bull. Biol. Soc. of Hiroshima University 38:2–12.

Nakano, T. 1971b. Some aerial and soil algae from the Ishizuchi mountains. Hikobia 6:139–52.

Nakano, H., T. Komiya, and S. Shibata. 1972. Anthraquinones of the lichens of *Xanthoria* and *Caloplaca* and their cultivated mycobionts. Phytochemistry 11:3505–8.

Paran, N., Y. Ben-Shaul, and M. Galun. 1971. Fine structure of the blue-green phycobiont and its relation to the mycobiont in two *Gonohymenia* lichens. Arch. Mikrobiol. 76:103–13.

Pearson, L. C., and S. Benson. 1977. Laboratory grown experiments with lichens based on distribution in nature. Bryologist 80: 317–27.

Peveling, E. 1969. Elektronenoptische Untersuchungen an Flechten. III. Cytologische Differenzierungen der Pilzzellen im Zusammenhang mit ihrer symbiontischen Lebensweise. Z. Pflanzenphysiol. 61:151–64.

Peveling, E. 1973. Fine structure. *In* V. Ahmadjian and M. E. Hale (eds.), The lichens, pp. 147–82. Academic Press, New York.

Piatelli, M., and G. de Nicola. 1968. Produzione di antrachinoni nel lichene *Xanthoria parietina* e in colture pure di *Xanthoriomyces parietinae*. Ric. Sci. (Roma) 38:850–54.

Poelt, J. 1970. Das Konzept der Artenpaare bei den Flechten. Deut. Bot. Ges., Neue Folge 4:187–98.

Santesson, J. 1969. Chemical Studies on Lichens. Acta Universitatis Upsaliensis, Abstracts of Uppsala Dissertations in Science 127, 28 pp.

Seaward, M. R. D. (ed.). 1977. Lichen ecology. Academic Press, London. 550 pp.

Shibata, S. 1974. Some aspects of lichen chemotaxonomy. *In* G. Benz and J. Santesson (eds.), Chemistry in botanical classification, pp. 241–49. Proceedings of the Twenty-fifth Nobel Symposium; Södergarn, Lidingö, Sweden, Academic Press, New York.

Smith, D. C. 1975. Symbiosis and the biology of lichenized fungi. *In* D. H. Jennings and D. L. Lee (eds.), Symbiosis, pp. 373–405. Symposia of the Soc. for Experimental Biology No. 29. Cambridge: At the University Press.

Spector, D. L., and T. E. Jensen. 1977. Fine structure of *Leptogium cynaescens* and its cultured phycobiont *Nostoc commune*. Bryologist 80:445–60.

Stahl, E. 1877. Beiträge zur Entwicklungsgeschichte der Flechte. II. Über die Bedeutung der Hymenialgonidien. Pp. 1–32. Felix, Leipzig.

Thomas E. A. 1939. Über die Biologie von Flechtenbildern. Beitr. Kryptogamenflora Schweiz, 9:1–208.

Tomaselli, R. 1957. Nuovo contributo alle richerche sulla presenza di "Fiscione" in colture pure di *Xanthoriomyces*. Atti Ist. Bot. Univ. Lab. Crittogam, Pavia, 14:1–18.

Tschermak-Woess, E. 1978. *Myrmecia reticulata* as a phycobiont and free-living—free-living *Trebouxia*: the problem of *Stenocybe septata*. Lichenologist 10:69–79.

Tu, J. C., and N. Colotelo. 1973. A new structure containing cyst-like bodies in apothecia-bearing sclerotia of *Sclerotinia borealis*. Can. J. Bot. 51:2249–50.

Umezawa, H., N. Shibamoto, H. Naganawa, S. Ayukawa, M. Matsuzaki, T. Takeuchi, K. Kono, and T. Sakamoto. 1974. Isolation of lecanoric acid, an inhibitor of histidine decarboxylase from a fungus. Journal of Antibiotics 27:587–96.

Vobis, G. 1977. Studies on the germination of lichen conidia. Lichenologist 9:131–36.

Warén, H. 1918–19. Reinkulturen von Flechtengonidien. Pp. 1–79. Öfversigt af Finska Vetenskaps-Societetens Förhandlingar 61, Afd. A, No. 14, 1–79.

Weise, R. 1936. Die Entstehung des Thallusmantels der *Cladonia*-Podetien. Hedwigia 76:179–88.

Werner, R. G. 1927. Recherches biologiques et expérimentales sur les Ascomycètes de Lichens. Thèses. Mulhouse. 88 pp.

Werner, R. G. 1958. La gonidie marocaine du *Parmelia tinctina* Mah. et Gill. Bull. Soc. Sci. Nancy, n. ser. pp. 262–74.

Werner, R. G. 1961. La gonidie marocaine du *Diploschistes scruposus* (Schreb.) Norm. Bull. Soc. Lorraines Sci., pp. 158–65.

Werner, R. G. 1965. La gonidie marocaine du *Ramalina usnea* (L.) R. H. Howe. Bull. Acad. Soc. Lorraines Sci. 5:3–16.

Werner, R. G. 1967. La gonidie marocaine du *Protoblastenia testacea* (Hoffm.) Clauz. et Rond. Bull. Acad. Soc. Lorraines Sci. 6:248–58.

Werner, R. G. 1968. La gonidie marocaine du *Ramalina subgeniculata* Nyl. Bull. Acad. Soc. Lorraines Sci. 7:228–39.

Figs. 1-3. Synthesis cultures on soil of a *Cladonia cristatella* mycobiont and its own phycobiont (*Trebouxia erici*). *Fig. 1.* Note squamules (arrow) and open mycelial growth outside of the area of squamule development. *Fig. 2.* Note the small podetia and the apothecial initials (arrow). *Fig. 3.* Note the mature pycnidia (arrows). *Fig. 4.* Synthesis culture on soil of a *Cladonia cristatella* mycobiont and the phycobiont (*Trebouxia glomerata*) of *Lecidea albocaerulescens*. Note the mature squamules and a beginning pycnidium (arrow).

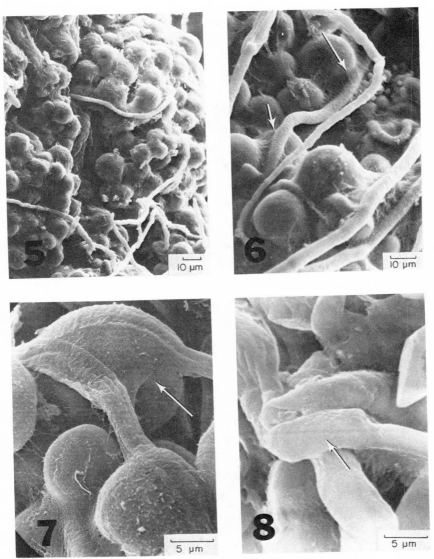

Figs. 5–8. Early stages of lichen synthesis viewed with a scanning electron microscope; synthesis culture on purified agar of a *Lecidea albocaerulescens* mycobiont and its own phycobiont (*Trebouxia glomerata*). *Fig. 5.* Fungal hyphae growing through several clusters of algal cells. *Fig. 6.* Note the thin veil of extracellular substance around the algal cells (arrows). This substance envelops parts of fungal hyphae that are in contact with the algal cells. *Fig. 7.* Two appressoria, flattened against an algal cell, are joined by an extracellular substance (arrow). *Fig. 8.* An appressorium (arrow) that has formed around a fungal hypha.

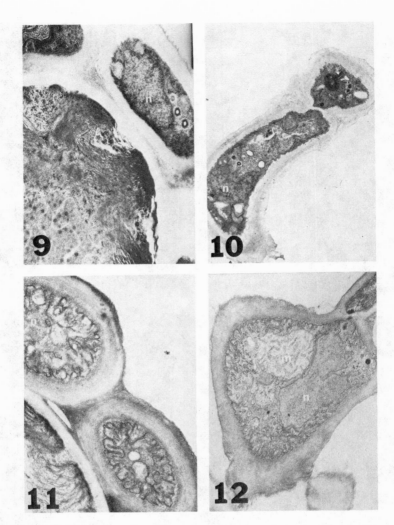

Figs. 9–10. *Cladonia cristatella* sections through a natural squamule. *Fig. 9.* Fungal appressorium against an algal cell. Note binding substance between the symbionts and the concentric bodies near the nucleus (n) of the fungal cell. × 26,750. *Fig. 10.* Segment of a fungal hypha; two concentric bodies are near the nucleus (n). × 19,875. Figs. 11–12. Sections through a synthesized squamule from a mixed culture on soil of a *Cladonia cristatella* mycobiont and the phycobiont (*Trebouxia* sp.) of *Lepraria zonata. Fig. 11.* Fungal appressoria against an algal cell. Note extracellular substance around the symbionts, nucleus (n), and invaginated plasmalemma of fungal cells. × 31,430. *Fig. 12.* Fungal cell; note large mitochondrion (m), nucleus, free ribosomes, invaginated plasmalemma, and absence of concentric bodies. × 26,080.

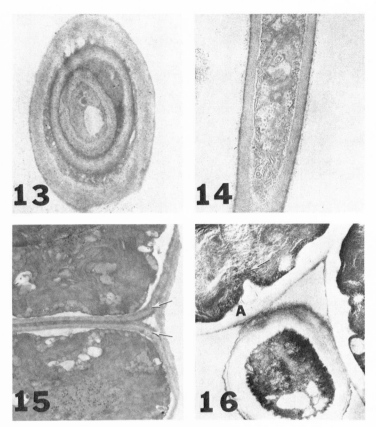

Figs. 13–14. *Cladonia cristatella* mycobiont grown in malt-yeast extract liquid medium. *Fig. 13.* Cross section of a hyphal cell showing two other hyphae within it. × 26,080. *Fig. 14.* Longitudinal section of a fungal hypha. A thin gelatinous matrix surrounds the cell wall, and concentric bodies are absent. × 30,763. *Fig. 15.* Algal symbiont (*Pseudotrebouxia* sp.) of *Diploschistes scruposus* showing vegetative cell division. Note the cell wall around each daughter cell (arrows). × 10,600. *Fig. 16.* Section through a natural squamule of *Cladonia cristatella*. Note cell (A) of the *Trebouxia* phycobiont undergoing cleavage of the chloroplast (arrow). × 30,763.

Host Cell Invasion
by Malarial Parasites

2

INTRODUCTION

Host-parasite interaction is of major importance to parasitic protozoa, since their survival depends on host cells that supply environmental and nutritional requirements. They cannot live apart from their host cells or host cell nutrients. Before the advent of electron microscopy, the description of the mode of invasion of protozoan parasites into host cells was mainly speculative. The information that has accumulated by electron microscopy in recent years now demonstrates that, with the exception of *Nosema*, all parasitic protozoa so far studied appear to enter cells by endocytosis. However, because a large number of variables may be operative, no generalization of specific mechanisms of endocytotic entry of parasitic protozoa can be made. For example, some infective stages of protozoa such as sporozoites of *Eimeria* and protomastigotes of *Leishmania* are actively motile, whereas others such as merozoites of *Plasmodium* and amastigotes of *Leishmania* are sluggish. On the other hand, some host cells such as erythrocytes are normally non-phagocytic, and others such as macrophages are professional phagocytes. In spite of these differences, there are three basic aspects of protozoan entry by endocytosis: (1) initial attachment of parasitic protozoa to the host cell membrane; (2) invagination of the host cell membrane around parasitic protozoa; and (3) sealing of the host cell membrane after completion of protozoan entry.

Erythrocyte entry by *Plasmodium* merozoites has been studied extensively since 1969 when Ladda et al. reported on the invasion of erythrocytes by merozoites of *P. berghei* and *P. gallinaceum*. They established that merozoites enter within an invagination of the erythrocyte membrane rather than by penetrating it.

Dvorak et al. (1975) studied by interference phase microscopy the invasion of erythrocytes by *P. knowlesi* and observed that the invasion consisted of attachment of the apical end of the parasite to the erythrocyte, deformation of the erythrocyte, and entry of the merozoite by invagination of the erythrocyte membrane.

Electron microscopic studies on erythrocyte invasion by *Plasmodium* were hindered by sampling problems. However, Dennis and his colleagues (1975) recently reported a new method for the collection of large quantities of free viable merozoites of *P. knowlesi*. By this method, Bannister et al. (1977) confirmed the previous electron microscopic observations on invasion. More recently Aikawa and his colleagues (1978) have used an approach similar to that of Bannister et al. and reported several new findings on the invasion process that have not been reported previously. In this chapter, we will focus our attention on (1) initial attachment, (2) invagination of the host cell membrane, and (3) sealing of the host cell membrane during erythrocyte entry by *Plasmodium*.

ULTRASTRUCTURE OF PLASMODIUM MEROZOITES

Since the understanding of the structure of merozoites of malarial parasites is essential for analysis of the interaction between erythrocytes and merozoites, a description of the merozoite is presented (fig. 1).

The merozoite is oval in shape and is surrounded by a pellicular complex composed of two membranes and a row of subpellicular microtubules (Aikawa and Sterling, 1974). The lateral side of the pellicle shows a small circular indentation called the cytostome. This organelle engages in the ingestion of the host cell cytoplasm in the intracellular trophozoite stage. The apical end (anterior end) is a truncated cone-shaped projection demarcated by two polar rings. Two teardrop-shaped electron-dense rhoptries and small micronemes are present in the apical end region of the merozoite. A ductule extends from each of the rhoptries and meets with the other forming a common ductule just proximal to the tip of the apical end. The common ductule then leads to the tip of the apical end. The nucleus is usually located in the midportion. Anterior to the nucleus is a cluster of a few vesicles and membranous lamellae that appears to be a Golgi complex. The posterior portion of the merozoite is occupied by a mitochondrion, electron-dense inclusions, and a spherical body of unknown function.

ENTRY INTO ERYTHROCYTES

Invasion of erythrocytes by merozoites of *Plasmodium* requires a number of distinct steps. They include (1) recognition and attachment of

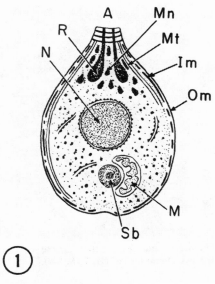

(The scale on all figures is measured in μm)

Fig. 1. Schematic diagram of the erythrocytic merozoites of *Plasmodium* showing the apical end (A), rhoptries (R), micronemes (Mn), a nucleus (N), a mitochondrion (M), a spherical body (Sb), a subpellicular microtubules (Mt), and outer (Om) and inner (Im) peculiar membranes. (From M. Aikawa, Exp. Parasitol. 30:284, 1971.)

the merozoite to the erythrocyte membrane, (2) invagination of the erythrocyte membrane around the merozoite to form a parasitophorous vacuole, and (3) sealing of the erythrocyte after completion of merozoite invasion. These aspects will be discussed.

Recognition and Initial Attachment

Evidence for the presence of host recognition sites on *Plasmodium* comes from observations that for successful endocytosis the parasite must come in contact with the host cell in a particular orientation (Miller et al., 1979). Endocytosis of a *Plasmodium* merozoite is initiated only when it attaches to the host cell by the apical end (fig. 2). However, it is not known what makes the exposed elements on the apical end different from the rest of the surface of the merozoite. Alteration of host plasma membrane by proteases has been shown to specifically abolish attachment of *Plasmodium*. Trypsinization of human erythrocytes reduces their susceptibility to penetration by *P. falciparum*, but does not alter penetration by a

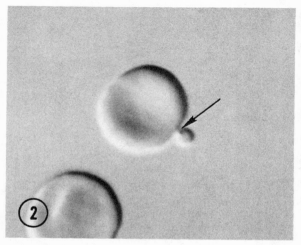

Fig. 2. Interference microscopy photograph showing the attachment (arrow between a merozoite and an erythrocyte. (Courtesy of Dr. L. Miller)

monkey parasite, *P. knowlesi.* Chymotrypsin treatment of erythrocytes blocks infection by *P. knowlesi* but has no effect on infection by *P. falciparum.* On the other hand, Miller and his associates (1975) reported that initial recognition and attachment between merozoites and erythrocytes are probably associated with Duffy blood group related antigens.

When a merozoite contacts the erythrocyte with the apical end, the erythrocyte membrane at the point of the interaction is slightly raised initially (fig. 3), but eventually a depression is created in the erythrocyte membrane (fig. 4). The erythrocyte membrane to which the merozoite is attached becomes thickened and forms a junction with the merozoite plasma membrane. However, no junction is formed between the merozoite and Duffy negative human erythrocytes, which are the only human erythrocytes refractory to invasion by malarial parasites. The Duffy negative erythrocyte is about 120 to 160 nm away from the merozoite, and is connected to the merozoite by thin filaments of 3–5 nm thickness (fig. 5). The filaments originate from the edge of the truncated, cone-shaped apical end, such that two distinct filaments can be seen in thin sections. This suggests that these filaments are arranged in a cylindrical fashion in three dimensions. Such filamentous attachments between merozoites and erythrocytes are not observed with Duffy positive human erythrocytes. Trypsin treatment of Duffy negative erythrocytes makes them susceptible to invasion by merozoites and permits the junction formation with merozoites (fig. 6).

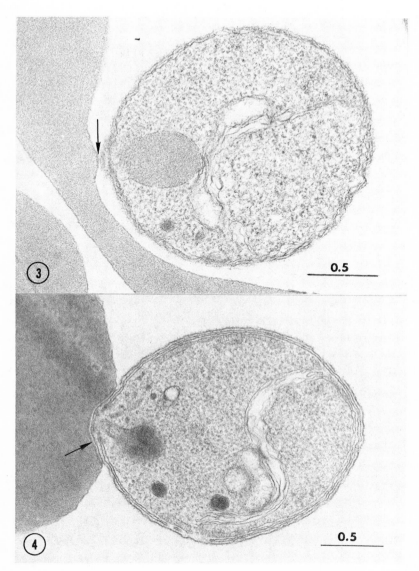

Fig. 3. Electron micrograph of a *Plasmodium* merozoite showing the initial contact with an erythrocyte. The erythrocyte membrane is slightly raised at the point of the interaction (arrow). (From Aikawa et al., J. Cell Biol. 77:72, 1978.)

Fig. 4. Electron micrograph of a merozoite contacting an erythrocyte. The erythrocyte membrane is slightly depressed and becomes thickened at the attachment site (arrow).

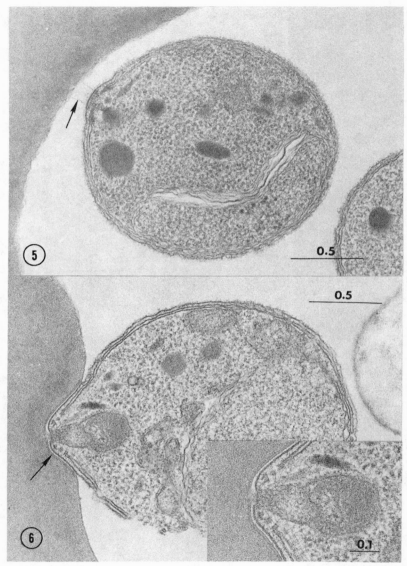

Fig. 5. Electron micrograph showing a merozoite connected with a Duffy negative human erythrocyte by two fine fibrils (arrow). (From Miller et al., J. Exp. Med. 1979.)

Fig. 6. Electron micrograph showing the attachment (arrow) between the apical end of a merozoite and a trypsin-treated Duffy negative human erythrocyte. ×51,000. Inset: High magnification showing the projection connecting the apical end and the erythrocyte. (From Miller et al., J. Exp. Med. 1979.)

Attachment of merozoites to erythrocytes is temperature sensitive (Miller et al., 1979). When *P. knowlesi* merozoites are mixed with rhesus erythrocytes at 4°C, 25°C, and 37°C for 4 minutes, the highest rate of attachment is at 37°C and few attach at 4°C. Attachment of merozoites to erythrocytes is specific in that merozoites do not attach to erythrocytes from nonsusceptible species.

Invagination of the Host Plasmalemma

There is increasing evidence that, following attachment, some product(s) from the apical organelles of the merozoite, such as rhoptries and micronemes, initiate the invagination of the host cell membrane and participate in the invagination of the host cell plasmalemma. A ductule runs from the rhoptries to the apical end, which is the point of initial attachment between the merozoite and erythrocytes (fig. 7). Throughout invasion, the apical end remains in contact with the erythrocyte membrane. An electron-dense band appears between the tip of the apical end of the merozoite and the erythrocyte and is continuous with the common duct of the rhoptries (fig. 7). The lower electron-density in the ductule during invasion suggests a release of rhoptry contents. Histochemical studies indicate the presence of protein and/or carbohydrate in rhoptries. Indirect experimental evidence suggests that a histidine-rich protein isolated from *Plasmodium lophurae* may be a component of the apical organelles of this malarial parasite (Kilejian, 1976); the protein was shown to cause invagination of erythrocyte membrane *in vitro*.

As invasion progresses, the depression in the erythrocyte deepens and conforms to the shape of the merozoite (fig. 8). At this time, the thickened, electron-dense zone on the erythrocyte is no longer observed at the point of the initial attachment but now appears at the orifice of the merozoite-induced invagination of the erythrocyte membrane (fig. 9). This thickened area of the erythrocyte membrane measures about 15 nm in thickness and about 250 nm in length, and appears to be a thickening of the inner leaflet of the bilayer. The thickened erythrocyte membrane forms a junction with the merozoite. The gap between these two membranes is about 10 nm, and fine fibrils extend between these two parallel membranes (fig. 9). The junction appears to form a circumferential interaction at the orifice of the invaginated erythrocyte membrane (Aikawa et al., 1978), since it is always located at each side of the orifice regardless of the plane through which the section passes.

The events occurring during invasion appear to relate to endocytotic processes by which phagocytic cells ingest inert particles, other cells, and microorganisms. In 1975, Griffin and his associates proposed a hypothesis

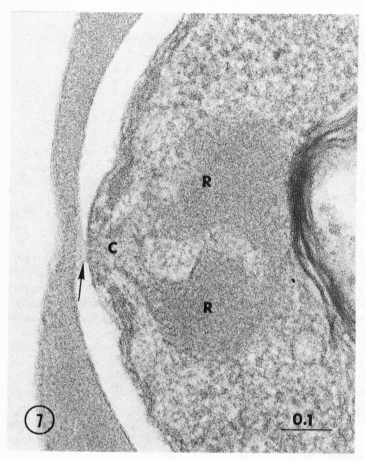

Fig. 7. Electron micrograph showing two rhoptries (R) at the apical end. The common duct (C) is formed by the meeting of ductules that lead from each rhoptry. An electron opaque projection connects the apical end and the erythrocyte membrane (arrow). (From Aikawa et al., J. Cell Biol. 77:72, 1978.)

for the endocytosis of inert particles, namely, specific attachment triggering endocytosis and zippering. Triggering requires specific receptors for attachment, but the ingestion is independent of receptors outside of the attachment. Zippering is the attachment to receptors around the circumference of the particles and may require a metabolically active cell such as macrophage, or it may be a passive process. However, this hypothesis appears not to explain fully the observation during invasion by *Plasmodium* merozoites. Aikawa et al. (1978), therefore, proposed the

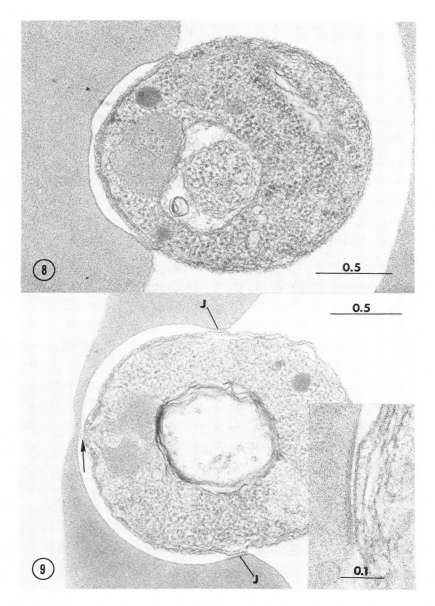

Fig. 8. Erythrocyte entry by a merozoite. The depression in the erythrocyte conforms to the shape of the merozoites. (From Aikawa et al., J. Cell Biol. 77:72, 1978.)

Fig. 9. An advanced stage of erythrocyte entry by a merozoite. The junction (J) is formed between the thickened membrane of the erythrocyte and merozoite plasma membrane. An electron-opaque projection connects the apical end and the erythrocyte membrane (arrow). Inset: Higher magnification of the junction. (From Aikawa et al., J. Cell Biol. J. Cell Biol. 77:72, 1978.)

following models for invasion of erythrocytes by *Plasmodium* merozoites. (1) Movement of the junction at the level of the membrane that may be related to the lateral displacement of the junction by the agency of membrane flow. However, it should be pointed out that this would necessitate simultaneous parallel flow of membrane components in both cells. (2) Attachment-detachment (modified zippering) model which prescribes that the junction itself may be capable of migrating on the surface of relatively stable plasma membrane. This would require that the leading edge of the moving junction becomes attached while the following edge becomes detached, perhaps by enzymatic cleavage.

Cytochalasin B is known to affect microfilaments (Wessells et al., 1971), glucose transport, and possibly other functions. When merozoites are treated with cytochalasin B, they can attach to erythrocyte membrane but do not invade the erythrocyte (fig. 10). Thus, invagination of the erythrocyte membrane, following attachment of the merozoite, is the endocytotic process.

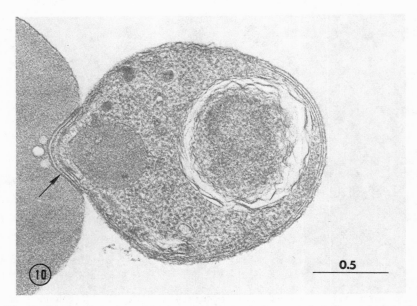

Fig. 10. The attachment (arrow) between the apical end of a cytochalasin B treated merozoite and an erythrocyte. The erythrocyte membrane is thickened at the attachment site. (From Miller et al., J. Exp. Med. 1979.)

As stated previously, Duffy negative human erythrocytes are refractory to merozoite invasion. Although the merozoite attaches to the erythrocyte membrane with filaments, no junction is formed between the merozoite plasmalemma and the erythrocyte plasmalemma. The absence of junction formation with Duffy negative erythrocytes may indicate that the Duffy associated antigen acts as a second receptor for junction formation or, alternatively, a determinant on Duffy negative erythrocytes blocks junction formation (Miller et al., 1979). This finding further supports the contention that the movement of the junction plays a major role in the mechanism by which the merozoite enters the erythrocyte.

The merozoite is covered with a uniform surface coat of 20 nm thickness (Aikawa et al., 1978; Miller et al., 1975) (fig. 11). During host cell entry, no surface coat is visible on the portion of the merozoite within the erythrocyte invagination, whereas the surface coat on that portion of the merozoite remaining outside the erythrocyte appears to be similar to that seen on free merozoites. However, no accumulation of the surface coat is seen at the orifice of the entry site of the erythrocyte by the merozoite (Aikawa et al., 1978). The surface coat beyond the junction appears to be unaltered in density (fig. 12). Therefore, simple capping does not seem to fit the observations. The significance of the surface coat in invasion must be further explored.

Sealing of Host Cell Membrane Following
Completion of Merozoite Entry

There are a few reports on the closure of the parasitophorus vacuole membrane after completion of interiorization of the merozoite. When entry is completed, the junction appears to fuse at the posterior end of the merozoite (fig. 13), closing the orifice in the fashion of an iris diaphragm (Aikawa et al., 1978). The merozoite still remains in close apposition to the thickened erythrocyte membrane at the point of the final closure. After completion of the host cell entry, the merozoite is surrounded by a membrane of the parasitophorous vacuole that originated from the erythrocyte membrane.

ALTERATION OF HOST CELL PLASMALEMMA
AFTER COMPLETION OF MEROZOITE INVASION

The parasitophorous vacuole membrane that originated from the erythrocyte membrane grows with the developing parasite and is retained until formation of the next generation of *Plasmodium* merozoites. Changes in the molecular organization of this parasitophorous vacuole membrane are apparent very early in development. Freeze-fracture studies (McLaren et al., 1977) have shown major differences in the distribution of intramembranous particles of erythrocyte membranes and the vacuole

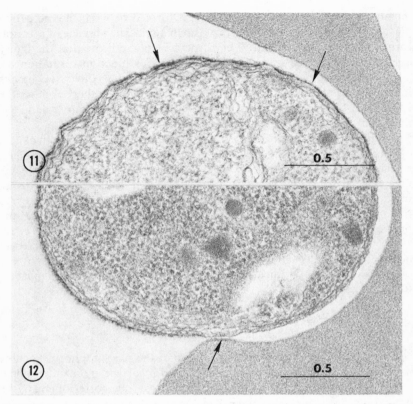

Fig. 11. A free merozoite showing a surface coat (arrow) covering the entire surface. (From Aikawa et al., J. Cell Biol. 77:72, 1978.)

Fig. 12. A merozoite entering an erythrocyte. No surface coat is visible on the portion of the merozoite surface that has invaginated the erythrocyte membrane, whereas the surface coat is visible behind the attachment site (arrow). (From Aikawa et al., J. Cell Biol., 77:72, 1978.)

membrane. Cytochemical studies (Langreth, 1977) also indicated differences in surface charge, glycoprotein, and enzyme distribution between these two membranes.

Intracellular malarial parasites appear to influence not only the parasitophorous vacuole membrane but also the host cell plasmalemma. Two types (Aikawa et al., 1975; Sterling et al., 1972) of erythrocyte membrane modifications can be seen in erythrocytes infected with certain species of *Plasmodium*: (1) electron-dense protrusions that have been described as excrescences (fig. 14) and (2) caveola-vesicle complexes along

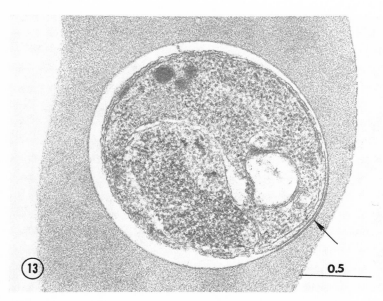

(13)   0.5

Fig. 13. A merozoite after completion of erythrocyte entry. The posterior end of the merozoite is still attached (arrow) to the thickened erythrocyte membrane. (From Aikawa et al.,
J. Cell Biol. 77:72, 1978.)

the erythrocyte membrane (figs. 15, 16). The chemical nature of these
structural components is not known, but they are immunologically
different from the rest of the erythrocyte membrane (Kilejian et al., 1977).

SUMMARY

In this chapter an attempt was made to review the interaction between
malarial *Plasmodium* and erythrocytes. Particular emphasis was placed on
four major topics that are basic for host-parasite interaction. They include:
(1) recognition and initial attachment, (2) invagination of the host
plasmalemma, (3) sealing of host cell membrane, and (4) alteration of host
cell membrane.

Plasmodium merozoites enter erythrocytes by endocytosis. However,
little is known of the mechanisms and factors that trigger the initiation of
endocytosis, although some experimental data indicate the possible
presence of a specific receptor for erythrocyte entry by the merozoite. A
distinct junction is formed between the erythrocyte membrane and the
merozoite membrane during the endocytotic process and moves along the
confronted membranes. The movement of this junction during entry may

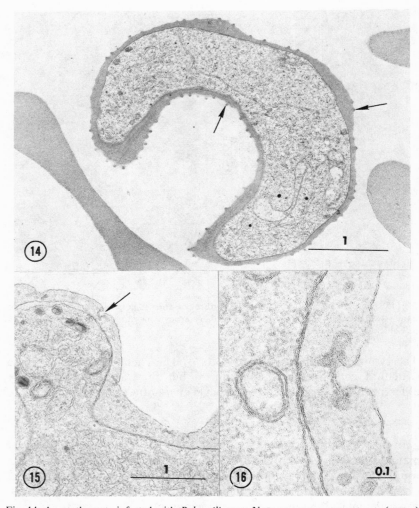

Fig. 14. An erythrocyte infected with *P. brasilianum*. Note numerous excrescences (arrow) on the erythrocyte plasma membrane. (From Aikawa, M., WHO Bull. 55:139, 1977.)

Fig. 15. *P. fieldi*-infected erythrocyte showing caveola-vesicle complexes (arrow) along the plasma membrane. (From Aikawa, et al., J. Parasit. 63:152, 1977.)

Fig. 16. Higher magnification micrograph of caveola-vesicle complex. (From Aikawa, et al., J. Parasit. 63:152, 1977.)

be an important component of the endocytotic mechanisms by which the merozoite enters the erythrocyte. The sealing of host cell membrane is accomplished by fusion on the junction in the fashion of an iris diaphragm.

After completion of host cell entry, the merozoite is surrounded by a parasitophorous vacuole membrane that originates from the erythrocyte membrane. This membrane becomes altered in its molecular organization. Such alteration of the erythrocyte-derived membrane may be one of the main determinants in assuring a safe and suitable intracellular environment for the intracellular *Plasmodium*.

ACKNOWLEDGMENTS

This work was supported in part by Research Grants (AI-10645 and AI-13366) from the U.S. Public Health Service and by the U.S. Army R&D Command (DADA 17-70-C-0006).

LITERATURE CITED

Aikawa, M., L. H. Miller, J. Johnson, and J. Rabbege. 1978. Erythrocyte entry by malarial parasites: a moving junction between erythrocyte and parasite. J. Cell Biol. 77:72–82.

Aikawa, M., L. H. Miller, and J. Rabbege. 1975. Caveola-vesicle complexes: the plasmalemma of erythrocytes infected by *Plasmodium vivax* and *Plasmodium cynomolgi*: Unique structure related to Schüffner's dots. Am. J. Path. 79:285–300.

Aikawa, M., and C. R. Sterling. 1974. Intracellular parasitic protozoa. Academic Press, New York.

Bannister, L. H., G. A. Butcher, and G. H. Mitchell. 1977. Recent advances in understanding the invasion of erythrocytes by merozoites of *Plasmodium knowlesi*. WHO Bull. 55:163–69.

Dennis, E. D., G. H. Mitchell, G. A. Butcher, and S. Cohen. 1975. *In vitro* isolation of *Plasmodium knowlesi* merozoite using polycarbone sieves. Parasitology 71:475–81.

Dvorak, J. A., L. H. Miller, W. C. Whitehouse, and T. Shiroishi. 1975. Invasion of erythrocytes by malaria merozoites. Science 187:748–50.

Griffin, F. M., J. A. Griffin, J. E. Leider, and S. C. Silverstein. 1975. Studies on the mechanism of phagocytosis. I. Requirements for circumferential attachment of particle-bound ligands to specific receptors on the macrophage plasma membrane. J. Exp. Med. 142:1263–82.

Kilejian, A. 1976. Does a histidine-rich protein from *Plasmodium lophurae* have a function in merozoite penetration? J. Protozool. 23:272–77.

Kilejian, A., A. Abati, and W. Trager. 1977. Immunogenicity of "knob-like protrusions" on infected erythrocyte membranes. Exp. Parasit. 42:157–64.

Ladda, R., M. Aikawa, and H. Sprinz. 1969. Penetration of erythrocytes by merozoites of mammalian and avian malarial parasites. J. Parasit. 55:633–44.

Langreth, S. G. 1977. Electron microscope cytochemistry of host-parasite membrane interactions with malaria. WHO Bull. 55:171–78.

McLaren, D. J., L. H. Bannister, P. I. Trigg, and G. A. Butcher. 1977. A freeze-fracture study on the parasite-erythrocyte interrelationship in *Plasmodium knowlesi* infections. WHO Bull. 55:199–203.

Miller, L. H., M. Aikawa, and J. A. Dvorak. 1975. Malaria (*Plasmodium knowlesi*) merozoites: immunity and the surface coat. J. Immunol. 114:1237–42.

Miller, L. H., M. Aikawa, J. Johnson, and T. Shiroishi. 1979. Interaction between cytochalasin B-treated malarial parasites and red cells: attachment and junction formation. J. Exp. Med. 149:172–84.

Miller L. H., S. J. Mason, M. Dvorak, H. McGinniss, and I. K. Rothman. 1975. Erythrocyte receptors for (*Plasmodium knowlesi*) malaria: Duffy blood group determinants. Science. 189:561–62.

Sterling, C. R., M. Aikawa, and R. S. Nussenzweig. 1972. Morphological divergence in a mammalian malarial parasites: the fine structure of *Plasmodium brasilianum*. Proc. Helminthol. Soc. Wash. 39:109–29.

Wessells, N. K., B. S. Spooner, J. F. Ash, M. O. Bradley, M. A. Luduena, E. L. Taylor, J. T. Wrenn, and K. M. Yamada. 1971. Micro-filaments in cellular and developmental processes. Science. 171:135.

Infection of Invertebrates with Algae

3

INTRODUCTION

Symbioses between invertebrates and symbiotic algae are characterized by a variety of cellular interactions. Previous reviews have summarized research on metabolite exchange between partners (Smith et al., 1969; Smith, 1974; Muscatine, 1974, 1979; Muscatine and Porter, 1977), cell surface interactions (Muscatine et al., 1975b), calcification (Vandermeulen and Muscatine, 1973), and integration between the partners (Taylor, 1973a; Trench, 1980). In this review I shall discuss recent research on the cellular events that occur in the establishment of these symbioses, and emphasize work on the two most intensively studied systems, *Paramecium bursaria* and *Hydra viridis*, in the hope that this discussion will stimulate additional work with other systems.

The cellular interactions that occur during the establishment of algae-invertebrate symbioses occur in three phases (Pardy and Muscatine, 1973; Muscatine, 1974). Initial events during contact between a host cell and potential symbionts include surface interactions involved with the "recognition" of partners. Following contact, appropriate algae are phagocytosed in instances that produce an intracellular association; algae are then enclosed by an endosome membrane. Once within cells, the algae must interact with other cellular components. Subsequent genetic adjustments of either partner may occur that affect functional integration; these latter events are considered elsewhere in this volume (Margulis, 1980; Trench, 1980).

In this review I frequently mention "infective" and "noninfective." I define a noninfective situation as one in which either the initial events involved in the formation of a symbiosis do not occur, or one in which an association is formed but does not persist. Clearly, the interpretation of "persistence" is somewhat arbitrary, and will vary from system to system. It would be more convenient to use functional criteria to define infection (or, perhaps more properly, reestablishment). However, the functional restitution of an algae-invertebrate symbiosis is rarely examined; exceptions are studies of *Paramecium bursaria* (Karakashian and Karakashian, 1965) and acoelous flatworms (Provasoli et al., 1968; Nozawa et al., 1972).

INFECTIVE ALGAE AND INFECTIBLE HOSTS: A SURVEY

Algae

The taxonomy and distribution of endosymbiotic algae have been most recently reviewed by Taylor (1973a). Table 1 lists many of the algae that have been used in infection studies. The list includes many strains that have been maintained in culture, as well as those that must be freshly obtained from host tissue. The symbiotic algae fit two general categories, "zoochlorellae" and "zooxanthellae." These terms have no taxonomic significance, but they are widely used in the literature. The primary distinction between them appears to be whether the cells are "green" or "brown".

Zoochlorellae include both the *Chlorella*-like symbionts of freshwater hosts (but see O'Brien, 1978) and the prasinophyte flagellates that are symbiotic with marine acoelous flatworms. The taxonomy of the *Chlorella*-like algae is confused; this is unfortunate because it is important to clearly identify strains used in infection studies. The situation is clouded since many of these algae are phenotypically plastic (see Karakashian, 1970, and Pardy, 1976a), and the infectivity of individual strains may be influenced by culture history (Pool, 1979). The symbiotic prasinophytes (e.g., *Platymonas convolutae*) can occur as free-living algae or as symbionts. The free-living and symbiotic stages have marked morphological differences (Oschman, 1966; Parke and Manton, 1967), and precise taxonomic descriptions of both forms are needed when these algae are used in infection studies (Taylor, 1973b).

The zooxanthellae used in infection studies are all dinoflagellates. *Symbiodinium* (=*Gymnodinium*) *microadriaticum*[1] has been isolated from various marine hosts; many of these isolates have been maintained in culture. *In situ* cells lack the flagella that are typical of most dinoflagellates, although evidence of flagellar apparatus is sometimes observed (Dodge, 1970; Schoenberg and Trench, 1979b). When cultured, these algae produce

TABLE 1

ALGAE THAT ESTABLISH SUCCESSFUL SYMBIOSES WITH
INVERTEBRATES IN EXPERIMENTAL STUDIES

I. "ZOOCHLORELLAE"

A. *Chlorella*-type algae

STRAIN/SPECIES	SOURCE	REFERENCE
*C. vulgaris (263, etc.)	Free-living	Karakashian and Karakashian, 1965 Hirshon, 1969
*NC64A	*Paramecium bursaria*	Karakashian, 1963
*32g	*P. bursaria*	Siegel, 1960
*Others	*P. bursaria*	Karakashian, 1963 Muscatine et al., 1975a
F/F	*Hydra viridis* (Florida)	Pardy and Muscatine 1973 Muscatine et al., 1975a
E/E	*H. viridis* (European)	Muscatine et al., 1975 Pardy, 1976a
*E/E	*H. viridis* (European)	Jolley and Smith, 1978
—	*H. viridis* (Carolina)	Jeon, 1977
—	*Chlorohydra hadleyi*	Park et al., 1967
O/O	*H. viridis* (Ohio)	Cook, unpublished
NC/F	NC64A resident in Florida hydra	Muscatine et al., 1975a Pool, 1979

B. Prasinophyceae

Platymonas convolutae	*Convoluta roscof-fensis*; also free-living	Parke and Manton 1967 Provasoli et al., 1968 Nozawa et al., 1972
Prasinocladus marinus	Free-living	Provasoli et al., 1968

II. "ZOOXANTHELLAE" (Dinophyceae)

Symbiodinium (Gymnodinium) microadriaticum	Various hosts**	Trench, 1971 Kinzie, 1974 Schoenberg and Trench, 1976 Schoenberg and Trench, 1979c
Amphidinium klebsii	*Amphiscolops langerhansi*	Taylor, 1971a
"Amphidinium sp."	Various hosts	Taylor, 1971b Kinzie, 1974

* Maintained in culture
** Sea anemones, corals, *Tridacna*, etc.

more typically flagellated forms. Formerly it had been thought that these algae are all similar (e.g., Taylor, 1973b), but evidence is accumulating that these isolates have morphological and genetic differences and differ in their infective properties (Schoenberg and Trench, 1976, 1979a–c).

The other infective dinoflagellates belong to the genus *Amphidinium. A. klebsii*[2] occurs as a symbiont with the acoelous flatworm *Amphiscolops langerhansi*. This alga appears to be unique among dinoflagellate symbionts in that it is an extracellular symbiont (Taylor, 1971a). *A. chattonii* and unidentified *Amphidinium* species have also been used in infection studies (Taylor, 1971b; Kinzie, 1974).

Hosts

Aposymbiotic (algae-free) individuals of normally symbiotic species occur naturally, as when uninfected gametes are produced (green hydra, Muscatine and Lenhoff, 1965; tridacnid clams, LaBarbera, 1975; convolutid flatworms, Keeble and Gamble, 1907; gorgonians, Kinzie, 1974). Aposymbionts can also be produced in the laboratory by a variety of methods, including: growth in darkness (paramecia, Siegel, 1960; sea anemones, Schoenberg and Trench, 1979a); high light intensities (hydra, Pardy, 1976b); high CO_2 concentrations (convolutid flatworms, Boyle and Smith, 1975); and the addition of glycerol (hydra, Muscatine, 1961). The addition of drugs has also been effective in removing algae. Successful "bleaching" has been achieved with the photosynthetic inhibitor DCMU (hydra, Pardy, 1976b; sea anemones, Schoenberg and Trench, 1979a), chloramphenicol (hydra, Fulton, 1961), and trimethoprim (Jeon, 1977). The susceptibility to these treatments varies from species to species and between individuals of the same species, so that often a combination of these treatments must be used.

Aposymbionts that have been successfully reinfected with algae are listed in table 2. Attempts have also been made to infect nonsymbiotic species (i.e., those not naturally symbiotic). These attempts essentially have been unsuccessful, even though such forms may be closely related to symbiotic species (e.g., subspecies, Theodor, 1969). The cellular basis of noninfection in these forms is not known, and is a problem for future research.

Aposymbiotic hosts vary in their suitability for infection studies. Polyps such as hydra and sea anemones are particularly useful for investigations of infection kinetics, because suspensions of algae can be readily pipetted into, or removed from, gut cavities where infection occurs; filter feeders such as *Paramecium* and bivalve larvae are somewhat less useful in this

TABLE 2

Species	Algae	Reference
Protozoa		
Paramecium bursaria	Various zoochlorellae from *P. bursaria, Hydra*; F/L *Chlorella*	Siegel, 1960 Karakashian, 1963 Karakashian and Karakashian, 1965 Hirshon, 1969 Reisser, 1976 Weis, 1976
Cnidaria		
Hydra viridis (Florida)	F/F, E/E, NC/F, NC64A	Pardy and Muscatine, 1973 Muscatine et al., 1975a Pardy, 1976a
H. viridis (European)	E/E, F/F	Pardy, 1976a Cooper and Margulis, 1977
H. viridis (Ohio)	O/O, F/F	Cook, unpublished
H. viridis (Carolina)	Native symbionts	Jeon, 1977.
Chlorohydra hadleyi	*C. hadleyi* symbionts	Park et al., 1967
Anthopleura elegantissima	*S. (G.) microadriaticum*	Trench, 1971
Aiptasia pulchella	*S. (G.) microadriaticum*	Kinzie et al., 1977
Aiptasia tagetes	*S. (G.) microadriaticum*	Schoenberg and Trench, 1976 Schoenberg and Trench, ms
Pseudopterogorgia bipinnata	*S. (G.) microadriaticum*	Kinzie, 1974
Platyhelminthes		
Convoluta roscoffensis	*P. convolutae*	Oschmann, 1966 Provasoli et al., 1968 Nozawa et al., 1972 Boyle and Smith, 1975.
Amphiscolops langerhansi	*A. klebsii*, *Amphidinium* sp.	Taylor, 1971a
Mollusca		
Tridacna crocea *T. gigas* *T. squamosa* *Hippopus hippopus*	*G. microadriaticum*	Jameson, 1976 LaBarbera 1975; Jameson 1976 LaBarbera, 1975 Jameson, 1976
(Various sacoglossans)	(Siphonaceous chloroplasts)	Muscatine and Green, 1973 McLean, 1976.

respect. The assay of infection is also more practical with some hosts than others: clones may be developed from individual paramecia, and the algal content of daughter cells easily counted. Similar counts may be made on cell maceration preparations from hydra (Pardy and Muscatine, 1973; Muscatine et al., 1975a) and hydroids (Trench, 1980). Chlorophyll fluorescence is a sensitive technique that can be used in direct microscopy (Muscatine et al., 1975a; Williamson, 1977) or for quantifying chlorophyll in pigment extracts (Pool, 1979; D'Elia et al., 1979). The latter technique is extremely rapid compared with direct cell counts and can detect very low levels of algae in single hydra.

Molluscs and Symbiotic Chloroplasts

In addition to associations between invertebrates and unicellular algae, there are symbioses between sacoglossan molluscs and choloroplasts that are obtained from feeding on siphonaceous algae such as *Codium* (Muscatine and Greene, 1973; Trench, 1975). These symbioses persist for variable periods of time, depending on the particular association. The cellular events involved with the uptake of chloroplasts from *Codium fragile* by the sea slug *Placida dendritica* have been described by McLean (1976), and will be discussed in the context of other infective systems.

ESTABLISHMENT OF CONTACT BETWEEN HOST CELL AND SYMBIONT

Contact between a potential symbiont and a potential host cell and host tissue initiates the infective process. In metazoans, cells that can be infected with algae generally are phagocytes that line a digestive cavity, so that ingestion (or artificial "force-feeding") precedes contact. A few observations indicate that cellular interactions between symbiont and host may potentiate the ingestion of potential symbionts. The classic observations of Keeble and Gamble (1907; see also Holligan and Gooday, 1975) indicate that free-living *Platymonas* actively swim toward the egg masses of *Convoluta* before the newly hatched worms are infected. Presumably this response is mediated by a chemotactic sense, but this has not been experimentally tested. Motile zooxanthellae are attracted to some hosts (Kinzie, 1974), but not to others (Trench, 1980); these behavioral differences may reflect genetic differences among isolates. Algal motility may elicit searching behavior by hosts. Taylor (1971a) has suggested that "noise," possibly the result of algal flagellar activity, attracts the acoel *Amphiscolops langerhansi* to *Amphidinium*. Algae might also increase the probability of infection by inducing host ingestive behavior. In coelenterates, feeding responses are mediated by surface chemoreceptors that respond to specific compounds released by captured prey (Lenhoff, 1968). Compounds released by potential symbionts could stimulate activities such as ciliary currents and mouth-opening that lead to ingestion (see Trench, 1980).

Most instances of the infection of invertebrates with algae involve the endocytosis of algae by phagocytes and the formation of an intracellular symbiosis. In convolutid flatworms (*Convoluta* and *Amphiscolops*) symbionts normally reside between parenchymal cells that underlie the epidermis. During infection algae are ingested and subsequently come to lie in their normal *in situ* position. The actual events that produce this arrangement are not clear despite ultrastructural studies of the infection of *Convoluta* (Oschman, 1966) and *Amphiscolops* (Taylor 1971a). Two scenarios are possible. First, the algae could be endocytosed by cells lining the digestive lumen; endocytosis would be followed by exocytosis to yield the extracellular position. The alternative is that algae are not phagocytosed but instead make their way between cells to reach their final position. The infective algae in each case (*Platymonas* and *Amphidinium klebsii*) are flagellated, and thus could be motile within host tissue. Both of these possibilities raise intriguing questions of cell-cell interactions that deserve further attention, as does the maintenance of these algae *in situ* between host cells.

Endocytosis of Algae: Morphological Aspects

In other instances of the infection of invertebrates with algae, the endocytosis of algae produces an intracellular symbiosis. Phagocytosis of algae results in an endocytic vacuole, bounded by a membrane of host origin (probably from the plasmalemma). Symbionts that are infective persist in individual vacuoles or "symbiosomes."[3] These endosomes have been reported to form in *Hydra viridis* (Park et al., 1967; Muscatine et al. 1975a,b; Cook et al., 1978), *Paramecium bursaria* (Karakashian and Karakashian, 1973; Reisser, 1976), *Anthopleura elegantissima* (Schoenberg and Trench, 1979b), and in the sea slug *Placida dendritica* (McLean, 1976).

The uptake of symbiotic algae by *H. viridis* appears to be a specialization of the digestive cells, which normally take up food materials from the gut cavity. Food uptake by these cells involves two distinct mechanisms.

Macromolecules such as ferritin and glycogen are endocytosed by discoidal-coated vesicles whose luminal surface has a distinctive coat of "pegs and globules" (Slautterback, 1967); the vesicles also participate in the uptake of larger food particles, but they do not appear to be involved with the endocytosis of infective algae. Larger food particles are also taken up by pseudopodial microvilli that lack the coat characteristic of the vesicles (Cook et al., 1978). Endocytosis is initiated by contact of algae with the digestive cell surface; this region of contact is marked by a zone of increased electron opacity of the digestive cell plasmalemma (Muscatine et al., 1975a). Contact is followed by the engulfment of algae by microvilli that possess a diffuse glycocalyx. Figures 1–3 show various stages in this process. During the early stages of endocytosis several microvilli are seen in the vicinity of an algal cell (fig. 1). Engulfment occurs as these microvilli enclose an alga; often several of these microvilli seem to fuse to form a "cytoplasmic veil" (fig. 3; Muscatine et al., 1975a). Newly formed endosomes contain single algae. The membranes that form these vacuoles show no trace of the coat that would be expected if coated vesicles were involved; in fact, the diffuse glycocalyx observed on the surface of endocytic microvilli is not seen on these membranes (fig. 4; Cook et al., 1978).

These observations are supported by observations made by scanning electron microscopy (P. McNeil, 1978). McNeil reports the formation and fusion of microvilli on the digestive cell surface, a process resulting in a "cytoplasmic tube" that actually effects the uptake of competent symbionts. Figures 2 and 3 may be sections through such "tubes." The microvilli are formed only when competent algae are introduced into the gut cavity. The injection of heated symbionts, which are taken up by digestive cells but not retained, does not produce microvillus formation. These dead algae are endocytosed by a cup-shaped extension of the plasmalemma that does not form from fused microvilli. McNeil has also observed that competent symbionts can be taken up by this mechanism. Thin sections indicate that microvilli also do not form when free-living *Chlorella* cells are injected into aposymbiotic hydra (Cook and Muscatine, unpublished); however, it is not known if these algae enter digestive cells through the cup-shaped cytoplasmic extensions.

The endocytosis of symbionts has been studied in detail in only one other system. The ultrastructural events associated with the endocytosis of chloroplasts by the sea slug *Placida dendritica* are similar to those described above for *H. viridis*. McLean (1976) has observed the formation of an electron-dense zone in regions in contact between cells of the digestive diverticula and the chloroplasts. Contact is followed by the formation of microvilli, and subsequently a "cytoplasmic veil" forms around the

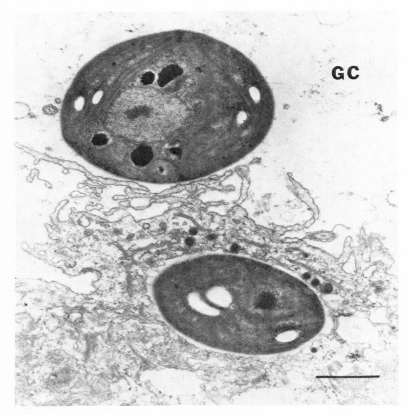

Fig. 1. The uptake of symbionts isolated from Florida strain *Hydra viridis* (F/ F algae) by a digestive cell of an aposymbiotic *H. viridis*; 10 min post-injection. One alga is surrounded by microvilli (mv), and a second is enclosed in an endosome and is surrounded by dense core vesicles (lysosomes?). GC = gut cavity. Scale bar = 1 μm.

chloroplasts, so that a membrane-bound endosome results. *P. dendritica* is a species that does not retain its chloroplasts for more than a few days, and it would be of interest to have comparable observations on slugs such as *Tridachia crispata* and *Elysia viridis* that retain their chloroplasts for much longer periods (Muscatine and Greene, 1973).

There have been no ultrastructural observations on the endocytosis of zooxanthellae by invertebrate cells. Kinzie and colleagues (1977) have reported preliminary observations on the phagocytosis of *S. microadriaticum* by the sea anemone *Aiptasia*, using light microscopy. Evidently the algae are taken up by gastrodermal cells; these cells may be in

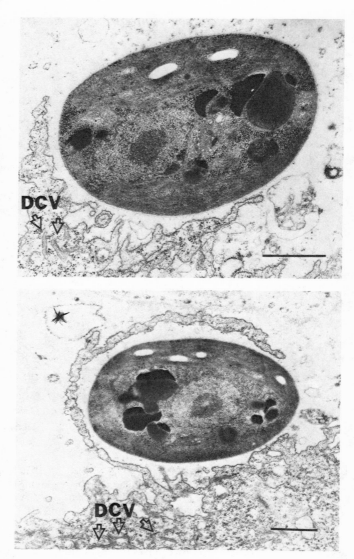

Figs. 2, 3. Further stages in the endocytosis of unheated F/F algae by *H. viridis*. Discoidal coated vesicles (DCV) are scattered in digestive cell cytoplasm. Scale bars: fig. 2: 1 μm; fig. 3: 0.5 μm. (Fig. 2 (top) from Cook et al., 1978; reprinted with permission of The Faculty Press; fig. 3 (bottom) from Muscatine et al., 1975a; reprinted with permission of Cambridge University Press and the Society for Experimental Biology.)

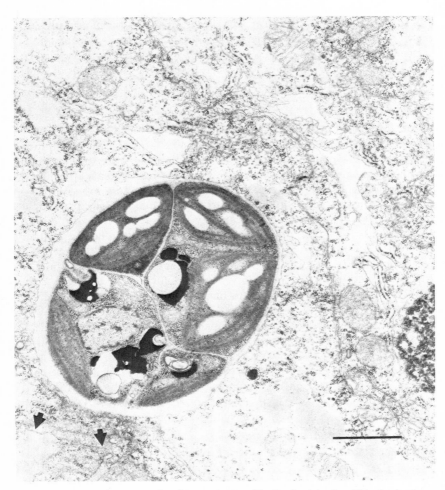

Fig. 4. An unheated F/F symbiont in an endosome 60 min after injection. This alga has been transported to the base of its host cell; the base is indicated by arrows. Scale bar: 1.0 μm. (From Cook et al., 1978; reprinted with permission of The Faculty Press.)

the absorptive zone of mesenterial filaments. This would be the region where food particles are endocytosed. Fine-structural studies of endocytic mechanisms in these cells have not been critically investigated.

Endocytosis of Algae: Quantitative Aspects

Kinetic studies. Measurements of the rate of endocytosis of algal cells during infection have been published only for *H. viridis*. The advantages of this system for such studies have been discussed above; the paucity of data from other systems may reflect technical difficulties. For example, measurements of uptake rates with *Paramecium bursaria* have yielded highly variable results, due in part to the digestion or exocytosis of some ingested algae (Karakashian and Karakashian, 1973).

The uptake of symbiotic hydra algae by the Florida strain of *H. viridis* follows kinetics that are typical of phagocytic systems (Pardy and Muscatine, 1973; Muscatine et al., 1975a; D'Elia et al., 1979). Uptake occurs at a maximum rate during the first hour after injection, and then decreases until uptake ceases 2 h after injection (fig. 5). Experiments in which hydra are reinjected with algae at successive intervals after initial injection indicate that the cessation of uptake is due to a loss of uptake capacity in hydra cells, rather than to any change in algae that might remain in the gut cavity; uptake capacity is restored over a 24 h period (D'Elia et al., 1979). The exhaustion of uptake capacity could result either from the consumption of appropriate membrane or from the energetic requirements of endocytosis. Observations that increased starvation of hydra does not decrease the ability of these cells to take up algae, even though starved aposymbiotic hydra have vastly depleted energy reserves (Kelty and Cook, in preparation), indicate that uptake is limited not by energy requirements but by suitable membrane for endocytosis. This is consistent with other phagocytic systems, such as macrophages (Werb and Cohn, 1973).

The endocytosis of other kinds of algae by *H. viridis* digestive cells generally follows the pattern of infective hydra symbionts. Cultured symbionts isolated from *Paramecium bursaria* (strain NC64A) are taken up in lesser amounts that are hydra symbionts, but the shape of the uptake curves are similar (Pardy and Muscatine, 1973). Free-living *Chlorella* are taken up only sparingly, but the available data suggest that the endocytosis of these algae also follows "exhaustion kinetics" (fig. 5). The exhaustion of uptake capability implies that membrane is limited for these algae as well as for the native Florida symbionts; since these algae are taken up in lesser amounts, there may be hydra cell membrane that is used for the endocytosis of hydra algae only, and other membrane used for the uptake of other algae. The properties of the digestive cell plasmalemma that could influence differential endocytosis could be the distribution of receptors on the digestive cell surface. By analogy with other endocytic systems, these receptors would probably be components of the diffuse glycocalyx which is seen on the digestive cell plasmalemma.

Studies of uptake specificity. Specificity is a characteristic of all algae-

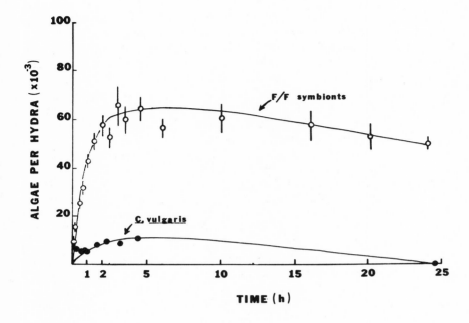

Fig. 5. The uptake of competent F/F symbionts and free-living *Chlorella vulgaris* by aposymbiotic *H. viridis*. Each point represents the mean of several replicates, ± 1 S.E.M. Algal content of individual hydra measured by fluorescence of chlorophyll extracts. (Data from D'Elia et al., 1979).

invertebrate infection systems. In some instances it is clear that discrimination between different kinds of algae takes place during the uptake process. In most systems infective and noninfective algae have differing fates following endocytosis. Post-uptake phenomena and their roles in effecting the specificity of infection will be discussed in a later section.

Hirshon (1969) examined the uptake of various algae by *Paramecium bursaria*. Symbionts from *P. bursaria* were taken up by greater percentages of paramecia than were free-living *Chlorella* cells, and other free-living algae were not taken up at all. Hirshon could not explain this discrimination, but he did find that autotrophic symbionts were taken up more than symbionts that had been grown in darkness. This observation suggests that released soluble carbohydrate, which is characteristic of autotrophic symbionts (Smith et al., 1969) may stimulate the uptake of algae by *P. bursaria*. Weis (1978) has also implicated carbohydrate release in the infection of *P. bursaria*, but his data do not indicate if the release influences endocytosis or subsequent events. Agglutination by the plant lectin Concanavalin A (Con A) is also characteristic of infective algae, but the relationship between Con A binding sites and infectivity, at least in *P. bursaria*, is not clear (Weis, 1978). Much work needs to be done to determine what factors stimulate endocytosis in these cells.

The Florida strain of *Hydra viridis* has been particularly useful for studies of uptake specificity, because uptake reaches a plateau following uptake exhaustion (Pool, 1979). As in *P. bursaria*, the number of algae taken up depends on the kind of algae. Symbiotic algae are taken up in the greatest numbers, whereas cultured symbionts from other hosts are taken up in lesser amounts, and very few free-living *Chlorella* cells are endocytosed (Pardy and Muscatine, 1973; fig. 5).

It seems clear that the recognition of algae by digestive cells during endocytosis depends in part on immunological properties of algal surfaces. The work of R. R. Pool, using a strain of algae from *P. bursaria*, has provided much information in this regard (tables 3 and 4; Pool, 1976, 1979; Muscatine et al., 1975a,b). When *H. viridis* is injected with native symbionts from Florida strain hydra (F/F algae), at least 90% of the digestive cells take up algae, and the mean number of algae in infected cells is between 5 and 6. In contrast, less than 10% of the digestive cells become infected when hydra are injected with cultured NC64A, and the mean number of algae in infected cells is only 2.7. However, if the infective NC64A algae are grown up in hydra tissue for many generations (NC/F algae), they become as infective as normal F/F symbionts (table 3). Pool has shown that this effect is not due to artifacts associated with contaminating host tissue but indeed represents changes in the infective properties of the algae.

Changes in the immunological properties of these algae are associated with this increased infectivity. Complement fixation studies demonstrate that the NC/F algae acquire surface characteristics that are more similar to those of F/F algae than they are to NC64A, and they lack some of the antigenic determinants of NC64A. If F/F and NC/F algae are treated with antisera prepared against F/F algae, their infectivity is decreased; this treatment does not affect the uptake of NC64A algae (table 4; Pool, 1979). These results indicate that specific protein or glycoprotein complexes on the surface of F/F and NC/F algae are recognized by receptors on the surface of digestive cells. (Presumably these sites on the algae are blocked

TABLE 3

Uptake of Zoochlorellae by
Digestive Cells of H. viridis

Algae	Algae per Cell* (N)	Percentage of Cells Infected
F/F	5.4 ± 0.8 (10)	90.0
NC/L (Cultured NC64A)	2.7 ± 0.4 (10)	8.5
NC/F	5.6 ± 1.2 (8)	80.5

Source: Data from Pool, 1979
*Algae per infected cell: $\bar{X}$ ± s.d.

TABLE 4

Effect of Incubation of Algae with Algal Antisera
on the Uptake of Algae by H. viridis
(Algal content estimated by fluorometry of single hydra)

Algae	Treatment	Relative Uptake	(Control)
F/F	Anti-F/F serum	25.5 ± 8.0 35.4 ± 14.1	73.4 ± 36.9
	Anti-NC/L serum	78.9 ± 32.2 82.3 ± 33.7	77.4 ± 31.3
NC/F	Anti-F/F serum	20.0 ± 8.0 14.5 ± 6.9	48.8 ± 28.9
	Anti-NC/L serum	10.5 ± 6.0	4.5 ± 2.5

Source: Data from Pool, 1979

by antiserum treatment.) The interaction between algal components and hydra cell receptors would then trigger endocytosis by the digestive cells.

The uptake of symbiotic algae can also be modified by treatment with Con A (Pool, 1978; Meints, 1978). Pool has found that Con A reduces the uptake of competent (unheated) F/F algae to levels similar to those of heated F/F algae, whereas the uptake of heated F/F algae is unaffected. It is tempting to speculate that the Con A-sensitive moieties of the algal surface stimulate the microvillus-mediated uptake of competent symbionts (see p. 54), and that these sites may also be the ones that are affected by treatment with antisera. Separate recognition systems may exist for the uptake of heated algae and symbionts treated with antisera. It is not known how these sites relate to those for competent algae or to the recognition of food particles. These questions clearly present opportunities for further investigation.

Digestion and Exocytosis

In addition to discrimination events that occur prior to or during endocytosis, post-uptake phenomena determine whether algae are retained as endosymbionts. Once they are endocytosed, algae may be degraded by host cell hydrolases or expelled from the cell by exocytosis. The exocytosis of endocytosed algae has not been examined, so it is not known how this process might be stimulated or if it discriminates between different algae. In established symbioses, selective expulsion of zooxanthellae has been suggested to occur during the regulation of algal numbers (Steele, 1976).

Interactions between host cell hydrolases and endocytosed algae have been examined in the *Chlorella* symbioses. In *P. bursaria* some endocytosed symbionts are incorporated into phagolysosomes and digested (Karakashian and Karakashian, 1965, 1973; Weis, 1976). The extent of digestion varies between strains of paramecia and also between paramecia from the same clone; Karakashian (1975) has suggested that symbiont digestion is influenced by a number of factors, including the nutritional state of the host. It is not clear that the digestion of some algae and the retention of others represents differences between algae in the same culture or differences in the properties of the membranes that form algal vacuoles.

Experiments using heat-killed algae provided some insight into the interactions between lysosomes and ingested algae in *P. bursaria* (Karakashian and Karakashian, 1973; Karakashian, 1975). Normally, acid phosphatase activity is not found in endosomes containing algae that are retained. When heat-killed algae are ingested, acid phosphatase activity occurs in endosomes with these algae much as it does in food vacuoles. If both heated and competent (unheated) algae are simultaneously ingested, endosomes that have both kinds of algae lack acid phosphatase activity. Karakashian and Karakashian (1973) offer two explanations for these results: either endosomes that contain competent algae do not fuse with lysosomes, or fusion occurs but competent algae inhibit acid phosphatase. The latter possibility could be tested by *in vitro* experiments in which algae are incubated with acid hydrolases and substrates, analogous to studies of the influence of parasitic helminths on host digestive enzymes (see Pappas, 1980, this volume).

In the Florida strain of *H. viridis*, it appears that "indigestibility" of ingested symbionts may be involved with the retention of algae. Unlike *P. bursaria* the digestive cells in these hydra do not show a decline in algal numbers following endocytosis (Muscatine et al., 1975a; Pool, 1979). Moribund algae, which might indicate the "digestion" of algae, are rarely observed following the uptake of competent algae (Cook, unpublished observations). Dense core vesicles often surround endosomes that contain competent algae (figs. 1, 6, 7) as well as heated algae (fig. 8). These vesicles appear to have acid phosphatase activity (Hohman, 1978, 1979), and acid phosphatase activity can be demonstrated in endosomes containing competent algae (Muscatine et al., 1975b; Cook and Muscatine, unpublished). Hohman (unpublished) has observed that F/F algae pass through the gut of *Daphnia* (an herbivore) functionally intact, whereas free-living *Chlorella* are readily digested. He also has evidence that the cell walls of the F/F algae possess sporopollenin, a polycarotenoid highly resistant to chemical hydrolysis. Unlike competent algae, heated algae appear to be broken down following endocytosis by green hydra digestive cells. Figure 9 shows such an alga 5 h after endocytosis by *H. viridis*; the disorganization of the cytoplasm and of the cell wall is evident. Whether this is an effect of the heat treatment, of host cell hydrolase activity, or both, is not clear at this time.

The above description of events does not apply to all strains of green hydra. In the "Jubilee" strains of *H. viridis*, the endocytosis of algae is followed by a significant decrease in the number of algae in digestive cells (D. C. Smith, pers. comm.). The fate of the lost symbionts, be it "digestion" or exocytosis, is not known.

Sequestering of Infective Algae

An asymmetrical distribution of algae within host cell cytoplasm is characteristic of several algae-invertebrate symbioses. In several coelenterates (hydra, Pardy and Muscatine, 1973; *Myrionema*, Trench, 1980), algae are confined to basal regions of gastrodermal cells. In *P. bursaria* symbionts are generally found in the peripheral cytoplasm (Weis, 1976). The significance of these distributions is not clear, but they may serve to isolate algal cells from sites of lysosomal activity, or to maintain algae away from regions of exocytosis.

At least in green hydra the asymmetrical distribution is restored during infection. The endocytosis of competent symbionts is followed by the transport of endosomes away from the site of uptake (Pardy and Muscatine, 1973; Muscatine et al., 1975a). The rate of algal transport to the base of digestive cells has been estimated to be 5 μm min^{-1}, and most algae are transported within 5 h of uptake (fig. 10; Muscatine et al., 1975a). This basal distribution of algae is retained by daughter cells in subsequent generations. In contrast, noninfective algae such as free-living *Chlorella* and heated hydra symbionts are not transported. These algae are retained at the apical tips of cells (figs. 11, 12) and subsequently disappear, as a result of digestion or exocytosis.

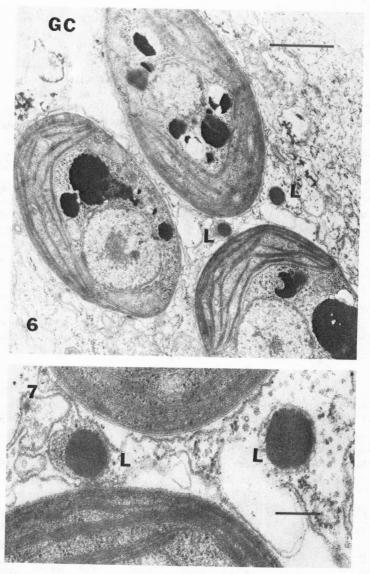

Figs. 6, 7. Interaction of lysosomes (L) with endosomes containing unheated F/F algae, 10 min after injection. The symbionts remain in apical cytoplasm; the gut cavity is noted (GC). Scale bars: fig. 6, 1 μm; fig. 7, 0.2 μm. (Figure 6 from Muscatine et al., 1975a; reprinted with permission of Cambridge University Press).

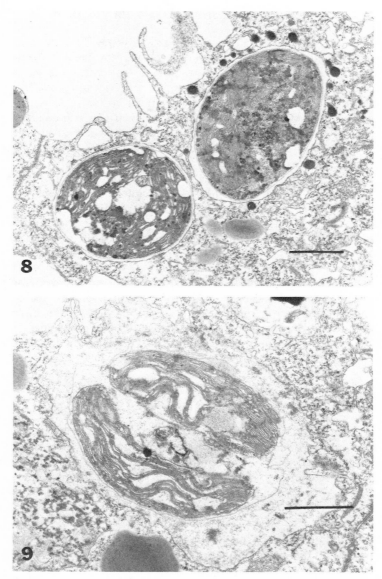

Figs. 8, 9. Endosomes containing heated F/F algae (10 min, 60°C) in *H. viridis* digestive cells. Figure 8 shows 2 algae 60 min after injection; note the dense core vesicles (presumably lysosomes) that surround one endosome. Figure 9 shows an endosome 5 h after injection. Disorganization of the alga is evident. Scale bars = 1 μm.

The mechanics of the transport system are not clear. Transport of infective algae can be reduced by the use of antimitotic drugs such as podophyllotoxin and vinblastine (Cooper and Margulis, 1977). These results implicate hydra cell microtubules in the transport process, although there are no morphological reports to confirm this. Oriented microfilaments do occur in these cells aligned along the route of transport (fig. 13). It is not known if these microfilaments are associated with the transport of algae or with general cytoplasmic circulation. Evidently, the cytoskeleton of hydra digestive cells functions in this aspect of infection, but whether the participation of host cell architecture is necessary for the infection of other invertebrate cells with algae remains to be investigated.

The basis of specificity in the transport system is an intriguing question whose answers are also unknown. Endocytosed infective algae may somehow activate or stabilize the transport system, possibly through factors that control microtubule assembly *in vitro*. These factors include Ca^{2+} concentration, microtubule-associated proteins and the availability of guanosine nucleotides (Dustin, 1978, p. 67). Alternatively, the operation of the transport system could depend on properties of the endosome membrane. In this case the fate of internalized algae would depend on the nature of the membrane used in endocytosis, and thus would be "decided" during the uptake process.

OVERVIEW

Invertebrates that can be infected with symbiotic algae represent several levels of cellular interactions. The establishment of intracellular symbioses in particular offers attractive opportunities for the study of phenomena such as cell recognition. Analyses of the infection of *Paramecium bursaria* and *Hydra viridis* with *Chlorella* indicate that specific sites on the surface of algae are recognized by receptors on host cell membranes. These receptors determine whether or not a particle will be endocytosed, and possibly control the nature of the membrane that forms around algae. The properties of this "symbiosome" membrane may influence subsequent events such as the fusion of endosomes and lysosomes, and the spatial distribution of algae within a host cell. Parallel studies are needed with other algae-invertebrate systems, as are studies of acceptance or rejection of algae by normally uninfected cells. Promising steps in this direction have been taken in studies of the infection of vertebrate tissue culture cells with zooxanthellae and other symbiotic flagellates (Taylor, 1978).

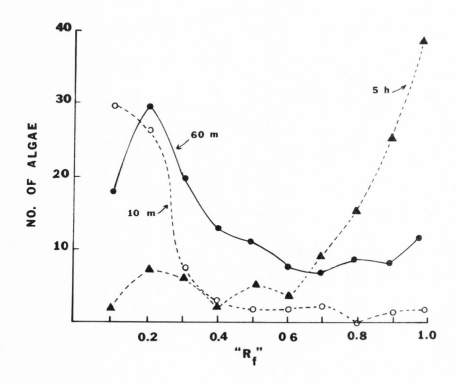

Fig. 10. The distribution of unheated F/F algae in digestive cells at various times after injection. "R_f" represents the relative distance of algae from the apical tips of digestive cells; an R_f of 0 indicates an alga at the tip, and an R_f of 1 indicates a cell that lies at the base of cells. (Data from Muscatine et al., 1975a).

ACKNOWLEDGMENTS

It is a pleasure to acknowledge the collaboration and advice of Dr. Leonard Muscatine and his coworkers at the University of California, Los Angeles. Much of the work reviewed here had its inception in his laboratory, and I am most grateful to C. F. D'Elia, T. C. Hohman, P. McNeil, and R. R. Pool for discussions of their work. Expert technical assistance with electron microscopy was provided by Bibbi Woloske (UCLA) and Jeanette

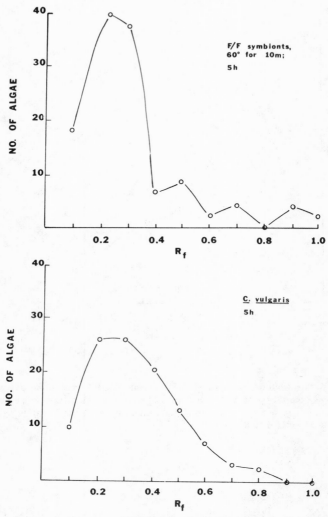

Figs. 11, 12. The distribution of noninfective algae in digestive cells 5 h post-injection; compare with figure 10. Figure 11 (top): heated F/F algae; figure 12 (bottom): *Chlorella vulgaris*. (figure 12 data from Muscatine et al., 1975a).

Pertuset (OSU). Some of the unpublished work was supported by an NSF grant to L.M. (PCM75–03380) and by grants from the Ohio State University Graduate School.

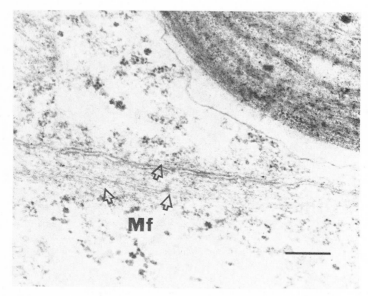

Fig. 13. Microfilaments (Mf) in peripheral cytoplasm of 2 adjacent digestive cells from a symbiotic green hydra. The filaments extend from basally located algae to the apical (luminal) end of the digestive cells. This axis parallels the direction of transport of infective algae. Scale bar is approximately 0.2 μm.

1. Strains of similar algae have recently been placed in the genus *Zooxanthella*, based on thecal morphology of motile and nonmotile forms (Loeblich and Sherley, 1979, J. Mar. Biol. Ass. U.K. 59:215–26). How this finding affects the taxonomy of other symbionts in the genus *Symbiodinium* is a question for future work.

2. Taylor (1971a) has assigned this species to the genus *Amphidinium*. Other authors retain it in the genus *Endodinium*; see discussion in Loeblich and Sherley (Note 1 above).

3. L. Muscatine (pers. comm.) has coined this term for endosomes that contain infective algae, and for vacuoles that normally contain algal symbionts. The term is most appropriate, and I would hope it gains acceptance.

LITERATURE CITED

Boyle, J. E., and D. C. Smith. 1975. Biochemical interactions between the symbionts of *Convoluta roscoffensis*. Proc. R. Soc. Lond. Ser. B. 189:121–35.

Cook, C. B., C. F. D'Elia, and L. Muscatine. 1978. Endocytic mechanisms of the digestive cells of *Hydra viridis*. I. Morphological aspects. Cytobios 23:17–31.

Cooper, C. G., and L. Margulis. 1978. Delay in migration of symbiotic algae in *Hydra viridis* by inhibitors of microtubule polymerization. Cytobios 19:7–20.

D'Elia, C. F., C. B. Cook and L. Muscatine. 1979. Endocytic mechanisms of the digestive cells of *Hydra viridis*. II. Kinetic aspects. (In preparation).

Dodge, J. D., and R. M. Crawford. 1970. A survey of thecal fine structure in the Dinophyceae. Bot. J. Linn. Soc. 63:53–67.

Dustin, P. 1978. Microtubules. Springer-Verlag, New York.

Fulton, C. 1961. Discussion. *In* H. M. Lenhoff and W. F. Loomis (eds.), The biology of hydra, p. 267. University of Miami Press, Miami.

Hirshon, J. B. 1969. The response of *Paramecium bursaria* to potential endocellular symbionts. Biol. Bull. 136:33–42.

Hohman, T. C. 1978. Intracellular digestion and symbiosis in *Hydra viridis*. Contributed paper, Fifth Annual Colloquium, College of Biological Sciences, Ohio State University, Columbus. 8 September 1978. (Unpublished).

Hohman, T. C. 1979. Ph.D. diss. University of California, Los Angeles.

Holligan, P. M., and G. W. Gooday. 1975. Symbiosis in *Convoluta roscoffensis*. Symp. Soc. Exp. Biol. 29:205–27.

Jameson, S. C. 1976. Early life history of the giant clams *Tridacna crocea* Lamarck, *Tridacna maxima* (Röding), and *Hippopus hippopus* (Linnaeus). Pac. Sci. 30:219–33.

Jeon, K. W. 1977. A new method for obtaining aposymbiotic hydra using trimethoprim as a bleaching agent. Devel. Biol. 59:255–58.

Jolley, E., and D. C. Smith. 1978. The green hydra symbiosis. I. Isolation, culture, and characteristics of the *Chlorella* symbiont of "European" green hydra. New Phytol. 81:637–46.

Karakashian, M. W. 1975. Symbiosis in *Paramecium bursaria*. Symp. Soc. Exp. Biol. 29:145–73.

Karakashian, S. J. 1970. Morphological plasticity and the evolution of algal symbionts. Ann. N.Y. Acad. Sci. 175:474–87.

Karakashian, M. W., and S. J. Karakashian. 1973. Intracellular digestion and symbiosis in *Paramecium bursaria*. Exp. Cell Res. 81:111–19.

Karakashian, S. J., and M. W. Karakashian. 1965. Evolution and algal symbiosis in the genus *Chlorella* and related algae. Evolution 19:368–77.

Keeble, F., and F. W. Gamble. 1907. The origin and nature of the green cells of *Convoluta roscoffensis*. Quart. J. Micros. Sci. 51:167–219.

Kinzie, R. A. 1974. Experimental infection of aposymbiotic gorgonian polyps with zooxanthellae. J. Exp. Mar. Biol. Ecol. 15:335–45.

Kinzie, R. A., G. Araki, and L. Kia. 1977. The specificity of symbiosis of zooxanthellae from a range of invertebrate hosts. Amer. Zool. 17:908 (Abstr.).

LaBarbera, M. 1975. Larval and post-larval development of the giant clams *Tridacna maxima* and *T. squamosa*. Malacologia 15:69–79.

Lenhoff, H. M. 1968. Chemical perspectives on the feeding response, digestion, and nutrition of selected coelenterates. *In* M. Florkin and B. T. Scheer (eds.), Chemical zoology, 2:157–221. Academic Press, New York.

Margulis, L. 1980. Symbiosis as parasexuality. *In* this volume.

McLean, N. 1976. Phagocytosis of chloroplasts in *Placida dendritica* (Gastropoda: Sacoglossa). J. Exp. Zool. 197:321–30.

Meints, R. H. 1978. Recognition and uptake of symbiotic algae by *Hydra*: an examination of host-symbiont cell surfaces. Contributed paper, Fifth Annual Colloquium, College of Biological Sciences, Ohio State University, Columbus. 8 September 1978. (Unpublished).

Muscatine, L. 1961. Symbiosis in marine and fresh water coelenterates. *In* H. M. Lenhoff and W. F. Loomis (eds.), The biology of Hydra: 1961, pp. 255–64. University of Miami Press, Miami.

Muscatine, L. 1974. Endosymbiosis of cnidarians and algae. pp. 359–395. *In* L. Muscatine and H. M. Lenhoff (eds.), Coelenterate biology: reviews and new perspectives, pp. 359–95. Academic Press, New York.

Muscatine, L. 1980. Uptake, retention, and release of dissolved inorganic nutrients by marine algae-invertebrate associations. *In* this volume.

Muscatine, L., and R. W. Greene. 1973. Chloroplasts and algae as symbionts in molluscs. Int. Rev. Cytol. 36:137–69.

Muscatine, L., and J. W. Porter. 1977. Reef corals: mutualistic symbioses adapted to nutrient-poor environments. Bioscience 27:454–60.

Muscatine, L., C. B. Cook, R. L. Pardy, and R. R. Pool. 1975a. Uptake, recognition, and maintenance of symbiotic *Chlorella* by *Hydra viridis*. Symp. Soc. Exp. Biol. 29:175–203.

Muscatine, L., R. R. Pool, and R. K. Trench. 1975b. Symbiosis of algae and invertebrates: aspects of the symbiont surface and the host-symbiont interface. Trans. Amer. Micros. Soc. 94:450–69.

Nozawa, K., D. L. Taylor, and L. Provasoli, 1972. Respiration and photosynthesis in *Convoluta roscoffensis* Graff, infected with various symbionts. Biol. Bull. 143:420–30.

O'Brien, T. L. 1978. An ultrastructural study of zoochlorellae in a marine coelenterate. Trans. Amer. Micros. Soc. 97:320–29.

Oschman, J. L. 1966. Development of the symbiosis of *Convoluta rescoffensis* Graff and *Platymonas* sp. J. Phycol. 2:105–11.

Pappas, P. W. 1980. Enzymological interactions at the host-parasite interface. *In* this volume.

Pardy, R. L. 1976a. The morphology of green hydra symbionts as influenced by host strain and environment. J. Cell Sci. 20:655–69.

Pardy, R. L. 1976b. The production of aposymbiotic hydra by the photo-destruction of green hydra zoochlorellae. Biol. Bull. 151:225–35.

Pardy, R. L., and L. Muscatine. 1973. Recognition of symbiotic algae by *Hydra viridis*: a quantitative study of the uptake of living algae by aposymbiotic *H. viridis*. Biol. Bull. 145:565–79.

Park, H. D., C. L. Greenblatt, C. F. T. Mattern, and C. R. Merrill. 1967. Some relationships between *Chlorohydra*, its symbionts, and some other chlorophyllous forms. J. Exp. Zool. 164:141–62.

Parke, M., and I. Manton. 1967. The specific identity of the algal symbiont in *Convoluta roscoffensis*. J. Mar. Biol. Ass. U. K. 47:445–64.

Pool, R. R. 1976. Symbiosis of *Chlorella* with *Chlorohydra viridissima*. Ph.D. diss., University of California, Los Angeles.

Pool, R. R. 1978. The role of the cell surface in the phagocytic recognition of algal symbionts by cells of *Chlorohydra*. Contributed paper, Fifth Annual Colloquium, College of Biological Sciences, Ohio State University, Columbus. 8 September 1978. (Unpublished).

Pool, R. R. 1979. The role of antigenic determinants in the recognition of potential algal symbionts by cells of *Chlorohydra*. J. Cell Sci. 35:367–79.

Provasoli, L., T. Yamasu, and I. Manton. 1968. Experiments on the resynthesis of symbiosis in *Convoluta roscoffensis* with different algal cultures. J. Mar. Biol. Ass. U. K. 48:465–79.

Reisser, W. 1976. Die stoffwechselphysiologischen Beziehungen zwischen *Paramecium bursaria* Ehrbg. und *Chlorella* sp in der *Paramecium bursaria*-Symbiose. II. Symbiose-spezifische Merkmale der Stoffwechelphysiologie und der Cytologie des Symbiose-Verbandes und ihre Regulation. Arch. Microbiol. 111:161–70.

Schoenberg, D. A., and R. K. Trench. 1976. Specificity of symbioses between marine cnidarians and zooxanthellae. *In* G. O. Mackie (ed.), Coelenterate ecology and behavior, pp. 423–32. Plenum Press, New York.

Schoenberg, D. A., and R. K. Trench. 1979a. Genetic variation in *Symbiodinium* (=*Gymnodinium*) *microadriaticum* Freudenthal, and specificity in its symbiosis with marine invertebrates. I. Isoenzyme and soluble protein patterns of axenic cultures of *S. microadriaticum*. Proc. R. Soc. Lond. (in press).

Schoenberg, D. A., and R. K. Trench. 1979b. Genetic variation in *Symbiodinium* (=*Gymnodinium*) *microadriaticum* Freudenthal, and specificity in its symbiosis with marine invertebrates. II. Morphological variation. Proc. R. Soc. Lond. (in press).

Schoenberg, D. A., and R. K. Trench. 1979c. Genetic variation in *Symbiodinium* (=*Gymnodinium*) *microadriaticum* Freudenthal, and specificity in its symbiosis with marine invertebrates. III. Specificity of infectivity in *S. microadriaticum*. Proc. R. Soc. Lond. (in press).

Siegel, R. W. 1960. Hereditary endosymbiosis in *Paramecium bursaria*. Exp. Cell Res. 19:239–52.

Slautterback, D. B. 1967. Coated vesicles in absorbtive cells of *Hydra*. J. Cell Sci. 2:563–72.

Smith, D. C. 1974. Transport from symbiotic algae and symbiotic chloroplasts to host cells. Symp. Soc. Exp. Biol. 28:485–520.

Smith, D. C., L. Muscatine, and D. H. Lewis. 1969. Carbohydrate movement from autotrophs to heterotrophs in parasitic and mutualistic symbiosis. Biol. Rev. 44:17–90.

Steele, R. D. 1976. Light intensity as a factor in the regulation of density of zooxanthellae in *Aiptasia tagetes* (Coelenterata: Anthozoa). J. Zool. 179:387–405.

Taylor, D. L. 1971a. On the symbiosis between *Amphidinium klebsii* (Dinophyceae) and *Amphiscolops langerhansi* (Turbellaria: Acoela). J. Mar. Biol. Ass. U.K. 51: 301–13.

Taylor, D. L. 1971b. Ultrastructure of the "zooxanthella" *Endodinium chattonii* in situ. J. Mar. Biol. Ass. U.K. 51:227–34.

Taylor, D. L. 1973a. The cellular interactions of algal-invertebrate symbiosis. Adv. Mar. Biol. 11:1–56.

Taylor, D. L. 1973b. Algal symbionts of invertebrates. Ann. Rev. Microbiol. 27:171–87.

Taylor, D. L. 1978. Artificially induced symbiosis between marine flagellates and vertebrate tissues in culture. J. Protozool. 25:77–81.

Theodor, J. 1969. Contribution à l'étude des Gorgones (VII): *Eunicella stricta aphyta* sous-espèce nouvelle sans Zooxanthelles, proche d'une espèce normalment infestée par ces algues. Vie Milieu Sér. A 20:635–37.

Trench, R. K. 1975. Of leaves that crawl: functional chloroplasts in animal cells. Symp. Soc. Exp. Biol. 29:229–65.

Trench, R. K. 1980. Integrative mechanisms in mutualistic symbioses. *In* this volume.

Vandermeulen, J. H., and L. Muscatine 1973. Influence of symbiotic algae on calcification in reef corals: Critique and progress report. *In* W. B. Vernberg (ed.), Symbiosis in the sea, pp. 1–19. University of South Carolina Press, Columbia.

Weis, D. S. 1976. Digestion of added homologous algae by *Chlorella*-bearing *Paramecium bursaria*. J. Protozool. 23:527–29.

Weis, D. S. 1978. Correlation of infectivity and Concanavalin A agglutinability of algae exsymbiotic from *Paramecium bursaria*. J. Protozool. 25:366–69.

Werb, Z., and Z. A. Cohn. 1972. Plasma membrane synthesis in the macrophage following phagocytosis of polystyrene latex particles. J. Biol. Chem. 247:2439–46.

Williamson, C. E. 1977. Fluorescence identification of zoochlorellae: a rapid method for investigating algal-invertebrate symbioses. J. Exp. Zool. 202:187–94.

DAVID H. S. RICHARDSON AND EVERT NIEBOER

Surface Binding and Accumulation of Metals in Lichens

4

INTRODUCTION

Lichens are the most generally recognized example of a symbiotic association in the plant kingdom. They result from the close association of an alga, usually green, with a fungus that is most frequently an Ascomycete. Aspects of the establishment of lichen symbioses and nutrient movement between lichen partners are discussed elsewhere (Ahmadjian, this volume; Smith, this volume). We have chosen to focus on the physiological effects on both partners of the surface accumulation of metals and the over-all mineral content. In this paper we shall consider the following aspects: (1) the physical and chemical characteristics of the lichen surface; (2) natural variations in lichen mineral content; (3) the influence of the uptake of metal ions on lichen sensitivity to sulphur dioxide; and (4) the ecological repercussions of metal accumulation by lichens. The need for additional research on each of these topics will become evident.

PHYSICAL CHARACTERISTICS OF THE LICHEN SURFACE

A macroscopic view of a lichen such as *Umbilicaria muhlenbergii* (fig. 1) shows a rather smooth upper surface and a lower surface covered with short lamellate outgrowths. In section, as in most lichens, there is a protective upper cortex, beneath which is an algal layer, then a less dense medulla, and finally a lower cortex (fig. 2). Detailed examinations of the upper cortex of lichens using the scanning-electron microscope (Hale, 1973) have shown considerable variation even in a single family, the Parmeliaceae. Thus several species of *Parmelia* have pores in the surface through which fine particles or gases could attain access to the interior (fig.

3), and others such as *Pseudoevernia furfuracea* have a very rough external surface with sizable interstices where particulates could also be trapped (fig. 4). Providing that there are ports of entry, there appears to be ample room for the accumulation of particulates within lichens. Collins and Farrar (1978) estimated the proportion of thallus volume occupied by the alga, fungus, extracellular matrix, and space in *Xanthoria parietina*. A value of 18% of the total volume was obtained for the unoccupied space in the thallus.

Particulate trapping as an accumulation mechanism for metals has been demonstrated by Nieboer et al. (1978). They showed a linear relationship between the iron and titanium contents (fig. 5) for lichens collected away from urban/industrial centers. The Ti and much of the Fe, Cr, V, and Ni contents of lichens from such rural areas could be assigned to particulates trapped within the thallus. Garty et al. (1977) has also concluded that the Ni content of lichens in urban areas is derived largely from fine particles released by abrasion of car components and from industrial sources; however, some Ni is also likely to be accumulated in ionic form. Evidence to support this conclusion, and also to confirm the importance of particulate trapping, is derived from studies around Sudbury, Ontario. Thus 20% of the Ni content of lichens growing within 8 km of Sudbury could be eluted with 1N HCl, and was therefore presumed to be bound to ion-exchange sites. In specimens collected at a greater distance (48 km), up to about 50% of the Ni could be eluted (Puckett et al., 1973a). Gough and Erdman (1977) have identified and photographed fly ash microspheres trapped on the surface of the rhizinae in *Parmelia chlorochroa* growing around a coal-fired power plant in Wyoming.

The possible magnitude of particulate trapping by lichens is emphasized by the observation that 5 g dry wt samples of *Cladonia rangiferina* that had been thoroughly washed following collection near (70 m) a uranium mine ventilation shaft had an ash content of ca. 300 mg. In contrast, comparable specimens collected 430 m away yielded only 40 mg of ash (fig. 6; Boileau et al., unpublished). The accumulation of large amounts of particulates would be expected to reduce a lichen's capacity for water uptake and perhaps the rate of gas diffusion in the thallus. Certainly lichens close to urban/industrial areas are often depauperate with few fruiting bodies. This aspect warrants investigation.

Little is known about the location of metals accumulated in lichen thalli either as particulates or in the ionic form. This will be discussed in some detail in the next section. Conspicuous deposits of iron oxides can, however, be observed, and are reported to be secreted on the surface of the upper cortex in a number of crustose lichens characteristic of iron-rich

Fig. 1. A macroscopic view of *Umbilicaria muhlenbergii*, the lichen used for many of the experiments summarized in this paper. Each thallus is approximately 5 cm in diameter. The darker spots on the upper surface are the fruiting bodies (ascocarps).

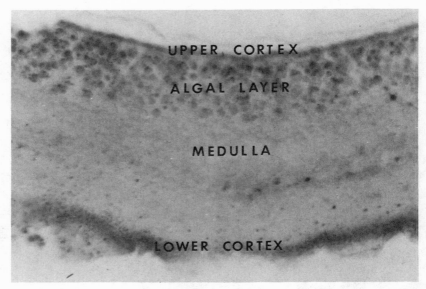

Fig. 2. A hand-cut vertical section of a living disc of *Umbilicaria muhlenbergii* showing the proportion of thallus occupied by the algae and the distribution of the various thallus zones.

rocks (Brightman and Seaward, 1977). In one species typical of this substrate, *Rhizocarpon oederi*, iron is also reported to be deposited on the hyphal cell walls within the thalli (Tuominen and Jaakkola, 1973).

The cortical layers of lichens appear to have a protective function. Evidence for this conclusion is deduced from a consideration of the SO_2 sensitivity of *Lobaria pulmonaria* and *Parmelia saxatilis*. Both lichens contain a green alga, have a foliose morphology, and often grow in the same habitat (tree boles). Tuerk and colleagues (1974) noted that the sensitive species, *Lobaria pulmonaria*, had a thin upper cortex and almost no lower cortex. *Parmelia saxatilis*, on the other hand, which is affected less by SO_2, has a robust cortex covered with a thick polysaccharide layer (Hale, 1976). Such structure could be a barrier against inward SO_2 diffusion or perhaps possess a barrage of cation binding sites (see below) that could also provide some protection.

CHEMICAL CHARACTERISTICS OF THE LICHEN SURFACE

Water Relations and Cell Wall Composition

Metal ions and SO_2 are accumulated by lichens from the aqueous phase.

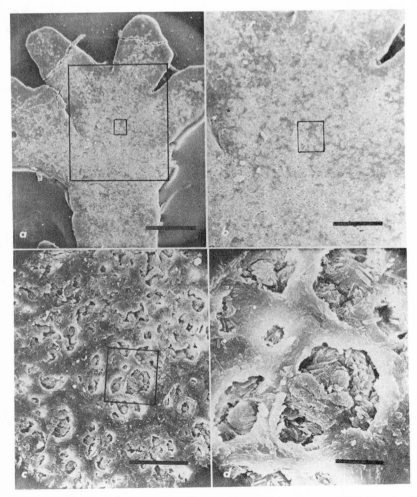

Fig. 3. A surface view of a lobe of *Parmeliopsis ambigua* at increasing magnification using the scanning electron microscope. Note the pores in the upper surface in this species. Reproduced with permission from Hale (1973). Scale bars: 3(a), 0.4mm; 3(b), 0.2mm; 3(c), 50μm; 3(d), 10μm.

Dry lichens are very resistant to SO_2 damage (Tuerk et al., 1974). Thus some consideration of lichen-water relations is essential. Many lichens need as little as 1–2 min to reach full saturation when placed in water and act like a hydrophilic gel taking up from 100 to 300% of their dry weight of

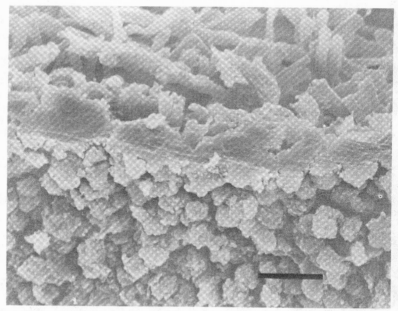

Fig. 4. A stereoview of the surface and interior of *Pseudevernia furfuracea* illustrating the rough upper cortex in this lichen. Reproduced with permission from Hale (1973). Scale bar = 20 μm.

water (Blum, 1973). Water moves rapidly into the free interstitial spaces in and between the walls of the fungal hyphae. Blum (1973) found that prolonged immersion of some species, e.g., *Ramalina farinacea*, resulted in greater net uptake due to increased ability of the thallus colloids to retain water.

Lichenin and isolichenin are no longer considered to be structural polysaccharides of lichens as was previously supposed (Mosbach, 1973). Lichenins and glucans are now thought to be components of the extracellular matrix of lichen thalli and fruiting bodies (Galun et al., 1976; Collins and Farrar, 1978) and contribute to the hydrophilic gel-like properties of these plants (Mittal and Seshadri, 1954). This extracellular matrix has been estimated to comprise 34% of the total thallus volume (including spaces) in *Xanthoria parietina* (Collins and Farrar, 1978).

Galun and colleagues (1976) suggest that chitin is an important structural component of lichens on the basis of incorporation of N- [^{3}H]-acetyl-D-glucosamine into their cell walls and on the binding of fluorescein isothiocyanate-conjugated lectins. D-[l-^{3}H]-mannitol and [^{3}H]-ribitol were also incorporated into the walls of germinating lichen ascospores, but in

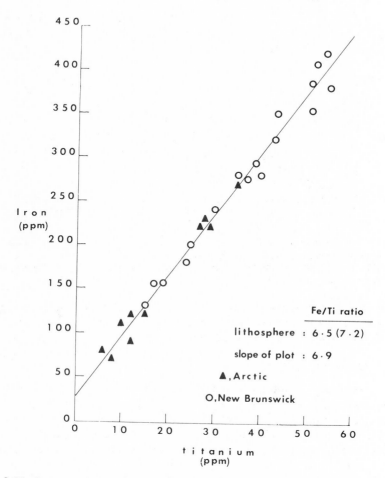

Fig. 5. The iron and titanium levels in *Cladonia* spp. from the Province of New Brunswick and the Northwest Territories, Canada. Modified from, and reproduced with permission of, Nieboer et al. (1978).

contrast to N-[³H]-acetyl-D-glucosamine, there was an almost complete lack of labeling in the hyphal septa. Mannitol and ribitol are, respectively, the main soluble carbohydrates of many lichen fungi and algae. This difference in incorporation of radioactive substrates suggests that in addition to chitin, other constituents are present in the cell wall material. Furthermore, it might be expected that mature hyphae in a differentiated thallus would be more complex than the very young hyphae studied by

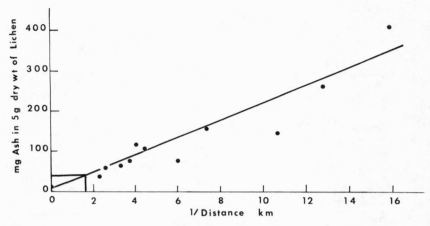

Fig. 6. The ash content of *Cladonia rangiferina* as a function of the reciprocal distance from a uranium mine ventilation shaft, Elliot Lake, Ontario. The lines close to the origin indicate the average ash content for samples collected in the general area but distant from obvious emission sources.

Galun and colleagues (1976). Indeed, the fungal hyphae in lichens frequently have very thick cell walls because of the secretion of a microfibrillar polysaccharide layer (see Hale, 1976).

Metal-Ion Uptake

In general, the uptake of metal ions by lichens is largely extracellular, passive, and rapid (Nieboer et al., 1978). The evidence to date, which is summarized below, indicates that metal ions are bound to carboxylic or hydroxycarboxylic acid functional groups, although the exact identification of the polymer on which these acidic groups are situated has not been achieved.

Metal ions are taken up from solution and bound reversibly on lichens through an ion-exchange process (Nieboer et al., 1976). The relative affinity order for the binding of different metal ions was Cu^{2+}, $Pb^{2+} > Zn^{2+} > Ni^{2+} > Mg^{2+} > Sr^{2+}$, which is diagnostic for the above-mentioned functional groups (see Nieboer et al., 1979). The pH measurements on hydrated lichen thalli covered with a thin film of water and titration curve studies (Tuominen, 1967) are consistent with this interpretation. Thus, Tuerk and colleagues (1974) observed thallus pH values of 4.2 ± 0.6 for members of the Parmeliaceae growing on tree bark. Higher values (5.9 ± 0.5) were only recorded for the nitrophilous species *Xanthoria parietina*. Tuominen (1967) and Nieboer and colleagues (1976) demonstrated the occurrence of functional groups of lichens suitable for metal ion binding with pK_a values between 2 and 6. Tuominen (1967) suggested that these functional groups were carboxylic acid moieties of pectin-like substances, and he also found a good linear relationship in 6 out of 12 lichens between the uronic acid content and the Sr^{2+} uptake capacity. The presence of uronic acids in lichens has yet to be proved conclusively (see Brown, 1976), although the fact that lichens readily stain reversibly with ruthenium red indicates at least that related kinds of molecules are present (Tuominen and Jaakkola, 1973).

The carboxylic acid groups in lichens to which metal ions bind reversibly are not those of oxalic acid, which is known to be secreted by lichen fungi (Syers and Iskandar, 1973), or those of lichen acids. Oxalic acid is a preferential trap for Ca^{2+} because of the relatively low solubility of calcium oxalate, and this is found abundantly in calcicolous lichens. Buck and Brown (personal communication) found that over 90% of the total calcium in samples of *Parmelia caperata* could not be displaced by water (hot or cold) or by $NiCl_2$ solution (an ion-exchange medium), but was released on heating in concentrated nitric acid. The latter digests calcium oxalate. Large fluctuations in total calcium and a high proportion in the acid-digestible fraction were also observed in *Usnea subfloridana* and *Ramalina siliquosa*. These workers believe that dryness of the habitat may be a factor in determining oxalate production by lichens. Hence, although oxalates are not involved in cation exchange in lichens, they may provide metabolically controlled sinks for metal ions especially in situations where particular metal ions are in great excess. Lawrey (1977) considers the possibility that lichen acids may be involved in cation binding but concludes on the basis of his electron probe studies that they are unimportant. Further evidence for this is deduced from the relatively low solubility of lichen acids in water (Iskandar and Syers, 1971), and also from the fact that acetone-treated lichens retain their cation binding capacity (Brown, 1976); both lichen acids and oxalic acid would be removed by this organic solvent.

The exact location of carboxylic acid functional groups within lichens requires clarification. The effects of the uptake of Cu^{2+} ions by *Umbilicaria muhlenbergii* (see below) would suggest that binding occurs to both symbionts. Although carboxylic acid containing polymers have not been unequivocally demonstrated for the fungal partner (though Schmidt et al., 1934, concluded that native lichenin chains consisted of 47 glucose residues and one carboxylic acid unit), pectic acids have been identified in the walls of green algae (Dainty et al., 1960; Percival, 1966). The electron probe study of Lawrey (1977) on small sections of the alga and fungus of *Cladonia cristatella* showed a greater concentration of Fe in or around the lichen algae than on adjacent fungal hyphae. This iron would likely be present as particulates as well as in ionic form. Even assuming that it is all bound ionically, the total amount of Fe associated with the fungal partner could still be greater than that bound to the algae because the latter makes up only a small proportion of the lichen. As mentioned earlier, Collins and Farrar (1978) estimated that the alga in *Xanthoria parietina* only comprised 7% of the total volume. The binding sites on lichens with pK_a values near 3 were assumed by Tuominen (1967) to be pectic carboxylic acids. In *Cladonia alpestris* these sites accounted for only 16% of the total Sr^{2+} uptake capacity. Because the proportion of Sr^{2+} uptake accounted for by the pectic carboxylic acid binding sites was similar to the proportion by weight of alga to fungus in the thallus, Tuominen (1967) considered that all the pectic carboxylic acid groups could be located on the alga in *Cladonia alpestris*. However, in other lichens, e.g., *Parmelia saxatilis*, groups with pK_a values near 3 accounted for more than 80% of the Sr^{2+} binding capacity. Thus, considering the small proportion of algae in lichens, the pectic carboxylic acids in these species would, accordingly, be present on both symbionts. A simple calculation on the basis of a total cation binding capacity in *Umbilicaria muhlenbergii* of 20 μ moles/g supports the concept of the localization of the cation binding sites on both symbionts.

NATURAL VARIATIONS IN LICHEN MINERAL CONTENT

Lichen physiologists have come to realize that in spite of the slow growth of lichens, collected material does not always behave alike when brought into the laboratory. Farrar (1976) comments that it is clear that physiological characteristics are dependent not only on the conditions in which the thallus is placed but also on the previous environmental history of the thallus. The importance of this will be examined in relation to short-term wetting and drying cycles and long-term seasonal variation.

Wetting and Drying Cycles

In the previous sections it has been indicated that metals can be accumulated as particulates (inorganic dusts and metal-ligand precipitates) or in ionic form through binding to ion exchange sites on the lichen. We have also noted the poikilohydrous nature of lichens; that is, they take up and lose water rapidly so that, over a period, a thallus will be subjected to cycles of wetting and drying. The continued metabolic activity of a lichen also requires that essential elements are accumulated within the cells of the fungus and alga. (Active uptake of cations may not be necessary if the electrochemical gradient is favorable: see Higinbotham, 1973a, b; Mierle and Stokes, 1976.) When a lichen dries, membranes are affected, and on rewetting a period of leakage ensues during which cations, particularly K^+, as well as phosphate and soluable sugars are lost. Farrar and Smith (1976) observed a phase of rapid leakage that was complete in 3 minutes for the lichen *Hypogymnia physodes*. A slower leakage phase continues for an hour or more, and this could be affected by inhibitors in contrast to the rapid phase. Farrar and Smith (1976) suggest that one function of the high sugar alcohol concentration in lichens was to provide a pool of energy-rich molecules that could be metabolized to repair cell membranes and hence curtail the slow-leakage phase.

Under natural conditions raindrops progressively wet different parts of the thallus. Larson and Kershaw (1976) have also shown that the rate of drying is related to the morphology and degree of clumping of particular lichen species. As a result of the rapid recovery mechanisms possessed by lichens, the areas wetted first may be able to absorb essential cations and nutrients from areas wetted later, so reducing the over-all nutrient loss from a thallus during wetting cycles (Farrar and Smith, 1976). On the basis of what is known for other plants (Higinbotham, 1973a,b), the most metabolically active areas of a thallus would be expected to develop a particularly favorable electrochemical (negative) gradient for intracellular cation uptake during the resaturation-respiration period. Thus, essential cations would tend to move and accumulate in the growing lichen apices where they are most required.

Seasonal Variation

Kershaw (1977) remarks that few studies make allowance for the time of year at which experimental material was collected or ensure that storage or pretreatment temperatures correspond to those found in the field at collection time. Considerable variation in the K^+ and Ca^{2+} contents have been observed by Kovács-Láng and Verseghy (1974) for lichens growing in the Great Hungarian Plain. Generally the highest values were in autumn and winter, although there was considerable variation within each season. Spring and summer levels were on an average 60% (Ca^{2+}) and 30% (K^+) lower. In *Umbilicaria muhlenbergii* seasonal variability has also been observed with low Ca levels of 165 $\mu g/g$ in November and high values of 760 $\mu g/g$ in early summer (Nieboer et al., 1979). In northern Ontario we find that this species can be collected and stored air-dry at ambient temperature in the summer without apparent harm, and in winter, thalli can be stored at $-10°C$ for 2–3 months. In spring or autumn either treatment is detrimental, and there is an indication that the lichen is less tolerant to stress situations when the mineral content is low following a rainy period. In a recent study we have shown that the reduction in ^{14}C fixation induced by 20–50 ppm aqueous SO_2 was 35% greater for samples collected in spring with a low mineral content (K^+, 1630; Ca^{2+}, 290; Mg^{2+}, 365 $\mu g/g$)than those collected a month later with a much higher content (K^+, 2,440; Ca^{2+}, 530; Mg^{2+}, 470 $\mu g/g$) (Richardson et al., 1979).

Kershaw (1977) was able to demonstrate acclimation of net photosynthesis to prevailing temperatures during his researches on seasonal variation in *Peltigera canina*. During late spring and early autumn, acclimation was restricted in direction and extent. It would be interesting to study whether the ability to acclimate in this species was in any way related to the mineral status of the lichen.

PHYSIOLOGICAL RESPONSE OF LICHENS TO METAL-ION BINDING

Response of Umbilicaria muhlenbergii to
Small Amounts of Various Metals

The effects of the uptake of known amounts (<20 μmoles/g) of a range of metal ions have been investigated by Nieboer and colleagues (1979) and Richardson and colleagues (1979). The effects on each symbiont were monitored using ^{14}C fixation and K^+ release as parameters. In addition, protection or exacerbation of SO_2-induced damage by the metal ion was also investigated. Significant but small K^+ losses ($\leqslant 10\%$ of the total K^+ content) were considered to be of both algal and fungal origin. However substantial K^+ release ($>10\%$) was accepted as a mark of damage to the fungal partner because the largest portion of the lichen is fungal and because, as discussed elsewhere (Nieboer et al., 1979), K^+ is largely located intracellularly. A change in the amount of photosynthetic ^{14}C fixation signified an effect on the algal partner, whereas a change in the ^{14}C leakage or proportion of photosynthate in ^{14}C-ribitol were more subtle effects on the alga.

The results of these studies are summarized in table 1. The metal ions used in these experiments can be classified as class A (having a preference for O-donors) or borderline (donor preference O$\approx$N$\approx$S). Puckett (1976) has shown that the uptake of class B metals such as Ag^+ and Hg^{2+} (which prefer S to N or O donors) severely inhibits the ability of lichen samples to photosynthesize and induce severe K^+ losses. (See Nieboer et al. [1979] and Nieboer and Richardson [1980] for a detailed discussion of metal-ion classification.)

The uptake of class A metal ions, Sr^{2+}, Ca^{2+}, and Mg^{2+}, had no

TABLE 1

SUMMARY OF THE INFLUENCE OF THE BINDING OF VARIOUS METAL IONS
(ALONE AND IN COMBINATION WITH A SUBSEQUENT SO_2 EXPOSURE) ON PHOTOSYNTHETIC ^{14}C
FIXATION AND K^+ RELEASE IN UMBILICARIA MUHLENBERGII

ION	CLASS	METAL ALONE		METAL + SO_2		LEAKAGE	CHANGE IN ^{14}C PRODUCTS
		K^+	^{14}C	*K^+	^{14}C		
Sr^{2+}	A	None	None	Protect[1]	Protect[1]	None	
Ca^{2+}	A	None	None	None	Protect[1]	None	More ribitol[1]
Mg^{2+}	A	None	None	None	Slight Decrease	None	
Zn^{2+}	Borderline	None	None	None	Slight Decrease	None	
Ni^{2+}	Borderline	None	None	None-protect	None-protect	Increase[2]	Less Ribitol[2]
Pb^{2+}	Borderline	Increase[2]	None	Increase[2]	None-protect	Increase[2]	Less ribitol[2]
Cu^{2+}	Borderline	Increase[2]	Decrease[2]	Increase[2]	None	Increase[2]	Less ribitol[2]
HSO_3		Increase[2]	Decrease[2]			Increase[2]	Less ribitol[2]

[1] Beneficial

[2] Toxic

* The potassium leakage due to Pb^{2+} and Cu^{2+} alone was 10% of the total while the increase induced by SO_2 · SO_2 + Cu^{2+}, and SO_2 + Pb^{2+} exceeded this amount.

Blanks indicate that parameter was not applicable or not measured.

SOURCE: Based on data of Nieboer et al. (1979) and Richardson et al. (1979).

detrimental effect on either partner as measured by the parameters of K^+ release and ^{14}C fixation. In addition, the binding of Sr^{2+} and Ca^{2+} provided some protection against the damaging effects of subsequent SO_2 exposure. In contrast, borderline metals tended to increase K^+ release and reduce the amount of ^{14}C fixation. Borderline metals can exhibit class A or class B properties, depending on conditions. In general, the uptake of those borderline metals with propensities to exhibit class B properties (Cu^{2+}, Pb^{2+}) resulted in marked physiological responses in lichen samples exposed to the metal alone and in those also given an SO_2 exposure. The uptake of Ni^{2+}, in contrast, induced less severe but subtle effects on the lichen samples.

Response of Umbilicaria muhlenbergii to the Uptake of Varying Amounts of Cu^{2+} Ion

Nieboer and colleagues (1976) identified two binding sites for Ni^{2+} on the lichen *Umbilicaria muhlenbergii* with capacities of 12–13 μmoles/g and 7–8 μmoles/g, giving a total uptake capacity under normal conditions of 20 μmoles/g. Recently, the maximum capacity for Ca^{2+} has also been shown to be 20 μ moles/g, and both binding sites also appear to be important for Cu^{2+} (Nieboer et al., 1979). The work of Puckett (1976) with this lichen indicates that a third binding site becomes available when very high external concentrations of Cu^{2+} are employed.

In the studies of Nieboer and colleagues (1979), the occupation of the first binding site by Cu^{2+} was accompanied by a decrease in ^{14}C fixation and small losses in K^+. It was therefore concluded that the functional groups associated with this binding site must in part be located on, near, or within the algal cell. However, the magnitude of the capacity of this binding site when compared with the small volume of the alga suggested binding at other locations as well. The occupation of the second binding site was associated with very significant K^+ losses (40% of the total K^+ content of the lichen). This loss implies binding on or near or within the fungal cells of the lichen. The occupation of the third binding site following incubation in solutions of very high Cu^{2+} concentrations (2M $CuCl_2$ for 3h) was associated with the loss of nearly all the K^+ (96%) from the lichen (Puckett, 1976). This large depletion suggests that intracellular binding of Cu^{2+} and complete loss of membrane integrity had occurred.

Quite high levels of free amino acids have been found in lichens; for example, up to 77 μmoles/g in *Pseudoevernia furfuracea* (Jager and Weigel, 1978). Cu^{2+} has a high affinity for these and other components such as nucleotides and proteins. Thus, it is not surprising that Cu^{2+} and other ions with significant class B character are taken up and become localized intracellularly.

ECOLOGICAL REPERCUSSION OF METAL ACCUMULATION

Accumulation of Class A Metals

Lichens growing on basic substrata such as limestone or asbestos-cement, when compared with lichens colonizing more acidic substrates, occur closer to urban centers where the ambient SO_2 levels are high (Brightman and Seaward, 1977). We have suggested previously, that this may be related to the capacity of the substrate to buffer SO_2 so that it is in the form of sulphite rather than the more toxic bisulphite ion (Puckett et al., 1973b). However, the results summarized in table 1 suggest that binding of class A metals such as Ca^{2+} and Sr^{2+} to sites on the lichen can help provide the lichen with protection against the effects of SO_2 exposure. This binding probably increases the stability of the cell membrane so that substances do not leak out of the lichen so readily when it is under stress. This role of Ca^{2+} in stabilizing membranes is known in both animals and plants (Salisbury and Ross, 1969; Ochiai, 1977; and especially Schroeder, 1965). Thus, if Ca^{2+} ions are readily available, cell membranes may be able to reestablish integrity faster following rewetting or a period of SO_2 exposure. Indeed, the observation that the distribution of particular lichen species often correlates best with the winter mean SO_2 level in the British Isles (Hawksworth and Rose, 1976) may be related to the fact that Ca contents tend to be lower during the rainy season.

The protective effect of Ca^{2+} in reducing the toxicity of other pollutants has been noted by Mierle and Stokes (1976), who established that the presence of Ca^{2+} in Cu^{2+} exposures reduced the K^+ loss by the green alga *Scenedesmus*, and Harding and Whitton (1977) found that it reduced the Zn^{2+} toxicity in *Stigeoclonium*. Indeed, though high levels of fluoride are usually toxic to lichens (see LeBlanc et al., 1971), *Cladonia alpestris* exhibited no symptoms of damage when transplanted to the vicinity of a fertilizer plant that emitted class A metals (K and Ca) as well as fluorides (Kauppi, 1976). All these examples suggest that class A metals have a protective effect. The most abundant naturally occurring example is Ca^{2+}.

Accumulation of Borderline and Class B Metal Ions

The experimental results summarized in table 1, as well as ecological observations, suggest that the uptake of borderline ions can affect the physiology of lichens and become harmful when certain threshold concentrations are exceeded. Damage in the field has been observed when the metal content is greater than the threshold values. These have been estimated by Rao and colleagues (1977) to be 3–5 μmoles/g for Pb and Cu, about 3 μmoles/g for Zn, and 0.05 μmoles/g for Cd in the lichen

Hypogymnia physodes. Nash (1975) found higher field threshold values for Zn in some lichens, e.g., 8 μmoles/g in *Parmelia taractica* and 39 μmoles/g in *Lasallia papulosa* (closely related to *Lasallia pustulata*, a known Zn accumulator; see Tuominen and Jaakkola, 1973, p. 192). However laboratory experiments revealed that the uptake of more than 3 μmoles/g of Zn^{2+} was harmful even for *Lasallia papulosa*. In the field much of the metal may have been in particulate form and accumulated slowly over many years. Thus, in field material the Zn^{2+} ions available for binding to sites on or within the lichen symbionts may have been insufficient to exceed the threshold value for damage. If we extrapolate from studies on *Umbilicaria muhlenbergii*, the continuous occupation of a significant number of sites at the first binding site can result in changes over a period of time that will upset the symbiotic association in a lichen and hence affect its distribution in the field. These laboratory experiments also suggest that simultaneous exposure to SO_2 could reduce metal threshold levels, particularly for borderline metals with significant class B character. Thus, around sources that emit Cd^{2+}, Cu^{2+}, or Pb^{2+}, lichens would be expected to be more than usually sensitive to SO_2. No data are available on the effects of class B ions on SO_2-induced damage. However, it is probably that, being themselves very toxic (Puckett, 1976), in combination with SO_2 they would be extremely harmful.

Finally, lichen physiologists have still to determine how the lichen fungus induces the alga to release the soluble mobile carbohydrate on which it depends for growth. It is possible that the fungus produces an extracellular enzyme or repressor that results in the accumulation and release of the mobile carbohydrate by the alga (but see Smith, 1975). In lichens containing blue-green algae, there is evidence that the fungus produces a substance that represses glutamine synthetase so that ammonia accumulates in the alga, is released, and subsequently is taken up by the fungus (Stewart, 1977). A further control mechanism has been demonstrated by Feige (1977) for *Peltigera aphthosa*. Here the fungal partner apparently produces a substance that inhibits the synthesis of the sulphate carrier protein in the lichen alga. As a result, sulphate uptake by algal cells within the thallus or by freshly isolated cells occurs at very low rates as compared with uptake by the cultured symbiotic alga. Recent studies by Wainwright and Woolhouse (1975, 1978) have shown that the borderline metals Zn^{2+} and Cu^{2+} can inhibit phosphatases located in the cell walls of the grass *Agrostis tenuis*. Class B ions such as Cu^+, Ag^+, Hg^{2+}, and Tl^{3+} would be expected to be even more inhibitory than borderline ions (Ochiai, 1977). Lichen enzymes are only now receiving deserved attention, but the production of extracellular enzymes by fungi is well known. It is evident that lichens that produce phosphatases or ureases (found particularly in species colonizing bird perching sites) could be adversely affected by exposure to toxic metal ions (Shapiro, 1975; Schmid and Kreeb, 1975).

Successful symbiosis in all lichens depends on the continued functioning of the control mechanisms mentioned above. Thus, the presence of low levels of toxic metal ions could, in the long term, also interfere significantly with these mechanisms. The ability of a few lichens to grow in environments enriched with high levels of potentially toxic metals is very interesting. For example, *Acarospora anomala*, *Lecanora vinetorum*, *Sarcogyne simplex*, and *Stereocaulon nanodes* have been recorded on vine supports in Austria that are regularly drenched with copper-containing fungicidal sprays (see Gilbert, 1977). This suggests that the threshold levels are species-dependent and that the factors which determine the damage threshold in particular lichens would be worthy of investigation.

CONCLUDING REMARKS

Recent research indicates that chemical interactions necessary for the continued functioning of the intact lichen are complex. An aim of our presentation has been to draw attention to the possible roles of the surface binding of both toxic and essential ions, as well as of the total mineral content, in the alga-fungus partnership under natural conditions. It is hoped that the interpretations and ideas presented in this paper will stimulate further experimental work and lead to an improved understanding of the various interactions in the lichen symbiotic association.

ACKNOWLEDGMENTS

The authors thank Ms. Pat Lavoie, Ms. Dilva Padovan, Mr. Luc Boileau, and Mr. Frank Tomassini for technical assistance. This research was supported by a team grant from the National Research Council of Canada.

REFERENCES

Blum, O. B. 1973. Water relations. *In* V. Ahmadjian and M. E. Hale (eds.), The lichens, pp. 381–400. Academic Press, New York.
Brightman, F. H., and M. R. D. Seaward. 1977. Lichens of man-made substrates. *In* M. R. D. Seaward (ed.), Lichen ecology, pp. 253–94. Academic Press, London.

Brown, D. H. 1976. Mineral uptake by lichens. *In* D. H. Brown, D. L. Hawksworth, and R. H. Bailey (eds.), Lichenology: progress and problems, pp. 419–40. Systematics Association special volume no. 8, Academic Press, London.

Collins, C. R., and J. F. Farrar. 1978. Structural resistance to mass transfer in the lichen *Xanthoria parietina.* New Phytol. 81:71–83.

Dainty, J., A. B. Hope, and C. Denby. 1960. Ionic relations of cells of *Chara australis.* II. Indiffusible anions of the cell wall. Aust. J. Biol. Sci. 13:267–76.

Farrar, J. F. 1976. Ecological physiology of the lichen *Hypogymnia physodes.* I. Some effects of constant water saturation. New Phytol. 77:93–103.

Farrar, J. F., and D. C. Smith. 1976. Ecological physiology of the lichen *Hypogymnia physodes.* III. The importance of the rewetting phase. New Phytol. 77:115–25.

Feige, G. B. 1977. Untersuchungen über die Aufnahme von Sulfat und Phoshat durch isolierte Phycobioten der Flechte *Peltigera aphthosa* (L.) Willd. Z. Pflanzenphysiol. 82:347–54.

Galun, M., A. Braun, A. Frensdorff, and E. Galun. 1976. Hyphal walls of isolated lichen fungi. Arch. Microbiol. 108:9–16.

Garty, J., M. Galun, C. Fuchs, and N. Zisapel. 1977. Heavy metals in the lichen *Caloplaca aurantia* from urban, suburban, and rural regions of Israel (a comparative study). Water, Air, Soil, Pollut. 8:171–88.

Gilbert, O. L. 1977. Lichen conservation in Britain. *In* M. R. D. Seaward (ed.), Lichen ecology, pp. 415–36. Academic Press, London.

Gough, L. P., and L. A. Erdman. 1977. Influence of a coal-fired power plant on the element content of *Parmelia chlorochroa.* Bryologist 80:492–501.

Hale, M. E. 1973. Fine structure of the cortex in the lichen family Parmeliaceae viewed with the scanning-electron microscope. Smithson. Contrib. Bot. 10:1–92.

Hale, M. E. 1976. Lichen structure viewed with the scanning electron microscope. *In* D. H. Brown, D. L. Hawksworth, and R. H. Bailey (eds.), Lichenology: progress and problems, pp. 1–16. Systematics Association Special Volume No. 8. Academic Press, London.

Harding, J. P. C., and B. A. Whitton. 1977. Environmental factors reducing the toxicity of zinc to *Stigeoclonium tenue.* Br. Phycol. J. 12:17–21.

Hawksworth, D. L., and F. Rose. 1976. Lichens as pollution monitors. Studies in Biology No. 66. Edward Arnold, London. 60 pp.

Higinbotham, N. 1973a. The mineral absorption process in plants. Bot. Rev. 39:15–69.

Higinbotham, N. 1973b. Electropotentials of plant cells. Annu. Rev. Plant Physiol. 24:25–46.

Iskandar, I. K., and J. K. Syers. 1971. Solubility of lichen compounds in water: pedogenetic implications. Lichenologist 5:45–50.

Jager, H. -J., and H. -J. Weigel. 1978. Amino acid metabolism in lichens. Bryologist 81:107–13.

Kauppi, M. 1976. Fruticose lichen transplant technique for air pollution experiments. Flora (Jena) 165:407–14.

Kershaw, K. A. 1977. Physiological-environmental interactions in lichens. III. The rate of net photosynthetic acclimation in *Peltigera canina* (L.) Willd. var. *Praetextata* (Floerke in Somm.) Hue, and *P. polydactyla* (Neck.) Hoffin. New Phytol. 79:391–402.

Kovács-Láng, E., and K. Verseghy. 1974. Seasonal changes in the K and Ca contents of terricolous xerophyton lichen species and their soils. Acta Biol. Acad. Sci. Hung. 23:325–33.

Larson, D. W., and K. A. Kershaw. 1976. Studies on lichen-dominated systems. XVIII. Morphological control of evaporation in lichens. Can. J. Bot. 54:2061–73.

Lawrey, J. D. 1977. X-ray emission microanalysis of *Cladonia cristatella* from a coal strip-mining area in Ohio. Mycologia 69:855–60.

LeBlanc, F., G. Comeau, and D. N. Rao. 1971. Fluoride injury symptoms in epiphytic lichens and mosses. Can. J. Bot. 49:1691–98.

Mierle, G., and P. M. Stokes. 1976. Heavy metal tolerance and metal accumulation by planktonic algae. *In* D. D. Hemphill (ed.), Trace substances in environmental Health-X, pp. 113–22. University of Missouri Press, Columbia.

Mittal, O. P., and T. R. Seshadri. 1954. Chemistry of lichenin and isolichenin. J. Sci. Industr. Res. (India) 13A:174–77.

Mosbach, K. 1973. Biosynthesis of lichen substances. *In* V. Ahmadjian and M. E. Hale (eds.), The lichens, pp. 523–46. Academic Press, New York.

Nash, T. H. 1975. Influence of effluents from a zinc factory on lichens. Ecol. Monogr. 45:183–98.

Nieboer, E., K. J. Puckett, and B. Grace. 1976. The uptake of nickel by *Umbilicaria muhlenbergii*: a physicochemical process. Can. J. Bot. 54:724–33.

Nieboer, E., D. H. S. Richardson, and F. D. Tomassini. 1978. Mineral uptake and release by lichens: an overview. Bryologist, 81:226–46.

Nieboer, E., D. H. S. Richardson, P. Lavoie, and D. Padovan. 1979. The role of metal-ion binding in modifying the toxic effects of sulphur dioxide on the lichen *Umbilicaria muhlenbergii*. I. Potassium efflux studies. New Phytol. 82:621–32.

Nieboer, E., and D. H. S. Richardson. 1980. The replacement of the nondescript term "heavy metals" by a biologically and chemically significant classification of metal ions. Environ. Pollut., Series B (in press).

Ochiai, E. I. 1977. Bioinorganic chemistry: an introduction. Allyn and Bacon, Boston.

Percival, E. 1966. The natural distribution of plant polysaccharides. *In* T. Swain (ed.), Comparative Phytochemistry, pp. 139–58. Academic Press, London.

Puckett, K. J., E. Nieboer, M. J. Gorzynski, and D. H. S. Richardson. 1973a. The uptake of metal ions by lichens a modified ion-exchange process. New Phytol. 72:329–42.

Puckett, K. J., E. Nieboer, W. P. Flora, and D. H. S. Richardson. 1973b. Sulphur dioxide: its effect on photosynthetic[14]C fixation in lichens and suggested mechanisms of phytotoxicity. New Phytol. 72:141–54.

Puckett, K. J. 1976. The effect of heavy metals on some aspects of lichen physiology. Can. J. Bot. 54:2695–2703.

Rao, D. N., G. Robitaille, and F. LeBlanc. 1977. Influence of heavy metal pollution on lichens and bryophytes. J. Hattori Bot. Lab. 42:213–39.

Richardson, D. H. S., E. Nieboer, P. Lavoie, and D. Padovan. 1979. The role of metal-ion binding in modifying the toxic effects of sulphur dioxide on the lichen *Umbilicaria muhlenbergii*. II. [14]C fixation studies. New Phytol. 82:633–43.

Salisbury, F. B., and C. Ross. 1969. Plant physiology, pp. 153–58. Wadsworth Publishing, Belmont, California.

Schmid, M. L., and K. Kreeb. 1975. Enzymatische Indikation gasgeschädigter Flechten. Angew Bot. 49:141–54.

Schmidt, E., R. Schnegg, and E. Wurzner. 1934. The chain length of lichenin of native composition. Naturwissenschaften 22:172.

Schroeder, H. A. 1965. The biological trace elements or peripatetics through the periodic table. J. Chronic. Dis. 18:217–28.

Shapiro, I. A. 1975. Determination of activity of urease in lichens by direct potentiometric method. Sov. Pl. Physiol. 23:844–46.

Smith, D. C. 1975. Symbiosis and the biology of Lichenised fungi. Symp. Soc. Expl. Biol. 29:373–405.

Stewart, W. D. P. 1977. A botanical ramble among the blue-green algae. Br. Phycol. J. 12:89–115.

Syers, J., and I. K. Iskandar. 1973. Pedogenic significance of lichens. *In* V. Ahmadjian and M. E. Hale (eds.), The lichens, pp. 225–48. Academic Press, New York.

Tuerk, R., V. Wirth, and O. L. Lange. 1974. Carbon dioxide exchange measurements for determination of sulphur dioxide resistance in lichens. Canadian Translation Bureau, Translation No. 366007; Oecologia (Berl.) 15:33–64.

Tuominen, Y., and Jaakkola, T. 1973. Absorption and accumulation of mineral elements and Bot. Fenn. 4:1–28.

Tuominen, Y., and Jaakola, T. 1973. Absorption and accumulation of mineral elements and radioactive nuclides. *In* V. Ahmadjian and M. E. Hale (eds.), The lichens, pp. 185–224. Academic Press, New York.

Wainwright, S. J., and H. W. Woolhouse. 1975. Physiological mechanisms of heavy metal tolerance in plants. *In* M. J. Chadwick and G. T. Goodman (eds.), The ecology of resource degradation and renewal, pp. 231–57. Br. Ecol. Soc. Symp. No. 15, Blackwell Scientific Publications, Oxford.

Wainwright, S. J., and H. W. Woolhouse. 1978. Inhibition by zinc of cell wall acid phosphatases from roots in zinc-tolerant and non-tolerant clones of *Argostis tenuis* Sibth. J. Exp. Bot. 29:525–31.

RICHARD D. LUMSDEN AND WILLIAM A. MURPHY

Morphological and Functional Aspects of the Cestode Surface

5

INTRODUCTION

Textbooks generally ascribe to the vertebrate intestine two principal functions: digestion and absorption of nutrients. Viewed in the broad context of Nature, however, there is a third important function, that being to provide a habitat for a large and diverse number of other organisms. The intestine as a place to live was monographically examined by Dr. Clark Read nearly thirty years ago (Read, 1950), and has since been readdressed several times by him and other investigators (e.g., Read, 1955, 1956, 1971; Read et al., 1963; Mettrick, 1970, 1971; Befus and Podesta, 1976). We will not dwell here on specifics of this particular topic, except to note that all workers agree that the intestine is a unique habitat as compared with those in which the majority of animals live. Clearly, the animals that utilize the intestine as such must be equally unique, as compared with their counterparts living in aquatic, marine, or terrestrial environments.

Many parasites of the vertebrate alimentary canal undoubtedly evolved from free-living organisms, or evolved from organisms that were previously parasitic in invertebrates. These organisms, after being accidentally or purposely ingested by their prospective vertebrate hosts, were able to survive the environment in which they found themselves, and were able to further adapt themselves to it. Representatives of some of the groups that were initially gut parasites of vertebrates have in the course of evolution extended their operations beyond the confines of the intestine, e.g., to the liver, lungs, and so forth, as we see in the case of certain nematodes and trematodes. Among the various metazoans that qualify as intestinal parasites, none surpass the tapeworms, as a group, in their dedication to the intestinal mode of existence as adults. With very few exceptions adult tapeworms live nowhere but the vertebrate intestinal tract or one of its major adjuncts. The only other group exhibiting this degree of environmental single-mindedness is the Acanthocephala.

Tapeworms, being denizens in the bowels of other animals, are regarded by lay people as among the most loathsome of creatures. On the other hand, biologists, some of whom may also find tapeworms loathsome, consider the gut itself to be a highly sophisticated organ, exemplarily specialized as it is for the acquisition of the nutrients that fuel the life processes of the entire organism. Tapeworms, we believe, have developed a comparable degree of sophistication, at least insofar as their structure and physiology for nutrient acquisition are concerned. Indeed, the degree to which certain aspects of tapeworm morphology and function mimic the host intestinal mucosa is truly remarkable. This has significance not only for parasitologists, who so often couch definitions of parasitism in nutritional terms, but also for transport physiologists and biologists in general. As remarked by Pappas and Read (1975), tapeworms possess inherent advantages over many other systems as a model for studying the unidirectional influx of solutes across biological membranes into intact organisms.

It is ironic that what has come to be regarded as the epitome of the intestinal metazoan parasite is denied a gut of its own; tapeworms have not even a vestige of an alimentary tract. Thus, absorption of nutrients is the province solely of the body surface. The over-all anatomical, histological, and fine structural organization of adult tapeworms has been reviewed recently by Lumsden and Specian (1979). Although unorthodox, this organization has proved to be a biologically highly successful one.

Tapeworms have what amounts to an insatiable appetite for carbohydrates, amino acids, purines, pyrimidines, and so on. Faced with the need to sustain a prodigous growth rate and reproductive capacity (see, e.g., Stunkard, 1962), adult tapeworms are at the same time severely limited in their capacity to utilize substances other than glucose for energy metabolism (Read, 1959, 1967; Jacobsen and Fairbairn, 1967; Saz, 1972). Due to the very limited capacity of these worms to carry out transamination (Wertheim et al., 1960), they utilize synthesis to only a minor extent in satisfying their amino acid requirements. Moreover, they are equally limited in their capacity for *de novo* synthesis of the carbon skeletons used as precursors to their lipids and nucleic acids (see von Brand, 1973, for review). Glucose, and virtually all the common amino acids, fatty acids, and nucleosides are dietarily essential for tapeworms. In their demands for these and other vital susbtances, tapeworms have a formidable competitor. The host's mucosal epithelium is highly specialized for, and most efficient in, removing from the gut lumen these same materials. Although the ability of tapeworms to compete successfully with their hosts is frequently exaggerated (see, e.g., fig. 20–1 in Schmidt and Roberts, 1977, and their discussion thereof), tapeworms are well suited to the task of fulfilling their nutritional requirements in the environment in which they find themselves.

Elucidating the nature of a given symbiosis is fundamentally a problem in ecology, but one that often requires detailed analysis of ultrastructure, physiology, biochemistry, and so on. This is especially true of endoparasite-host relationships. The present review addresses one facet of a highly complex host-parasite relationship—the surface at which chemical interchange between a parasite, in this case a tapeworm, and its microenvironment occurs. In order to put the tapeworm surface in proper perspective, it is appropriate to first consider some general features of absorptive (transport) epithelia.

ABSORPTIVE SURFACES

Tissues dedicated to the function of absorption can be so identified by a number of morphological, physiological, and biochemical hallmarks. Among those morphological are structures that amplify the area of the surface exposed to the surrounding milieu. For example, where the vertebrate intestinal mucosa is concerned (fig. 1), the chief cell type is of the columnar variety, bearing at the lumenal pole a distinctive brush border of microvilli. These digitiform structures, measuring on the order of 1.5 μm in length and some 80 nm in diameter, amplify the surface exposed to the lumen by as much as thirtyfold. The so-amplified plasma membrane is further adorned with a calyx of mucopolysaccharides and glycoproteins, whose physicochemical properties and distribution enhance the *ad*sorptive (or binding) characteristics of the lumenal cell surface. Elsewhere, the plasma membrane may be amplified by infoldings or plications. For details concerning these and other features of the cytoarchitecture of the mucosal epithelium, see Bloom and Fawcett (1975).

Within the membrane proper of the brush border are the molecular portals through which sugars, amino acids, and so on, pass from the gut lumen into the mucosal cell cytoplasm. To a significant extent these are not passive channels, but their subunits, in the fashion of enzymes, catalyze the cross-membrane transport functions, perhaps according to the model suggested by Singer (1974). This model (fig. 2) calls for gates formed by amphipathic proteins spanning the lipid bilayer. In the "unloaded" state

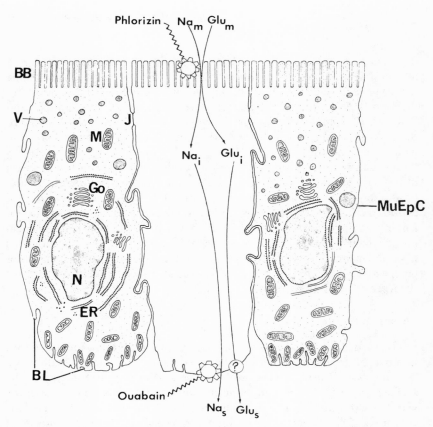

Fig. 1. Cytoarchitecture of the mammalian intestinal mucosal epithelium. Fine structural features of the absorptive mucosal epithelial cells (MuEpC) include a lumenal surface brush border of microvilli (BB), subjacent vesicles (V), mitochondria (M), intercellular junctional complexes (J), Golgi (Go), endoplasmic reticulum (ER), a basally located nucleus (N), and a somewhat infolded basolateral membrane (BL). Organelles have been omitted from the cell in the center of the drawing, to present schematically a model (after Schultz and Zalusky, 1964) for sodium-dependent transport in this system. Phlorizin-sensitive absorption of glucose from the medium (Glu_m) is accumulative and coupled to the uptake of ambient sodium (Na_m). Sodium uptake is, in effect, downhill, due to a low internal concentration of sodium (Na_i) maintained by active (ouabain-sensitive) extrusion of sodium at the "serosal" surface (Na_s). Transcellular diffusion of glucose, toward the serosal surface is downhill (Glu_s). Passage of glucose across the basolateral membrane is accomplished by a yet unknown (?) mechanism— i.e., it might be carrier-facilliated, or by passive diffusion, driven by an osmotic gradient, etc.—but at this site is not coupled directly to the sodium efflux pump.

OUTSIDE

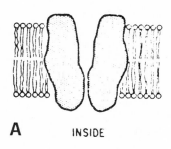

A   INSIDE

OUTSIDE   OUTSIDE

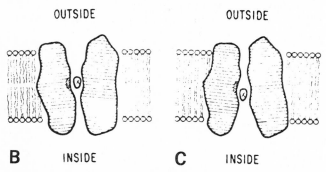

B   INSIDE   **C**   INSIDE

Fig. 2. A model for a transport carrier, according to Singer (1974). Integral proteins, spanning the lipid bilayer, form gates, which undergo alternations in configuration and relative position to one another with the binding of the solute. *A*: "unloaded carrier"; the pore is effectively closed. *B*: when solute attaches to specific binding site(s) on "carrier" protein(s), the pore opens. *C*: subsequent dissociation of the solute from the carrier results in its translocation to the inside compartment. Diagrams fron Singer (1974), with the permission of Annual Reviews, Inc.

their conformation is such that the gap between them is effectively closed. With the binding of solute, however, the conformation of these membrane proteins is altered, opening a channel through which the solute moves to the cytoplasm. Not infrequently, transport of an organic solute—e.g., hexose sugar, amino acid, and so on—is enhanced by the prior attachment and subsequent cotransport of an inorganic cation, especially sodium (Crane, 1965). At this locus the effect of the inorganic species may be to increase the affinity of the "gate" proteins for binding the organic solute (i.e., affect K_t) or otherwise accelerate the nonelectrolyte's traversal of the membrane (i.e., increase V_{max}). Under such circumstances influx of the organic solute species may occur against sizable concentration differences existing between the lumenal and cytoplasmic compartments. This situation creates a thermodynamically feasible movement of the organic solute from the lumenal compartment, across the cytoplasmic compartment, to the vascular compartment (i.e., collecting blood vessels of the lamina propria underlying the mucosal epithelium). One theory, formulated by Dr. Robert Crane (1960) and subsequently extended by him and others (see, e.g., Schultz and Curran, 1970; Schultz et al., 1974; Crane, 1977), is that sodium-dependent organic molecule transport is driven, without the intervention of covalent intermediates, by an electrochemical gradient established when sodium, entering concomitantly with the organic solute, is selectively extruded at a site separate from the one of influx. It is a misinterpretation, apparently, by some authors (e.g., Podesta et al., 1977) that adherents to this hypothesis consider the sodium gradient the *only* source of energy driving nonelectrolytes, through sodium-coupled accumulative transport, across epithelial tissues. Since the outpumping of sodium, which establishes the gradient, is accomplished at the expense of ATP, "direct" metabolic functions by which ATP is generated (e.g., Mitchell, 1961) are obviously important to the over-all process as well. This, clearly, is not at odds with the findings of Crane et al. (1961), which dissociated the sodium pump *per se* (efflux) from the membrane transport carrier for the nonelectrolyte absorbates.

Whether all features of the Gradient hypothesis prove correct, at least to the satisfaction of all its critics (see reviews of Mitchell [1973] and Crane [1977] for discussion), remains to be seen. For the systems under review here, however, we believe the supportive "hard" data on hand outweigh the so far largely hypothetical disadvantages of the Gradient theory.

One characteristic of the sodium ATPase pump is its selective inhibition by the glycoside ouabain (Glynn, 1964; Hoffman, 1973). In some systems ouabain may also inhibit some "direct" sources of chemical energy generation (see, e.g., Newey et al., 1968); this strikes us as a less than compelling argument for discarding the Gradient hypothesis (Podesta et al., 1977). In any event, important to the establishment of a transepithelial gradient is the nonuniform distribution of the enzyme—i.e., in the intestinal mucosa, ouabain-sensitive ATPase is found in the basolateral regions of the plasma membrane, but is essentially absent in the brush border (Stirling, 1972; Fujita and Nakao, 1973). In this way a lumenal surface to ablumenal surface electrochemical gradient that can drive the transport of nonelectrolyte absorbates across the mucosal epithelium is feasible. Placing the sodium efflux pump close to (i.e., in the same region of the membrane as) the entry site of the nonelectrolyte had theoretical advantages at the time this disposition was proposed (Crane, 1962), but has not proved to be the case.

In transport epithelia the lumenal (apical) surface faces an environment different from that at the ablumenal (basal) surface. This imposes a geometrical, or spatial, asymmetry on the component cells of such tissues, reflected by differences in morphology, biochemistry, and physiology of the membrane where it interfaces with these different environments. These features should be kept in mind when mechanisms for transport by epithelia are compared with absorption by nonepithelial, or isolated, cells. We are reminded that in erythrocytes, for example, where sites for sugar entry and sodium extrusion are equidistant from one another and from the intracellular compartment, the uptake of sugar is nonaccumulative.

Features of the apical and basal surfaces of transport epithelia are appropriately emphasized by many investigators, though the lateral associations of the cytological elements comprising such tissues merit equal attention. In discretely cellular epithelial tissues, the adjacent cells cohere tightly to one another, a function of a variety of junctions. At the *zonula/macula adherens*, the apposed membranes are separated by a space some 15 to 20 nm wide, but forces generated at such junctions are sufficient to provide relatively strong attachment of the cells. In transport epithelia, e.g., the intestinal mucosa, there are, in addition to desmosomes, sites along apposing lateral cell surfaces where the membranes apparently fuse—the *zonula occludens* or so-called tight junctions. While also providing for mechanical cell-to-cell adhesion, the obliteration of significant intercellular space at these sites impedes passive leakage of substances between cells, such that the physiological mediation of solute traversal of the epithelium is not unduly compromised. However, inorganic ions may be shunted back intercellularly to the lumen through the tight junctions. At the *nexus*, or "gap" junction, the intercellular space is not totally obliterated, but is reduced to some 2–4 nm in width. Although this is again a site of relatively firm cohesion between cells, it is also believed (see, e.g., Lowenstein, 1972) that the nexus mediates the flow of ionic currents between cells; i.e., at a nexus junction, cells are electrically coupled, providing coordination of function. Well-studied examples are cardiac and smooth muscle of vertebrates (see Bloom and Fawcett, 1975).

Only rarely among the various kinds of epithelia developed in animals is syncytial organization encountered; i.e., where there is direct cytoplasmic continuity established between the cellular elements. Nonetheless, syncytial cytoarchitecture serves as an alternative to the above-mentioned junctions, insofar as intercellular leakage is prevented and the functions of structural and chemical coherence are clearly accomplished.

STRUCTURE OF THE TAPEWORM SURFACE

At the risk of oversimplification, the body plan of tapeworms (at least that pertaining to the cortex) may be viewed as a gut turned inside out, with the integument serving the absorptive functions just attributed to the intestinal mucosa. In some respects, its structure is unique, in others, quite comparable. It is first of all syncytial (fig. 3). The ectocytoplasm[1] forms a continuous band over the entire surface of the worm, corresponding to what was formerly termed (in the pre-1960's literature) the cuticle. While bound apically and basally by plasma membranes and containing within it mitochondria and numerous membrane-bound vesicles, the ectocytoplasm is devoid of nuclei, endoplasmic reticulum, and Golgi bodies. The latter organelles are found within underlying cytons, which are confluent with the ectocytoplasm via tendrillar processes. Resembling more conventional absorptive epithelia, the ectocytoplasm bears at its free surface a brush border of digitiform processes and rests on a basement lamina of connective tissue, which is in turn underlain by longitudinal and circular muscle fibers. A structural and functional analogy between tapeworm integumentary cytoarchitecture and that of the intestinal mucosa can be drawn, by comparing figures 1 and 3 (see also Beguin, 1966). Original micrographs, on which figure 3 is based, can be found in many previous publications (see, e.g., Lumsden, 1966a, 1975a,b; Lumsden and Specian, 1979).

In most respects the digitiform processes of the adult tapeworm's brush border, at least of the strobilar integument (fig. 4), are reminiscent of the microvilli adorning the host's mucosal cells, and likewise provide a significant amplification of the free surface area. Otherwise, they differ from microvilli, *sensu stricto*, in the presence of a densely fibrillar apical cap, which is set off from the remainder of the shaft by a trilaminate baseplate. Rothman (1963) coined for these structures the term "microthrix" (from the Greek *thrix*, meaning hair), though, to our way of thinking, those associated with the strobila are typically more "villus"-like than "hair"-like. However, elsewhere than the strobila, "microtriches" may be modified to form spine- or hooklike structures (e.g., Mount, 1970).

Like the luminal surface plasma membrane of the mucosal epithelial cells, the free surface plasma membrane of the tapeworm integument is coated with carbohydrate-rich marcromolecules. This does not represent an adsorption of host intestinal mucins (Lumsden, 1974). Rather, the

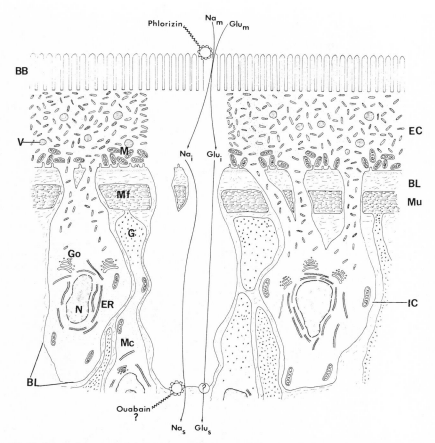

Fig. 3. Cytoarchitecture of the tapeworm integument. The syncytial ectocytoplasm (EC) rests on a connective tissue basement lamina (BL) and underlying muscle fibers (Mu). The ectocytoplasm is interconnected with nucleated cell bodies, or integumentary cytons (IC). Surficially, the ectocytoplasm bears a brush border (BB) of "microtriches," contains numerous vesicles (V), and mitochondria. Constituents of the integumentary cytons include a Golgi apparatus (Go), endoplasmic reticulum (ER), nucleus (N), and basolateral membrane (BL). Over-all, the arrangement is analogous, in most respects, to that of the host mucosal epithelial cells (see fig. 1). Organelles, etc., have been omitted from the central part of the drawing, to present schematically a model of glucose transport comparable to that of the host's mucosa (see fig. 1, and text for evidence and discussion). Also figured are muscle elements that interdisperse with integumentary elements. Mf is the myofibrilar region of a muscle cell, Mc, the nucleus-containing region or myocyton, which typically contains numerous glycogen particles (G). Junctional complexes elsewhere described in this paper are not figured in the drawing.

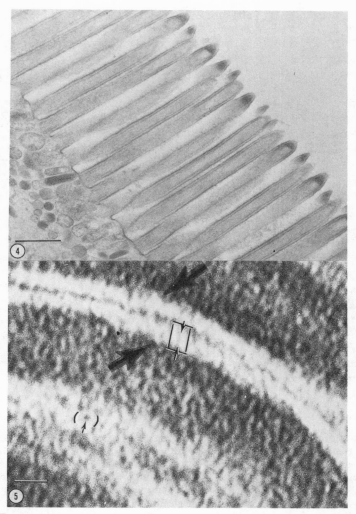

Fig. 4. Tapeworm (*Lacistorhynchus tenuis*) integumentary brush border. Scale marker = 0.5 µm.

Fig. 5. Ultrastructure of tapeworm (*Hymenolepis diminuta*) integumentary ("microthrix") surface membrane, as seen in chemically unfixed preparations. The sample was frozen in liquid nitrogen-cooled isopentane, freeze-dried, embedded in Epon, and then sectioned. Sections were subsequently stained with uranyl acetate. The "white" lines (at large arrows) represent two very closely approximated membrane profiles ("negative images thereof), the "black" background the internal "microthrix" cytoplasm (magnification here is on the order of 1.5 million ×). Within the area demarcated by the rectangle is what might be interpreted as a longitudinal section through a "pore complex" (see fig. 2), comprised of a pair of particulate subunits, separated by a narrow gap (small arrows). At the lower left of this micrograph, the plane of section is tangential to the membrane surface; within the brackets appears what may be a cross-sectioned pore complex. Its putative aperture is noted at the small arrow. Scale marker = 10 nm.

tapeworm glycocalyx is produced essentially by the synthetic activities of the integument itself (Lumsden, 1966b; Oaks and Lumsden, 1971). This is not to deny the findings of Befus (1977) that host antibodies (those of a glycoprotein nature) may attach to the tapeworm surface, a point emphasized by Podesta (1977). However, there are abundant data, omitted in Podesta's (1977) discussion but elsewhere reviewed by Lumsden (1975a,b), that the majority of the topochemical features exhibited by tapeworms are patently intrinsic to the worms. Curtis's (1972) suggestion that glycocalyces generally may be artifacts of microtechnical preparation has been contravened by numerous subsequent findings (reviewed by Lumsden, 1975a).

Like that of the host mucosa, the glycocalyx of the tapeworm surface membrane contains a preponderance of acidic groups that invest the membrane with a net electronegative fixed surface charge. Among the plasma membrane proteins isolated from the tapeworm brush border and examined by polyacrylamide electrophoresis (Knowles and Oaks, 1979) are some that have been tentatively identified as phosphoproteins. The presence of neuraminic acid-containing glycans, suggested by earlier cytochemically based observations (Lumsden et al., 1970), has not been confirmed, however, by biochemical analyses (Knowles and Oaks, 1979). Otherwise, of some 28 surface membrane proteins separable by electrophoresis, 10 are glycoproteins, with molecular weights ranging from 12,000 to over 200,000 daltons (Knowles and Oaks, 1979). The electrophoretic patterns obtained for brush border membrane proteins and those from a vesicle fraction believed to be elaborated by the Golgi bodies of the integumentary cytons (Oaks and Lumsden, 1971), are quite similar (Knowles and Oaks, 1979). These data are supportive of the thesis that formation and maintenance of the brush border and glycocalyx of the tapeworm integument are predicated on the same mechanisms operable in the vertebrate mucosa (see Bonneville and Weinstock, 1968; Lumsden, 1975a,b; Lumsden et al., 1974). However, it is indicated from the study of Knowles and Oaks (1979) that the phosphorylation of certain proteins characteristic of the brush border (vs. vesicle) fraction takes place only after fusion of the vesicle and brush border plasma membranes. Thus, modification of chemistry may be afforded at sites where the membrane components function—a topic to which we will return in a following section.

A current view of cell membrane structure generally is that a substantial amount of the membrane proteins are embedded within the lipid bilayer (Singer and Nicolson, 1972). Freeze fracture preparations (e.g., Belton, 1977) indicate that the surface membrane of the tapeworm integument, like that of the host mucosal cell plasma membrane (see, e.g., images of freeze-fractured brush borders by Friend and Gilula, 1972; Swift and Mukherjee, 1976), is structured along the lines of the Singer fluid-mosaic model. Particulate substructure within the tapeworm surface membrane is also suggested by an alternate technique to freeze-fracturing/etching. Figure 5 is of material (adult *Hymenolepis diminuta* from the rat) frozen in liquid nitrogen-cooled isopentane, freeze-dried, embedded in anhydrous epoxy resin, thin-sectioned, and examined after "negative" staining with uranyl acetate. Such images reveal what we would interpret as particles, many effectively spanning the membrane proper. Certain of these particles are arranged in a manner that might further be interpreted, fancifully perhaps, as "carriers" or pore complexes, after the model depicted in figure 2. In any event, as they appear in longitudinal section, each of the particles in such a complex measures approximately 2×10 nm, with an intervening slit about 1–3 nm wide; in cross section, the complex has the appearance of a doughnut, on the order of 7 nm diameter, with an apparent aperture of approximately 3 nm diameter. A schematic summary is presented in figure 6.

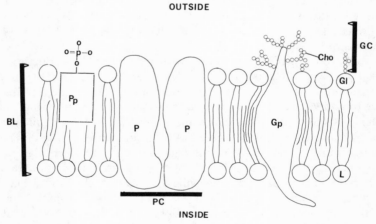

Fig. 6. A model for the hypothetical structure of the tapeworm integumentary surface plasmalemma. Following Singer's "fluid-mosiac' concept, it includes a lipid bilayer, into which particulate proteins are inserted. A surficial glycocalyx (GC) is present, the carbohydrate (Cho) moieties being attached to glycoprotein (Gp) and, possibly, glycolipids (Gl), though chemical analyses indicate the majority of the membrane lipids to be phospholipids (L). The presence of phosphoproteins (Pp) has also been suggested. Included is a "carrier", based on the "pore complex" model (PC) (see fig. 2); the structures labeled P are the protein subunits.

The basal membrane of the ectocytoplasm rests on a lamina of connective tissue, where adhesion thereto is provided by structures

reminiscent of half-desmosomes (Lumsden, 1966a; fig. 7). Not infrequently, pleat-like and/or tubular invaginations of this membrane are also
encountered (Threadgold and Read, 1970; fig. 8). The basal membrane
continues inwardly (i.e., into the cortical parenchyma) as the plasmalemma
of the integumentary cytons (fig. 9). Two features of the cyton surface are
noteworthy (figs. 10—12). The first is the occurence of what appear to be
gap, or nexus, junctions between adjacent cytons and, especially noteworthy, between cytons and adjacent muscle cells. The second is the
consistent proximity of ER cisternae to the plasmalemma. The intermem-
brane distances are on the order of 10 nm or less. Such complexes,
elsewhere observed in a variety of vertebrate and invertebrate nerve and
muscle cells (e.g., Rosenbluth, 1962; Lumsden and Byram, 1967; Lumsden
and Foor, 1968; Walker et al., 1970), have been suggested as having the
function of lowering membrane impedence. Subsurface cisterns of ER are
not especially common, however, in "nonexcitable" cells (see, e.g.,
Kumegawa et al., 1968); thus their presence in cestode integumentary
cytons is of interest.

Space does not permit a review of other aspects of the structure of the
tapeworm integument, but we would be remiss in not at least mentioning
one more. This concerns the sensory aspect. Interdicting the otherwise
syncytial ectocytoplasm are uniciliated dentrites whose function may be
that of chemoreception. Although tapeworms may be gutless wonders and
otherwise loathsome creatures, they are, perhaps, not without taste. For
more discourse on surface sensory structures in tapeworms, the reader is
invited to consult Lumsden and Specian (1979).

PHYSIOLOGICAL PROPERTIES OF THE TAPEWORM SURFACE

Functional correlations for the fine structural and topochemical features
of the tapeworm integument described in the preceding paragraphs remain
to an extent matters of speculation, but such speculation is not lacking in at
least some supportive data.

Adsorption (Surface Binding) and the Glycocalyx

Given its composition predominately of polyionic glycoproteins, and
such, the surface glycocalyx might be expected to influence electrostatic
interactions between constituents of the external milieu and the surface
membrane. Among these would be the concentration of inorganic cations
of importance to enzymatic and other activities. Using autoradiography,
coupled with liquid scintillation spectrometry and cytochemical staining
methods, Lumsden and colleagues (Lumsden, 1972, 1973; Lumsden and
Berger, 1974) have examined the adsorption of calcium to the tapeworm

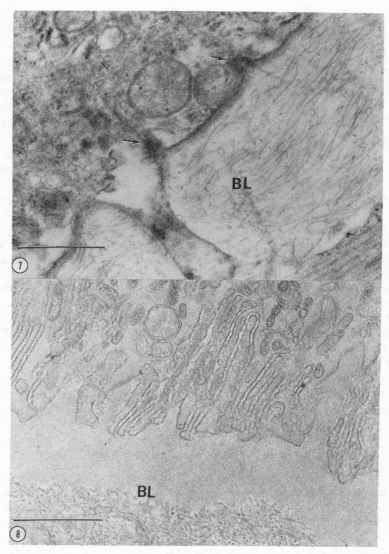

Fig. 7. Basal surface of tapeworm (*Hymenolepis diminuta*) integumentary ectocytoplasm, illustrating hemidesmosome-like junctions (arrows) with connective tissue basement lamina (BL). Scale marker = 0.5 μm. From Lumsden (1966a.)

Fig. 8. Infoldings of tapeworm basal integumentary (ectocytoplasmic) membrane (*Gyrocotyle urna*; BL, basement lamina. Scale marker = 0.5 μm.

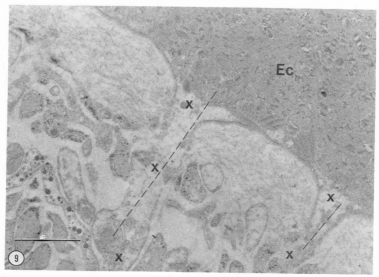

Fig. 9. Basal region of tapeworm (*Hymenolepis diminuta*) integumentary ectocytoplasm (Ec), illustrating cytoplasmic connections (X—X) with underlying cytons. Scale marker = 1.0 μm.

brush border. Figure 13 is of an autoradiograph of a worm breifly exposed to radiocalcium (^{45}Ca), showing labeling of the integumentary surface. This binding of calcium is sensitive to pH and to reagents that react with, and thereby neutralize, anionic components of the glycocalyx, but is not materially affected by reagents (e.g., lectins) that react nonionically with neutral sugars of the glycocalyx. The binding sites in the tapeworm surface that so adsorb calcium are apparently nonspecific for this cation, as other metal ions (e.g., zinc and magnesium) are also adsorbed with comparable affinity. In the presence of zinc, magnesium, lanthanum, or nonradioactive calcium, ^{45}Ca adsorption decreases nonlinearly, as a function of the magnitude of the concentration difference between the ambient ^{45}Ca concentration and the concentration of the other cation species (fig. 14); the kinetics are suggestive of a competition between radiocalcium and the other cations for the same membrane binding sites. In a similar fashion, these and other cations, including sodium and potassium, will displace previously adsorbed radiocalcium. The observed properties of the surface glycocalyx are, accordingly, not unlike those of an ion exchange resin.

The functional contributions of this system are undoubtedly manifold. For example, the brush border phosphohydrolase activity has a demonstrable requirement for adsorbed divalent cations (Lumsden and

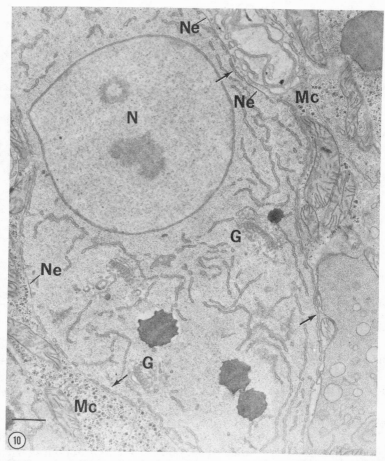

Fig. 10. Tapeworm (*Lacistorhynchus tenuis*) integumentary cyton. N, nucleus; G, Golgi; subsurface cisterns of endoplasmic reticulum at arrows; Ne, nexus-like junctions with adjacent myocytons (Mc). Scale marker = 1.0 μm.

Berger, 1974); several transport functions described later in this chapter exhibit sensitivity to monovalent ions, notably sodium; cation binding (especially of calcium) could also play a role in controlling tegument membrane structure (see, e.g., Williams, 1975). Additionally, it would be expected that the hydration layer that results from ion concentration gradients at the lumen/membrane interface materially influences a variety of surface functions in this system.

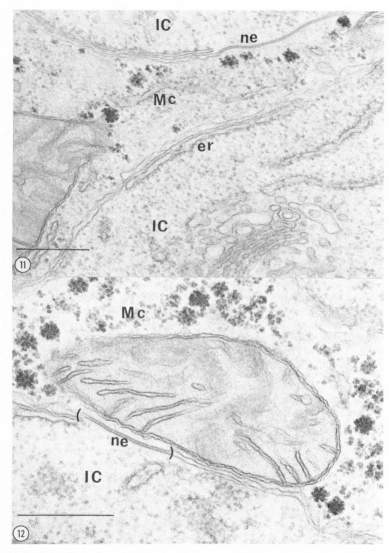

Fig. 11. Regions of tapeworm (*Lacistorhynchus tenuis*) integumentary cytons (IC) interfacing with a myocyton (Mc), illustrating subsurface cisternae of endoplasmic retiulum (er) and a nexus-like junction (ne). Scale marker = 0.5 μm.

Fig. 12. Higher magnification image of a junctional complex (ne) between a tapeworm (*Lacistorhynchus tenuis*) integumentary cyton (IC) and mycocyton (Mc). Scale marker = 0.5 μm.

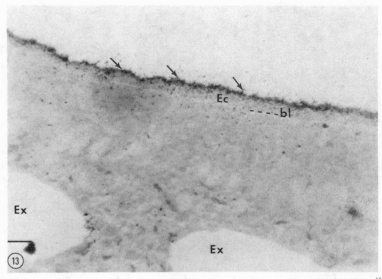

Fig. 13. Autoradiograph of tapeworm (*Hymenolepis diminuta*) incubated with ^{45}Ca. Radioactivity is localized in the region of the integumentary brush border (arrows). Ec, ectocytoplasm; bl, basal lamina; Ex, major excretory canals. Scale marker = 10 μm.

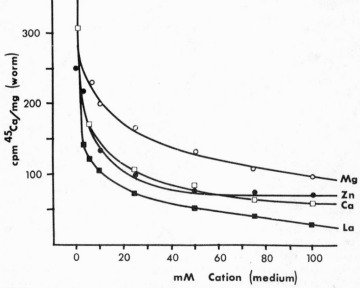

Fig. 14. Effect of metal ions on calcium adsorption by *Hymenolepis diminuta*. Worms were incubated in media containing a fixed amount of radioactive calcium (2.6 mM) and concentrations of nonradioactive calcium, magnesium, zinc, or lanthanum at the concentrations indicated on the abscissa. Ordinal values represent radioactivity adsorbed by the worms. (From Lumsden and Berger, 1974.)

TABLE 1

DIFFERENTIAL ADSORPTION OF ^{3}H-LABELED POLYPEPTIDES
BY HYMENOLEPIS DIMINUTA INTEGUMENTARY SURFACE MEMBRANE

Preincubation	Incubation polypeptide (polyelectrolyte)	cpm/mg bound
KRT	polylysine (+)	57,162
Diazomethane	polylysine (+)	19,638
KRT	polyglutamate (−)	17,114
Diazomethane	polyglutamate (−)	27,075

SOURCE: Lumsden, 1972

The glycocalyx also serves as an adsorptive site for organic substances, including proteins. In some cases the attachment is loose and transitory, apparently by relatively weak interactions such as van der Waals forces (Ruff and Read, 1973). In other cases peptides attach as a function of electrostatic charge (Lumsden, 1972; Read, 1973). Table 1 presents some data on the binding of two peptides, polylysine and polyglutamate, of comparable molecular weight (100,000 and 80,000, respectively). With preincubation only in saline (KRT of Read et al., 1963), the polycationic species (polylysine) is preferentially adsorbed to the net negatively charged surface, but when the electronegative charges of the glycocalyx are first neutralized (as in the case exhibited, by diazomethane pretreatment), the converse obtains. Possibly germane to these findings is the observation of Revel and Ito (1967) that most amphoteric proteins in the alkaline mammalian digestive tract will bear a net negative charge; certain of these proteins, potentially deleterious to the parasites, might thereby be excluded from establishing intimate contact with the tapeworm surface. On the other hand, there is evidence that some proteins of the digestive tract *are* adsorbed to the tapeworm surface (Read, 1973; Pappas and Read, 1972a,b; Ruff and Read, 1973). This function has potential physiological relevance to contact enhancement as well as inactivation of extrinsic (host) enzymes, a subject addressed elsewhere in this volume by Dr. Pappas (Pappas, 1979).

Absorption (Permeability and Transport)

The transport functions of the tapeworm integument have been the subject of study for many years, originating in the laboratory of the late Dr. Clark Read (see, e.g., Phifer, 1960; Read, 1961; Read et al., 1960). Periodic reviews have included those of Read and colleagues (1963), Read (1966, 1971), Podesta and Mettrick (1974), Arme (1975), and Read and Pappas (1975).

As with the vertebrate intestinal mucosa, the substances that classify as nutrients are acquired by tapeworms in the form of low molecular weight organic compounds, whose entry from the gut lumen involves passage across a membrane interface. Adult tapeworms are impermeable to macromolecules. No pinocytotic or other mechanisms of bulk transport have been identified (Lumsden et al., 1970). Relatively few organic solutes of physiological significance enter tapeworms by passive diffusion alone; rather, entry of the greatest majority of organic solutes to which the worms are permeable appears to be mediated by so-called carriers. Thus, the uptake of hexose sugars, amino acids, nucleosides, and so on, by tapeworms is characterized by stereospecificity; this is reflected in competitive inhibitions by chemically similar compounds, and a rate of movement that follows saturation kinetics.

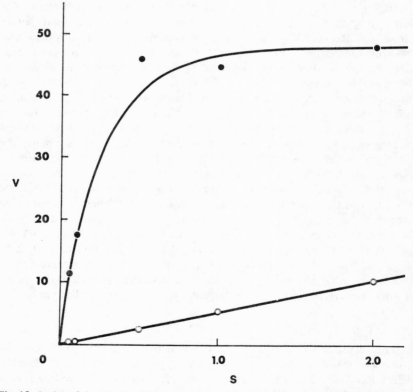

Fig. 15. A plot of the velocity of uptake (V, μmoles/g ethanol-extracted dry wt/hr) of ³H-thymidine by *Hymenolepis diminuta* as a function of solute concentration (S, mM). The curves represent experiments conducted at sodium concentrations of 120 mM (closed circles) and 0 mM (open circles). The deleted sodium chloride was replaced isosmotically with choline chloride. Each point is the mean of 3 replicates. (From McCracken et al., 1975.)

For certain solutes, e.g., hexose sugars, some amino acids, and nucleosides, net influx can be accomplished against an apparent chemical concentration difference. Accumulative transport in these instances is inhibited by poisons of energy metabolism and, at least for sugar and pyrimidine nucleoside uptake, is sensitive to the inorganic ion composition of the incubation medium, notably the sodium composition. As reported by McCracken and colleagues (1975), when sodium is isosmotically replaced in the incubation medium with choline, uridine, and thymidine influx reverts to a linear function of solute concentration, i.e., from an "active" transport to a passive diffusion system (fig. 15). The sodium requirement for mediated nucleoside transport is highly specific, satisfied by neither lithium nor potassium. Kinetic analysis indicates that the primary effect of sodium deletion is to decrease the affinity of the carrier for the nucleoside solute; there is little or no effect, however, on the maximal influx rate (V_{max}). For the glucose transport system, on the other hand, sodium effects principally a change in V_{max} rather than K_t (Read et al., 1974). In both systems transport of the organic solute species appears to be coupled to sodium transport, and the sodium deletion effects are readily reversible.

Glucose influx also exhibits a chloride sensitivity, in that replacing chloride in the incubation medium with other anions effects, primarily, a net decrease in V_{max} (Pappas et al., 1974; Pappas and Hansen, 1977). Chloride appears to closely follow sodium entry during glucose absorption, serving to maintain electrical neutrality relative to the influxing cation population and, thereby, the native surface electrical potential. The failure of other permeating anions (e.g., bicarbonate or acetate) to compensate fully for a chloride deletion may be due to their slower diffusion rates. The effect would be a decrease in the electrochemical gradient as sodium crosses the membrane in the absence of, or in significant excess of, an electrically compensating anion or countertransport of cations; as the transmembrane potential is so reduced, the rate of glucose uptake would be likewise reduced.

Localization of the Carriers and Pumps

Intuitively, the specific transport loci mediating the initial carrier-driven uptake processes would be placed in the plasma membrane of the integumentary brush border, and the data available do not so far belie such intuition. Surface active, nonpermeating compounds have been localized in the brush border that have the physiological effect of blocking carrier-mediated absorptive functions. Particularly instructive have been studies employing phlorizin and concanavalin A (McCracken and Lumsden, 1975a,b).

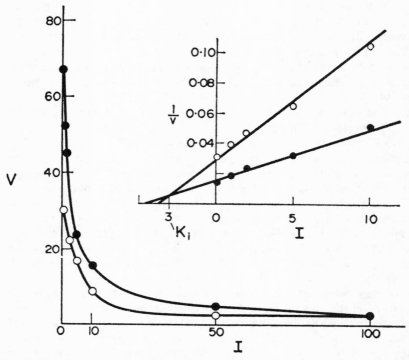

Fig. 16. The velocity (V, μmoles/g ethanol-extracted dry wt per hr) of uptake of 0.05 mM (open circles) and 0.1 mM (closed circles) ^{3}H-glucose by *Hymenolepis diminuta* in the presence of increasing concentrations of phlorizin (I, μmoles). Each point is the mean of three replicates. Insert: the data replotted as the reciprocal of influx (1/V) vs. I. Lines were fitted by regression analyses. (From McCracken and Lumsden, 1975a.)

Phlorizin, a biphenolic glycoside, proves to be a fully competitive inhibitor of glucose uptake by tapeworms. The affinity of the hexose carrier for phlorizin is quite high, indeed some 250 times higher than that for glucose itself. The activity of phlorizin as a competitive inhibitor of hexose transport derives from the fact that phlorizin contains a glucose moiety, in addition to the aglycone (biphenolic) moiety. Figure 16 illustrates the kinetics of glucose uptake and the effect of phlorizin. The apparent "uptake" of phlorizin, like that of glucose, exhibits a typical adsorption isotherm (fig. 17), but that this represents surface binding rather than actual permeation of phlorizin is indicated by the following observations: concentrations of phlorizin on the order of 100 μm inhibit mediated glucose assimilation by as much as 98%, with glucose uptake in the presence of such levels of phlorizin reverting to a passive diffusion

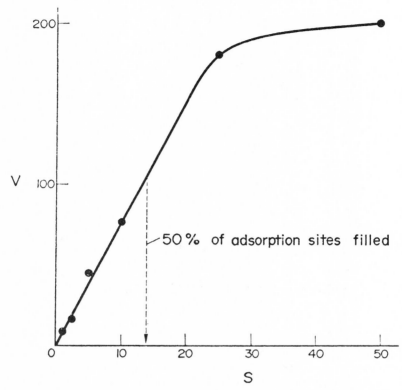

Fig. 17. Velocity of phlorizin adsorption (V, nmoles/g ethanol-extracted dry wt per hr) by *Hymenolepis diminuta* as a function of the ambient phlorizin concentration (S, μmoles). Each point is the mean of 3 replicates. (From McCracken and Lumsden, 1975a.)

mechanism (fig. 18); however, the inhibitory effect of phlorizin is immediately reversed by brief washing in saline; and autoradiographs (fig. 19) of worms incubated in ^{3}H-phlorizin, frozen in isopentane-liquid nitrogen, freeze-dried, and sectioned (technique of Stirling, 1967), indicate only surficial binding of the inhibitor (to the brush border).

That phlorizin localization may be taken as concomitantly localizing the hexose carrier loci follows the reasoning of Caspary and colleagues (1969) and others that phlorizin combines with the carrier through its glucose moiety but is not transported because of the formation of secondary bonds between the aglycone portion of the phlorizin molecule and regions of the membrane proximal to the transport locus. Analysis of these interactions for tapeworms include those of Murphy and colleagues (1977, 1979). In comparing the tapeworm tegument to the host mucosa, it is

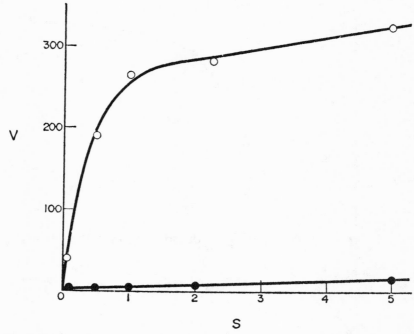

Fig. 18. Velocity of glucose uptake (V, μmoles/ gethanol-extracted dry wt per hr) as a function of the ambient glucose concentration (S, mM) by *Hymenolepis diminuta*, in the presence (closed circles) or absence (open circles) of 500 μM phlorizin. Each point is the mean of 3 replicates. (From McCracken and Lumsden, 1975a.)

instructive that ^{3}H-phlorizin similarly localizes in the brush border of that tissue, as demonstrated autoradiographically by Stirling (1967).

Concanavalin A (Con A) is a lectin (molecular weight 110,000) that reacts with glycans containing multiple terminal nonreducing glucopyranosyl residues and close analogues thereof (Lis and Sharon, 1973). Con A does not permeate the tapeworm integument, but binds surficially to the glycocalyx (McCracken and Lumsden, 1975b; Knowles and Oaks, 1979). Interaction of Con A with cell surfaces has been found to affect a variety of functions of plasma membranes, including transport of ions and metabolites; in certain cell types, the Con A binding sites appear to be closely approximated to transport sites for particular solutes (e.g., Inbar et al., 1971). Utilizing ferritin-conjugated Con A and electron microscopy, McCracken and Lumsden (1975b) determined that the Con A binding sites on the tapeworm surface are not uniformly distributed, but occur in clusters separated by distances of 100 nm or more. Con A adsorption was

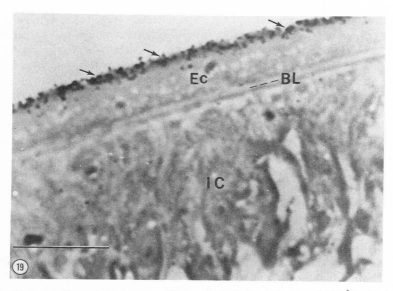

Fig. 19. Autoradiograph of tapeworm (*Hymenolepis diminuta*) incubated with ^{3}H-phlorizin. Radioactivity is localized in the region of the integumentary brush border (arrows). Ec, ectocytoplasm; BL, basal lamina; IC, integumentary cytons. White vacuolar areas are artifacts produced by ice crystal formation. Scale marker = 10 μM.

found in this study to have no effect on test solutes entering solely by free diffusion, but did inhibit certain others entering by a carrier-mediated process; among the latter were the hexose sugars glucose and galactose, the amino acid lysine, and the pyrimidine nucleoside uridine. The inhibition exerted by Con A was found to be of the noncompetitive type and to affect V_{max} rather than K_t. McCracken and Lumsden (1975b) suggested that Con A interferes with solute translocation across the membrane after the solute is bound to the transport locus, binding of the lectin possibly altering the spacial distribution and/or numerical availability of the carriers.

As noted previously for the vertebrate intestinal mucosa, sodium-dependent active transport systems invariably have associated with them a ouabain-sensitive ATPase. For some tapeworms (adult *Calliobothrium* from the shark spiral valve), the addition of ouabain to the incubation medium reportedly inhibits accumulative hexose sugar transport (Fisher and Read, 1971). However, for the "model" species on which the greatest amount of data on tapeworm transport has been gathered—the rat intestinal tapeworm *Hymenolepis diminuta*—ouabain is without effect (Dike and Read, 1971a; Podesta et al., 1977). However, that ouabain, so employed, does not inhibit the sodium-dependent transport mechanisms of intact *H. diminuta* should not be misconstrued as evidence against the involvement of Na-ATPase in sodium-dependent integumentary transport functions by this species. As it turns out, the sites of ouabain-sensitive activity are probably not at the free surface of the worm, and intact *H. diminuta* are apparently impermeable to ouabain.

Gallogly (1972) identified in homogenates of *H. diminuta* significant amounts of ouabain-sensitive ATPase, though from the design of his experiments, the location(s) of this activity cannot be ascertained. Podesta et al. (1977) found that if worms are sliced, and corrections made for leakage of solute through the wound, the net effect of ouabain is to reduce intracellular accumulation of glucose with a concomitant elevation of intracellular sodium. In table 2 we compare the binding of ^{3}H-ouabain to intact worms and to worms briefly pretreated with triton x-100 according to the method of Knowles and Oaks (1979); this treatment effects removal of the brush border, but leaves the remaining carcass (with the basal ectocytoplasmic plasmalemma and cytons, etc.) relatively intact (see also Oaks et al., 1977). The specific activities from table 2 translate to essentially 0 nm bound to intact worms, versus nearly 90 nm/g wet weight to the brush border-stripped carcass. It would therefore appear that ouabain exerts its effect on transport at a level below, or internal to, the brush border of the integument.

TABLE 2

HYMENOLEPIS DIMINUTA INCUBATED IN 1 mM OUABAIN (^{3}H, 5 MIN PULSE, SPECIFIC ACTIVITY 2.5×10^2 cpm/nmole)

Intact worms	1,000 cpm/g tissue	4.0 ± 3.6 nmole ouabain/g wet wt
Worms with integumentary brush border removed by triton	22,600 cpm/g tissue	88.0 ± 8.9 nmole ouabain/g wet wt

Paradoxically, the integumentary brush border does contain abundant hydrolytic activity against ATP and a variety of other phosphorylated substrates (Dike and Read, 1971b; Pappas and Read, 1974; Lumsden and Berger, 1974), but this is probably not related to the possible ATPase-mediated sodium extrusion pump associated with the sugar and other transport systems. This brush border phosphatase activity, shown in figure 20 by the histochemical lead method, is insensitive to sodium deletion, and is not inhibited by ouabain (Dike and Read, 1971a,b; Lumsden, unpublished observations).

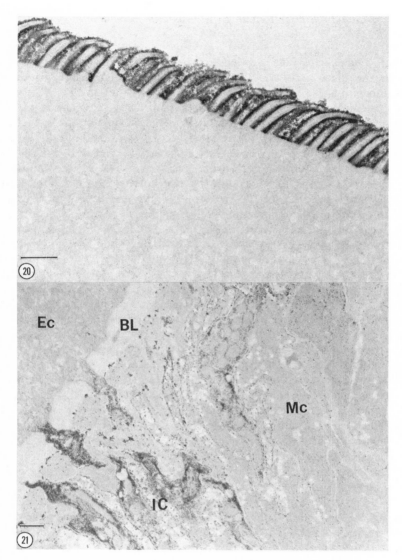

Fig. 20. Tapeworm (*Hymenolepis diminuta*) integumentary brush border, following lead method for localization of phosphohydrolase activity (ATP substrate). Section otherwise unstained. Scale marker = 0.5 μm.

Fig. 21. Tapeworm (*Hymenolepis diminuta*) integumentary cytons (IC) exhibiting phosphohydrolase activity via the lead method (ATP substrate). Ec, ectocytoplasm; BL, basal lamina; Mc, myocyton. Section otherwise unstained. Scale marker = 1 μm.

Intuition and previous knowledge of the intestinal mucosal system would favor the location of the tapeworm ATPase-mediated integumentary sodium pump, if present, in the basal membrane system—i.e., the basal plasmalemma of the ectocytoplasm and/or the plasmalemma of the attached cytons. Figure 21 illustrates a cytochemical reaction product for ostensible ATPase activity in the integumentary cytons. That this in fact might represent the Na-ATPase must, however, be qualified. It should be pointed out that the localization of Na-ATPase is not readily accomplished by cytochemistry, since its activity is often inhibited by the fixatives and metal-trapping agents usually employed for phosphatases (see Ernst, 1972a,b, for detailed discussion). Even when a positive reaction for "ATPase" is obtained, as in this case, a convincing demonstration of substrate specificity or the effect of activators and inhibitors is difficult to achieve by cytochemical staining.

An alternative approach, we believe, will be to employ ^{3}H-ouabain and visualize its location by autoradiography, in concert, of course, with the appropriate biochemical assays. Autoradiographic localization of ^{3}H-ouabain binding has proved successful in identifying sites of Na-ATPase in other systems (Stirling, 1972; Quinton et al., 1973).

A remaining question dealing with morphological correlates of tapeworm Na-sensitive transport concerns the site(s) of sodium efflux from the worms (i.e., its return to the bathing medium). In the intestinal mucosa a significant amount of the sodium extruded at the basolateral surfaces of the epithelial cells is shunted back into the lumen via the intercellular spaces and leakage between the tight junctions. This shunt accounts for much of the total ion conductance of the tissue (Frizzell and Schultz, 1972). A comparable shunt is precluded for tapeworms by the syncytial structure of the integument, prompting Uglem (1976) and others to propose a backflow of sodium directly across the brush border. It occurs to us, however, that, under normal physiological conditions (i.e., *in vivo*), a likely site of sodium efflux would be the excretory system, the structure and physiological features of which have been reviewed by Lumsden and Specian (1979). Thus, sodium (and chloride, etc.) initially extruded at the parenchymal (basal) surface of the tegument (possibly via an ATPase pump) might diffuse intercellularly toward the collecting ducts of the excretory system; the apparent concentrations of these ions in the excretory fluid (Webster and Wilson, 1970) would make this thermodynamically feasible. Movement of ions across the *entire* integument (i.e., not just the brush border) would have the advantage as well of providing an osmotic gradient for water movement toward the basal integumentary surface. This, in turn, would facilitate the ultimate delivery of absorbed sugars and such to the internal tissues. As described and discussed elsewhere (Lumsden and Specian, 1979), the intercellular parenchymal space probably serves as an important route for the distribution and exchange of metabolites with the cells it surrounds and the tissues it pervades.

Intercellular Communication

A most interesting feature of the tapeworm integument, which may or may not bear on the subject of electrolyte fluxes discussed in the preceding paragraphs, is the high frequency of nexus-like junctions between integumentary cytons and muscle cells. This appears to be an exception to the rule that such junctions do not ordinarily occur between cells of different types. However, the tapeworm integument may be otherwise exceptional, in that it may be an example of an "epidermal" tissue exerting a modulating effect on muscle tissue. In a related system, the digenetic trematode *Schistosoma mansoni*, a transintegumental electrical potential of some –30 mv has been recorded (Fetterer et al., 1978). Experimentally manipulated changes in ambient ionic concentrations materially alter this potential, and these changes have concommitant effects on the contraction of schistosome musculature. Fetterer et al. (1978) tentatively concluded that certain drugs affecting this parasite's motility owe their action at least in part to their influence on the transintegumentary potential. The significance of morphologically apparent electrical coupling between tapeworm integument and muscle elements (through nexus-like junctions and, perhaps subsurface ER cisterns) has not been experimentally examined, though these interactions may have relevance to the control of contractile activity in these worms (see further discussion in Lumsden and Specian, 1979).

CONCLUSIONS

For organisms that are literally bathed in a nutrient-rich milieu, the development of a body surface versus an internal, alimentary absorptive mechanism is clearly advantageous. It is of interest to note that this phenomenon is not restricted to parasitic forms, but exhibited also among some free-living helminths (see, e.g., Fisher and Oaks, 1978). However, at least among those invertebrates commonly referred to as "worms," the greatest development of the body surface route of nutrient acquisition has been realized by the enteroparasites, exemplified by tapeworms and acanthocephalans (see Lumsden, 1975a, and Starling, 1975, for a review of the latter).

Flatworms are phylogenetically primitive when compared with the majority of other metazoan groups, even those comprised by "worms." As noted by Pappas and Read (1972c), the occurrence of coupled organic solute-sodium transport systems in the brush border of the tapeworm integument, on one hand, and the brush border of the mammalian intestinal mucosa, on the other, indicate that the evolution of such systems has been very conservative.

It is a rather well adhered to principle that tissues of like function in otherwise widely diverse organisms will display common features of cytoarchitecture. Among those exhibited in common by the tapeworm integument and vertebrate intestinal mucosa are features that amplify surface area. Moreover, the membranes are arranged both in series and in parallel. Transport by epithelia further depends on a polarization, or geometrical asymmetry, of the component cells such that the inward- and outward-facing membranes exhibit different properties, functionally as well as structurally. This same feature is likewise exhibited in the otherwise less conventional integument of tapeworms.

It has been said that in her design of tapeworm anatomy, Mother Nature followed a somewhat unique blueprint (i.e., one calling for placement of the "gut" on the outside instead of the inside, etc.). However, at least insofar as the functional morphology for transport epithelia is concerned, wherever they are located, it would appear that she has been rather conservative after all.

ACKNOWLEDGMENTS

The senior author thanks the National Science Foundation and National Institutes of Health for their generous support of his and his students' research over the years. His students' work has also been aided by grants and such from the Graduate School of Tulane University. Pilot investigations reported in this paper were supported in part by funds provided from NIH Biomedical Research Support Grant 5 SO7-RR07040-11.

For his gift of precious time and artistic talents, Mr. Michael Hildreth has our gratitude. We are likewise indebted to Drs. John Oaks, George Cain, and William Knowles for providing us previews of their manuscripts and permission to discuss their findings in this paper.

---

1. In the helminthological literature, what we term here "integument" and "ectocytoplasm" are frequently called the "tegument" and "distal cytoplasm," respectively.

REFERENCES

Arme, C. 1975. Tapeworm-host interactions. Symp. Soc. Exp. Biol. 24:505–32.

Befus, A. D. 1977. *Hymenolepis diminuta* and *Hymenolepis microstoma*: mouse immunoglobulins binding to the tegument surface. Exp. Parasit. 41:242–51.

Befus, A. D., and R. B. Podesta. 1976. Intestine. *In* C. R. Kennedy (ed.), Ecological aspects of parasitology. North Holland Publ. Co., Amsterdam.

Beguin, F. 1966. Etude au microscope électronique de la cuticle et ses structures associés chez quelques cestodes. Essai d'histologie comparée. Zeit. f. Parasitenk. 72:30–46.

Belton, C. M. 1977. Freeze-fracture study of the tegument of larval *Taenia crassiceps*. J. Parasit. 63:306–13.

Bloom, W., and D. W. Fawcett. 1975. A textbook of histology. W. B. Saunders Co., Philadelphia.

Bonneville, M., and M. Weinstock. 1970. Brush border development in the intestinal absorptive cells of *Xenopus* during metamorphosis. J. Cell Biol. 44:151–71.

Brand, T. von 1973. Biochemistry of parasites. Academic Press, New York.

Caspary, W. F., N. R. Stevens, and R. K. Crane. 1969. Evidence for an intermediate step in carrier-mediated sugar translocation across the brush border of hamster small intestine. Biochim. Biophys. Acta 193:168–78.

Crane, R. K. 1960. Intestinal absorption of sugars. Physiol. Rev. 40:789–825.

Crane, R. K. 1962. Hypothesis of mechanism of intestinal active transport of sugars. Fed. Proc. 21:891–95.

Crane, R. K. 1965. Na^+-dependent transport in the intestine and other tissues. Fed. Proc. 24:1000–1006.

Crane, R. K. 1968. Absorption of sugars. *In* Handbook of physiology. Vol. 3. American Physiological Society, Washington.

Crane, R. K. 1977. The Gradient Hypothesis and other models of carrier-mediated active transport. Rev. Physiol. Biochem. Pharmacol. 78:99–159.

Crane, R. K., D. Miller, and I. Bihler. 1961. The restriction on the possible mechanism of intestinal active transport of sugars. *In* A. Kotyk (ed.), Membrane transport and metabolism. Czechoslovak Academy of Science Press, Prague.

Curtis, A. 1972. Adhesive interactions between organisms. Symp. Br. Soc. Parasit. 10:1–21.

Dike, S. C., and C. P. Read. 1971a. Relation of tegumentary phosphohydrolase and sugar transport in *Hymenolepis diminuta*. J. Parasit. 57:1251–55.

Dike, S. C., and C. P. Read. 1971b. Tegumentary phosphohydrolases of *Hymenolepis diminuta*. J. Parasit. 57:81–87.

Ernst, S. A. 1972a. Transport adenosine triphosphatase cytochemistry I. Biochemical characterization of a cytochemical medium for the ultrastructural localization of ouabain-sensitive, potassium-dependent phosphatase activity in the avian salt gland. J. Histochem. Cytochem. 20:13–22.

Ernst, S. A. 1972b. Transport adenosine triphosphatase cytochemistry. II. Cytochemical localization of ouabain-sensitive, potassium-dependent phosphatase activity in the secretory epithelia of the avian salt gland. J. Histochem. Cytochem. 20:23–38.

Fetterer, R. H., R. A. Pax, and J. L. Bennett. 1978. The effect of putative neurotransmitters and praziquantel on membrane potential from male *Schistosoma mansoni*. Prog. Abst. 53d Ann. Meet. Amer. Soc. Parasit. P. 71.

Fisher, F. M., Jr., and J. A. Oaks. 1978. Evidence for a nonintestinal nutritional mechanism in the rhynchocoelan, *Lineus ruber*. Biol. Bull. 154:213–25.

Fisher, F. M., Jr., and C. P. Read. 1971. Transport of sugars in the tapeworm *Calliobothrium verticillatum*. Biol. Bull. 140:40–62.

Friend, D. S., and N. B. Gilula. 1972. Variations in tight and gap junctions in mammalian tissues. J. Cell Biol. 53:758–76.

Frizzell, R. A., and S. G. Schultz. 1972. Ionic conductances of extracellular shunt pathway in rabbit ileum: influence of shunt on transmural sodium transport and electrical potential differences. J. Gen. Physiol. 59:318–46.

Fujita, M., and M. Nakao. 1973. Localization of ouabain-sensitive ATPase in intestinal mucosa. *In* M. Nakao and L. Packer (eds.) Organization of energy-transducing membranes. University Park Press, Baltimore.

Gallogly, R. L. 1972. ATPase and membrane transport in the rat tapeworm *Hymenolepis diminuta*. Ph.D. diss., Rice Universty, Houston.

Glynn, I. M. 1964. The action of cardiac glycosides on ion movements. Pharm. Rev. 16:381–407.

Hoffman, J. F. 1973. Molecular aspects of the Na^+, K^+-pump in red blood cells. Nakao and L. Packer (eds.) Organization of energy-transducing membranes. University Park Press, Baltimore.

Inbar, M., H. Ben-Bassat, and L. Sachs. 1971. Location of amino acid and carbohydrate transport sites in the surface membrane of normal and transformed mammalian cells. J. Membr. Biol. 6:195–209.

Jacobsen, N., and D. Fairbairn. 1967. Lipid metabolism in helminth parasites. II. Biosynthesis and interconversion of fatty acids by *Hymenolepis diminuta* (Cestoda). J. Parasit. 53:355–61

Knowles, W. J., and J. A. Oaks. 1979. Isolation and partial biochemical characterization of the brush border plasma membrane from the cestode, *Hymenolepis diminuta*. J. Parasit. (in press).

Kumegawa, M., M. Cattoni, and G. Rose. 1968. Electron microscopy of oral cells in vitro. II. Subsurface and intracytoplasmic confronting cisternae in strain KB cells. J. Cell Biol. 36:443–52.

Lis, H., and N. Sharon, 1972. Lectins; cell-agglutinating and sugar specific proteins. Science 177:949–59.

Lowenstein, W. R. 1972. Cellular communication through membrane junctions. Arch. Int. Med. 129:299–309.

Lumsden, R. D. 1966a. Cytological studies on the absorptive surfaces of cestodes. I. The fine structure of the strobilar integument. Zeit. f. Parasitenk. 27:355–82.

Lumsden, R. D. 1966b. Cytological studies on the absorptive surfaces of cestodes. II. The synthesis and intracellular transport of protein in the strobilar integument. Zeit. f. Parasitenk. 28:1–13.

Lumsden, R. D. 1972. Cytological studies on the absorptive surfaces of cestodes. VI. Cytochemical evaluation of electrostatic charge. J. Parasit. 58:229–34.

Lumsden, R. D. 1973. Cytological studies on the absorptive surfaces of cestodes. VII. Evidence for the function of the tegument glycocalyx in cation binding by *Hymenolepis diminuta*. J. Parasit. 59:1021–30.

Lumsden, R. D. 1974. Relationship of extrinsic polysaccharides to the tegument glycocalyx of cestodes. J. Parasit. 60:374–75.

Lumsden, R. D. 1975a. Surface ultrastructure and cytochemistry of parasitic helminths. Exp. Parasit. 37:267–339.

Lumsden, R. D. 1975b. The tapeworm tegument: a model system for studies on membrane structure and function in host-parasite relationships. Trans. Amer. Micros. Soc. 94:501–7.

Lumsden, R. D., and B. Berger. 1974. Cytological studies on the absorptive surfaces of cestodes. VIII. Phosphohydrolase activity and cation adsorption by the glycocalyx of *Hymenolepis diminuta*. J. Parasit. 60:744–51.

Lumsden, R. D., and J. E. Byram, III. 1967. The ultrastructure of cestode muscle. J. Parasit. 53:326–42.

Lumsden, R. D., and W. E. Foor. 1968. Electron microscopy of schistosome cercarial muscle. J. Parasit. 54:780–94.

Lumsden, R.D., J. A. Oaks, and W. Alworth. 1970. Cytological studies on the absorptive surfaces of cestodes. IV. Localization and cytochemical properties of membrane fixed cation binding sites. J. Parasit. 56:736–47.

Lumsden, R. D., J. A. Oaks, and J. F. Mueller. 1974. Brush border development in the tegument of the tapeworm *Spirometra mansonoides*. J. Parasit. 60:209–26.

Lumsden, R. D., and R. D. Specian. 1979. The morphology, histology, and fine structure of the adult stage of the cyclophyllidean tapeworm *Hymenolepis diminuta*. *In* H. Arai (ed.), The biology of *Hymenolepis diminuta*. Academic Press, New York. In press.

Lumsden, R. D., L. T. Threadgold, J. A. Oaks, and C. Arme. 1970. On the permeability of cestodes to colloids: an evaluation of the transmembranosis hypothesis. Parasitology 60:185–93.

McCracken, R. O.,and R. D. Lumsden. 1975a. Structure and function of parasite surface membranes. I. Mechanism of phlorizin inhibition of hexose transport by the cestode *Hymenolepis diminuta*. Comp. Biochem. Physiol. 50B:153–57.

McCracken, R. O., and R. D. Lumsden. 1975b. Structure and function of parasite surface membranes. II. Concanavalin A adsorption by the cestode *Hymenolepis diminuta*. Comp. Biochem. Physiol. 52B:331–37.

McCracken, R. O., R. D. Lumsden, and C. R. Page, III. 1975. Sodium-sensitive nucleoside transport by *Hymenolepis diminuta*. J. Parasit. 61:999–1005.

Mettrick, D. F. 1970. Protein nitrogen, amino acid and carbohydrate gradients in the rat intestine. Comp. Biochem. Physiol. 37:517–41.

Mettrick, D. F. 1971. *Hymenolepis diminuta*: the microbial fauna, nutritional gradients, and physicochemical characteristics of the small intestine of uninfected and parasitized rats. Can. J. Physiol. Pharm. 49:972–84.

Mitchell, P. 1961. Coupling of phosphorylation to electron and hydrogen transfer by a chemiosmotic type of mechanism. Nature (Lond.) 191:144–48.

Mitchell, P. 1973. Performance and conservation of osmotic work by proton-coupled solute porte systems. Bioenergetics 3:63–91.

Mount, P. 1970. Histogenesis of the rostellar books of *Taenia crassiceps* (Zeder, 1800) (Cestoda). J. Parasit. 56:47–61.

Murphy, W. A., J. R. Devoll, and R. D. Lumsden. 1977. Phloretin and phloretin/phlorizin effects on glucose transport in *Hymenolepis diminuta*. Prog. Abst. 57th Ann. Meet. Amer. Soc. Parasit. p. 37.

Murphy, W. A., J. R. Devoll, and R. D. Lumsden. 1979. The combined effects of phlorizin and phloretin on glucose transport by a tapeworm, *Hymenolepis diminuta*. Comp. Biochem. Physiol. (submitted).

Newey, H., P. A. Sanford, and D. H. Smyth. 1968. Some effects of ouabain and potassium on transport and metabolism in rat small intestine. J. Physiol. 194:237-48.

Oaks, J. A., W. Knowles, and G. Cain. 1977. A simple method of obtaining an enriched fraction of tegumental brush border from *Hymenolepis diminuta*. J. Parasit. 63:476-85.

Oaks, J. A. and R. D. Lumsden. 1971. Cytological studies on the absorptive surfaces of cestodes. V. Incorporation of carbohydrate-containing macromolecules into tegument membranes. J. Parasit. 57:1256-68.

Pappas, P. W. 1979. Enzyme interactions at the host-parasite interface. *In* this volume.

Pappas, P. W., and B. D. Hansen. 1977. Chloride-sensitive glucose transport in *Hymenolepis diminuta*. J. Parasit. 63:800-804.

Pappas, P. W., and C. P. Read. 1972a. Trypsin inactivation by intact *Hymenolepis diminuta*. J. Parasit. 58:864-71.

Pappas, P. W., and C. P. Read. 1972b. Inactivation of α- and β-chymotrypsin by intact *Hymenolepis diminuta*. Biol. Bull. 143:605-16.

Pappas, P. W., and C. P. Read. 1972c. Sodium and glucose fluxes across the brush border of a flatworm (*Calliobothrium verticillatum*, Cestoda). J. Comp. Physiol. 81:215-28.

Pappas, P. W., and C. P. Read. 1974. Relation of nucleoside transport and surface phosphohydrolase activity in *Hymenolepis diminuta*. J. Parasit. 60:447-52.

Pappas, P. W., and C. P. Read. 1975. Membrane transport in helminth parasites. Exp. Parasit. 37:469-530.

Pappas, P. W., G. L. Uglem, and C. P. Read. 1974. Anion and cation requirements for glucose and methionine accumulation by *Hymenolepis diminuta* (Cestoda). Biol. Bull. 146:56-66.

Phifer, K. O. 1960. Permeation and membrane transport in animal parasites: the absorption of glucose by *Hymenolepis diminuta*. J. Parasit. 46:51-62.

Podesta, R. B. 1977. *Hymenolepis diminuta*: unstirred layer thickness and effects on active and passive transport kinetics. Exp. Parasit. 43:12-24.

Podesta, R. B., W. S. Evans, and E. Stallard. 1977. *Hymenolepis diminuta* and *Hymenolipis microstoma*: effect of ouabain on active nonelectrolyte uptake across the "epithelial" syncytium. Exp. Parasit. 43:25-38.

Podesta, R. B., and D. F. Mettrick. 1974. Components of glucose transport in the host parasite system, *Hymenolepis diminuta* (Cestoda) and the rat intestine. Can. J. Physiol. Pharm. 52:183-97.

Quinton, P. M., E. M. Wright, and J. McD. Tormey. 1973. Localization of sodium pumps in the choroid plexus epithelium. J. Cell Biol. 58:724-30.

Read, C. P. 1950. The vertebrate small intestine as an environment for parasitic helminths. Rice Institute Pamphlet 35:1-94.

Read, C. P. 1955. Intestinal physiology and the host- parasite relationship. *In* W. H. Cole (ed.), Some physiological aspects and consequences of parasitism. Rutgers University Press, New Brunswick, N.J.

Read, C. P. 1959. The role of carbohydrates in the biology of cestodes. VIII. Some conclusions and hypothesis. Exp. Parasit. 8:365-82.

Read, C. P. 1961. Competitions between sugars in their absorption by tapeworms. J. Parasit. 47:1015-16.

Read, C. P. 1966. Nutrition of intestinal helminths. *In* E. Soulsby (ed.), Biology of parasites. Academic Press, New York.

Read, C. P. 1967. Carbohydrate metabolism in *Hymenolepis* (Cestoda). J. Parasit. 53:1023–29.

Read, C. P. 1971. The microcosm of intestinal helminths. *In* A. M. Fallis (ed.), Ecology and Physiology of Parasites. University of Toronto Press, Toronto.

Read, C. P. 1973. Contact digestion in tapeworms. J. Parasit. 59:672–77.

Read, C. P., A. Rothman, and J. E. Simmons, Jr. 1963. Studies on membrane transport, with special reference to parasite-host integration. Ann. N.Y. Acad. Sci. 113:154–205.

Read, C. P., J. E. Simmons, Jr., J. W. Campbell, and A. Rothman. 1960. Permeation and membrane transport in parasitism: studies on a tapeworm-elasmobranch symbiosis. Biol. Bull. 119:120–33.

Read, C. P., G. L. Stewart, and P. W. Pappas. 1974. Glucose and sodium fluxes across the brush border of *Hymenolepis diminuta* (Cestoda). Biol. Bull. 147:146–72.

Revel, J.-P. and S. Ito. 1967. The surface components of cells. In B. Davis and L. Warren (eds.), The specificity of cell surfaces. Prentice-Hall, Engelwood Cliffs, N.J.

Rosenbluth, J. 1962. Subsurface cisterns and their relationship to the neuronal plasma membrane. J. Cell Biol. 13:405–21.

Rothman, A. 1963. Electron microscopy studies of tapeworms: the surface structures of *Hymenolepis diminuta* (Rudolphi, 1819) Blanchard, 1891. Trans. Amer. Micros. Soc. 82:22–30.

Ruff, M., and C. P. Read. 1973. Inhibition of pancreatic lipase by *Hymenolepis diminuta*. J. Parasit. 59:105–11.

Saz, H. J. 1972. Comparative biochemistry of carbohydrates in nematodes and cestodes. *In* H. van den Bossche (ed.), Comparative biochemistry of Parasites. Academic Press, New York.

Schmidt, G. D., and L. S. Roberts. 1977. Foundations of parasitology. C. V. Mosby Co., St. Louis.

Schultz, S. G., and P. F. Curran. 1970. Coupled transport of sodium and organic solutes. Physiol. Rev. 50:637–718.

Schultz, S. G., R. A. Frizzell, and N. Nellans, II. 1974. Ion transport by mammalian small intestine. Ann. Rev. Physiol. 36:51–91.

Schultz, S. G., and R. Zalusky. 1964. Transport of sodium and sugar in rabbit ileum. J. Gen. Physiol. 47:1043–59.

Singer, S. J. 1974. The molecular organization of membranes. Ann. Rev. Biochem. 43:805–33.

Singer, S. J., and G. Nicolson. 1972. The fluid mosiac model of the structure of cell membranes. Science 175:720–31.

Starling, J. A. 1975. Tegumental carbohydrate transport in intestinal helminths: correlation between mechanisms of membrane transport and the biochemical environment of absorptive surfaces. Trans. Am. Micros. Soc. 94:508–23.

Stirling, C. E. 1967. High-resolution radioautography of phlorizin-^{3}H in rings of hamster intestine. J. Cell Biol. 35:605–18.

Stirling, C. E. 1972. Radioautographic localization of sodium pump sites in rabbit intestine. J. Cell Biol. 53:704–14.

Stunkard, H. W. 1962. The organization, ontogeny, and orientation of the Cestoda. Quart. Rev. Biol. 37:23–34.

Swift, J. G., and T. M. Mukherjee. 1976. Demonstration of the fuzzy surface coat of rat intestinal microvilli by freeze etching. J. Cell Biol. 69:491–96.

Threadgold, L. T., and C. P. Read. 1970. *Hymenolepis diminuta*: ultrastructure of a unique membrane specialization in tegument. Exp. Parasit. 28:246–52.

Uglem, G. L. 1976. Evidence for a sodium ion exchange carrier linked with glucose transport across the brush border of a flatworm (*Hymenolepis diminuta*, Cestoda). Biochim. Biophys. Acta 443:126–36.

Walker, S. M., G. R. Schrodt, and M. B. Edge. 1970. Electron dense material within sarcoplasmic reticulum apposed to transverse tubules and to the sarcolemma in dog papillary muscle fibers. Amer. J. Anat. 128:33–44.

Webster, L. A., and R. A. Wilson. 1970. The chemical composition of protonephridial canal fluid from the cestode *Hymenolepis diminuta*. Comp. Biochem. Physiol. 35:201–9.

Wertheim, G., R. Zeledon, and C. P. Read. 1960. Transaminases of tapeworms. J. Parasit. 46:497–99.

Williams, R. J. P. 1975. The binding of metal ions to membranes and its consequences. *In* D. S. Parsons (ed.), Biological membranes. Clarendon Press, Oxford.

JOHN R. SEED, MARY S. BOGUCKI,
AND STEPHEN C. MERRITT

Interactions between Immunoglobulins and the Trypanosome Cell Surface

6

The glycocalyx (fig. 1) of certain clones of African trypanosomes appears to consist of both host serum components and trypanosome antigens (Ag). These constituents are thought to be involved in the parasites' immune escape mechanisms under appropriate immunological conditions (deRaadt and Seed, 1977; Seed, 1974). For example, new variant specific antigens (VSA) are expressed by the trypanosomes following the production and exposure of the variant specific antibody (Ab) to the previous serotype. Host Ags, on the other hand, are believed to be associated with trypanosomes harvested from hosts that do not manifest a detectable variant specific immune response.

HOST SERUM COMPONENTS

By quantitative immunofluorescence, it has been shown that the accretion of plasma components by rodent-harvested *Trypanosoma congolense* is a rapid process *in vivo* and *in vitro* (Diffley and Honigberg, 1977). Both immunoglobulins (Ig) IgG and IgM are associated with the parasites, but only heterospecific Ab activity is recovered in acid eluates of *Trypanosoma brucei gambiense* (*T.b.g.*) (Bogucki, 1978; Bogucki and Seed, 1978). The relative proportions of IgG and IgM in acid eluates of the trypanosomes depends upon the relative concentrations of the Ig classes in the host serum at the time of infection. A significant amount of the heterospecific Ab appears to be adsorbed by the parasite with the Ag-combining region of the molecule facing away from the parasite membrane, free to bind homologous Ag (Bogucki, 1978; Bogucki and

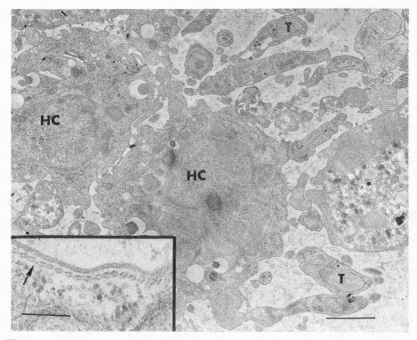

Fig. 1. A transmission electron micrograph of trypanosomes (*Trypanosoma brucei gam-biense) in vivo*. Note the intimate relationship between host cells (HC) and the trypanosomes (T) at this extravascular site; scale bar = 2 μm. The insert is a higher-magnification picture of a portion of the glycocalyx, cell membrane, associated microtubules, and interconnecting microfilaments (arrows). The diameter of the glycocalyx is approximately 12.0 nm and the cell membrane 9.4 nm.

Seed, 1978). Furthermore, different clones of the same trypanosome strain acquire different amounts of heterospecific Ab during rodent passage (Bogucki, 1978). This suggests that either (1) host Igs are bound by VSA and the structural variability of VSA from different clones affects their Ig-binding capacity; or (2) trypanosomes adsorb more host Ig with increased time in serial passage.

If VSA has particular sites that are serving as Ig receptors, it is probable that the receptors are light chain specific. This would explain the parasites' apparent indiscrimination between IgG and IgM-class heterospecific Ab. A specific VSA-Ig light chain interaction would also explain the ability of trypanosome-bound heterospecific Ab to combine with homologous Ag *in situ* (see fig. 2). Since complement has also been shown to be bound to the surface of the trypanosome (Diffley and Honigberg, 1978), it is also conceivable that host Igs are bound to this cell-associated complement through the Fc portion of the molecule. Since the VSA has been shown to activate the complement (Musoke and Barbet, 1977), it would be assumed that complement is bound to the trypanosome by the VSA forming a VSA-complement-Ig complex on the cell surface. In any case, if the parasites accumulate progressively more host Ig with increased time in rodent passage, it may reflect an increased density or altered orientation of VSA (Ig receptors) under these conditions. Alternatively, the density of the surface VSA may be decreased in the absence of host immunity. Host Ig adsorption could then be a compensatory mechanism to preserve the integrity of the glycocalyx. The Igs may bind either to unoccupied membrane receptors for VSA or to other receptors that are exposed only when diminished quantities of VSA are present. If most Igs do indeed bind to VSA or VSA-receptor molecules, it should be possible to induce modulation of the Igs, and to observe co-capping of host and parasite antigens. It should be noted that host serum proteins other than host Igs are also bound to the African trypanosome (Bogucki and Seed, 1978; Diffley and Honigberg, 1977), however, we have not made any attempt at this time to incorporate them into our topographic model (fig. 2).

Trypanosomes harvested from rodent hosts that are known to have host Igs associated with their cell surface are agglutinated (and neutralized) by rabbit anti-blood trypanosome (VSA) sera of either the IgM or IgG class. Therefore, the heterospecific Igs do not appear to block or sterically inhibit specific host Igs from combining with the VSA. However, albino mice and rats do not produce detectable Ab to an infection with our clones of *T.b.g.*, and a single cell will kill a mouse in 5.6 days with a maximum parasitemia of greater than 1.0×10^9 trypanosomes per ml of blood. The trypanosomes appear to grow logarithmically without any sign of a host immune response. It has therefore been suggested that this reduced immune response to our clone TTrT-1 of *T.b.g.* is due to the coating of the trypanosome surface with host Igs. We have hypothesized therefore that these antigens effectively mask the VSA from being recognized as foreign.

The observation (Bogucki, 1978) that late variants from a chronically infected rabbit are difficult to isolate by subpassage in rodents is also consistent with either hypothesis. The VSA expressed by late-appearing variants may have limited regions complementary to host Ig. It is also possible that with an increased number of generations in an immunologically responsive host, the parasites' capacity for binding host Igs through one of the mechanisms proposed above may be impaired.

VARIANT SPECIFIC ANTIBODY

The phenomenon of antigenic variation (fig. 3) has been recently and

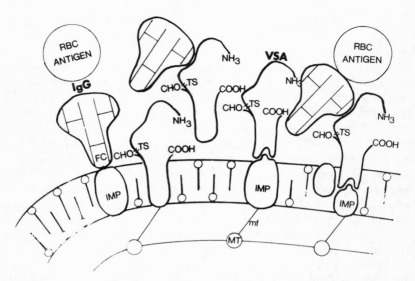

Fig. 2. Diagrammatic representation of the variant specific antigen and host components within the cell membrane of the blood trypanosomes. The VSA is shown as either extending completely through the cell membrane or associated with intramembranous particles (IMP). If the VSA extended through the cell membrane (CM), it could be in contact with both the interior and exterior of the cell and would presumably be in direct contact with the underlying microfilaments (mf) and/or microtubules (mt). The other models show that the VSA extends only partially into the cell membrane and would only be indirectly in contact with the submembranous cytoskeleton.

The molecule is split into 2 portions by trypsin (TS-Trypsin sensitive site). The intact trypanosome has been found to be concanavalin A (Con A) negative, whereas following trypsin treatment the trypanosomes become con A positive (Cross, 1975; Seed et al., 1976). The small 17,000 mw piece of the VSA containing the carbohydrate (CHO) moiety (con A positive piece) would appear to be the portion of the molecule associated with the cell membrane and is hidden in the intact molecule by the larger portion of the molecule. This small piece of the molecule is assumed to contain regions that are similar (if not identical) between the different VSA, since it is this portion of the molecule which presumably interacts with the IMP and/or the mf (or mt). This model is basically identical to that proposed earlier by Cross and Johnson (1976), and by Cross (1978), except that it incorporates the association of the VSA directly or indirectly with the mfs (and/or mt) elements of the cell.

The host immunoglobulins (Ig) that are associated with the cell membrane are shown as being either bound through their Fc portion of the molecule to a receptor IMP or by the variable region of the light chain to the VSA. It is somewhat difficult to envision the host Ig combining directly with an Fc receptor site since (1) con A is unable to penetrate the glycocalyx and combine with the CHO portion of the VSA molecule, and (2) other investigators have suggested that the cell surface is covered with an almost continuous layer of VSA molecules. However, in both models the bound Igs are shown with their variable regions facing outward with at least a portion of the molecules sufficiently exposed to the exterior environment to be able to combine with large particular antigens (Ag-RBC).

extensively reviewed (Cross, 1978; Cross and Johnson, 1976; deRaadt and Seed, 1977; Vickerman, 1978). It is assumed for the purposes of this discussion that the phenomenon is similar to the Ab-induced phenotypic change observed in the immobilization antigens of *Paramecium aurelia* (Baele, 1974). This is similar to models previously discussed by other investigators (Cross, 1978; Cross and Johnson, 1976; Vickerman, 1978). A very recent report has suggested that a single clone of trypanosomes contains at least two different genes coding for different antigens (Williams et al., 1978). This work would eliminate the prior suggestions that frameshift and/or suppressor mutations are the predominate mechanism for antigenic variation (Seed et al., 1977). There have been several reports where antibody has been suggested to induce a phenotypic change in the VSA (Inoki et al., 1956; Takayanagi and Enriquez, 1973). However, in one case the experiments in our opinions have not been successfully repeated, and in the other the data were obtained in systems that did not totally rule out phenotypic change with selection by antibody as the possible mechanism. We wish to note that Le Ray and colleagues (1978) have obtained evidence that would suggest that VSA-specific antibody is not required for antigenic variation to occur. Therefore, in our opinion, there is not yet any definite proof that Ab can act as an inducer.

In the paramecium-like model, Ab is hypothesized to induce a phenotypic change in the trypanosome's serotype through an interaction with VSA on the cell surface. It has been stated therefore that antibody plays two roles in animals chronically infected with *T.b.g.* Antibody may act (1) as an inducer of antigenic variation at low concentrations and (2) as a lytic or a selective agent at higher concentrations (fig. 4). It has also been proposed (Takayanagi and Enriquez, 1973) that IgM is the more effective class of Ab at induction of phenotypic change (fig. 4). Furthermore, the concentration of Ab required to induce a change or to cause cell lysis may depend not only on concentration but also on the host niche (extravascular vs. intravascular) in which the interaction between the trypanosome and Ab occurs. The concentration of complement is also assumed to be important in these reactions. Therefore, the curve shown in figure 4 could presumably be displaced to either the right or left depending on environmental conditions (fig. 5). It may be significant that changes in temperature and nutritional status of *Paramecium* also induce changes in the immobilization antigens (Beale, 1974).

We believe that Ab has been clearly and repeatedly demonstrated to play a role in killing the trypanosomes (Seed, 1977). It has been found to increase the attachment of the trypanosomes to macrophages, and thereby increase the rate of phagocytosis (Seed, 1964; Stevens and Moulton, 1978;

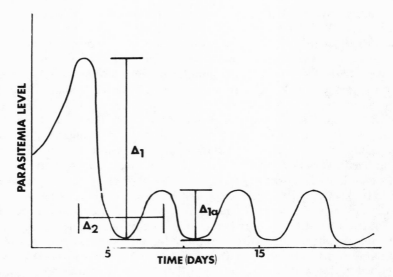

Fig. 3. Diagrammatic representation of parasitemia levels in the blood of animals infected with *Trypanosoma brucei gambiense*. The decrease in cell numbers at Δ1 and Δ1a is presumably due to antibody (Ab)-mediated destruction of the trypanosomes. The increase in cell number (Δ2) at the second relapse is suggested to be the result of an early induction of a new varient type by Ab (or other environmental factors) followed by the selection of this new variant by specific antibody. Relapses in the blood of rabbits are known to occur every 3 to 5 days.

Takayanagi et al., 1974). These studies are continuations of earlier ones on immune adherence (Lamana, 1957). This phagocytic process has been found by some authors to be complement-dependent (Stevens and Moulton, 1978). In addition, it has been shown that antibody can be both lytic and/or cytotoxic depending upon the presence of either the intact or the alternate complement pathway (Flemming and Diggs, 1978). Cytotoxicity was measured by determining the reduction in the rate of incorporation of leucine into protein (Flemming and Diggs, 1978).

The VSA has been shown by both surface-labeling techniques and immunoelectron microscopy to be located within the surface coat (glycocalyx) of the African trypanosome (Cross, 1975; Fruit et al., 1977; Vickerman and Luckins, 1969). It is unclear at present whether the Ag is buried within the cell membrane or is external to the outer lipid layer. Our laboratory has shown (Merritt and Seed, 1977) that when low (sub-lytic) doses of Ab are incubated with the homologous trypanosomes, the Ag-Ab complexes are displaced to either the anterior or posterior end of the

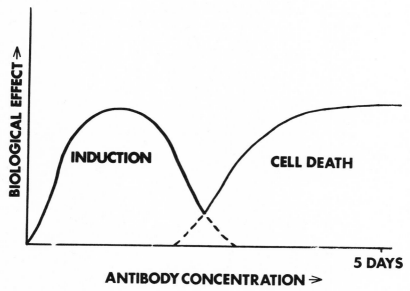

Fig. 4. The proposed roles of varying concentrations of IgM type antibody to *Trypanosoma brucei gambiense*. At low concentrations of variant specific IgM type antibody, it is hypothesized that this Ab combines with the VSA in the glycocalyx and initiates in a limited number of cells a series of events resulting in gene activation. There is a critical concentration of antibody required for this phenomena that when exceeded results in cell death. This concentration could vary and be shifted to the left or right depending upon the environmental parameters (see fig. 5).

trypanosome and are rapidly lost from the cell surface (fig. 6). Similar results have been reported by others (Barry and Vickerman, 1977; Doyle, 1977). This loss of Ag has been shown to be energy-dependent, and appears analogous to the capping of lymphocyte surface antigens. It has also been shown to be cytochalasin B and D sensitive, and is, therefore, assumed to be microfilament (MF)-dependent (Merritt and Seed, unpublished). Thus, it is likely that the membrane-associated VSA is linked either directly or indirectly to a submembranous cytoskeleton. If VSA is directly linked to the MF, it must extend through the lipid bilayer and would require the presence of a substantial hydrophobic moiety in the numerous different VSAs. Alternatively, there may be a membrane receptor protein specific for a common region on the VSA. This integral membrane, or transmembranous protein, would presumably be attached to the submembranous MF (fig. 1). In a related organism, the microtubules (MT)

138   *Cellular Interactions in Symbiosis and Parasitism*

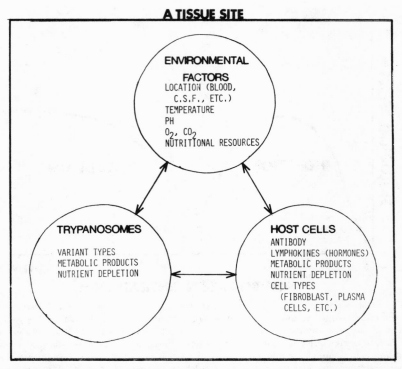

Fig. 5. A diagrammatic representation of the complex interaction at a localized tissue site between the host, the trypanosomes, and some of the factors involved.

plus interconnecting filaments were shown to be linked to the cell membrane (Linder and Staehelin, 1977).

Cross (1975) has approximated the size of a VSA molecule at $4.0 \times 4.0 \times 10.0$ nm. Based on the molecular size of globular proteins (i.e., IgG) (Humphrey and While, 1970), we have estimated the dimensions of the VSA at 4.0 nm in diameter $\times$ 6.0 nm in depth. Since the VSA is exposed at the cell surface, and the cell membrane is approximately 9.4 nm in width (fig. 1), a transmembranous receptor protein appears necessary to provide the attachment of VSA to the cellular MF. In fact, trypsin treatment of live trypanosomes, by releasing a 48,000 MW fragment of the VSA into the supernatant (Cross, 1975), causes the cells to become serologically negative (Cross, 1975; Seed et al., 1976, unpublished). It would therefore appear that very little of the VSA (a 17,000 MW piece) is bound within the cell membrane. Barbet and McGuire (1978) recently presented substantial evidence for the presence of a common antigenic site between serologically different VSAs. This common region could be the portion of the VSA

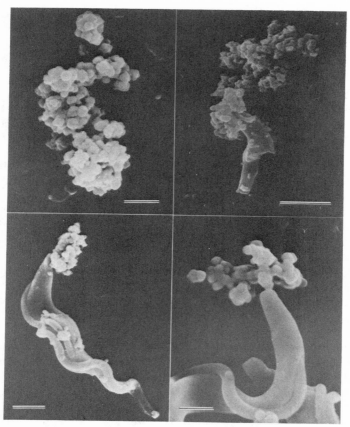

Fig. 6. Antibody-mediated redistribution of variant specific antigen on *Trypanosoma brucei gambiense* cell surface. Goat anti-rabbit IgG antiserum was fractionated by DEAE-cellulose and the IgG fraction was attached to methacylate latex beads by the method of Molday et al. (1975). Cells were labeled with the IgG fraction of rabbit anti-TTrt-1 at 0°C, washed repeatedly, and labeled with the bead cell surface marker at 0°C. After various times intervals at 37°C, the cells were fixed and processed for scanning electron microscopy and the redistribution of antigen was monitored. No binding was observed in controls where the specific pre-label was omitted or where a heterologous (anti-TTrT-17) antiserum was substituted. Movement was sensitive to sodium fluoride, 2-deoxy-D-glucose and cytochalasin B and D. No change in redistribution was noted with colchicine or sodium azide.

Upper left. One min after incubation at 37°C; bar represents 1.5 μm. *Upper right.* Three min; bar represents 3 μm. *Lower left.* Five min; bar represents 1.5 μm. *Lower right.* Ten min; bar represents 0.8 μm.

molecule that binds to the membrane receptor proteins. Cross (1978) has noted that the amino acid composition of the smaller 17,000 MW piece is very polar and therefore suggested that the VSA molecule is not deeply embedded within the cell membrane. He hypothesized that the VSA is bound to the surface of the cell membrane. We conclude, therefore, that the VSA is bound to a transmembranous receptor site that connects the VSA to the cytoskeletal elements beneath the cell membrane.

Capping of the VSA determinants may be the first phase of induction of phenotypic variation. In our laboratory we have hypothesized that if Ab is truly the inducer, a change in metabolic parameters should be observed following treatment of the cells with the appropriate concentrations of Ab, and under the proper environmental conditions. To test this hypothesis, cells have been incubated with various concentrations of Ab and radioactive precursors of DNA, RNA, and protein. The rates of incorporation of these labeled compounds into macromolecules (a VSA-containing fraction) have been measured. Preliminary results have shown variable rates of incorporation of label, with the slight suggestion that at low (inducer) doses of homologous Ab there appears to be a very slight stimulation of nucleic acid and protein synthesis. This stimulation may reflect an increased incorporation of amino acids into VSA, or of nucleotides into relevant message. As anticipated, the preliminary results suggest that homologous Ab has extremely different effects upon the cells, varying from possible metabolic stimulation to producing cell death (fig. 4).

Both the class and concentration of Ab may vary greatly at different sites within the host, e.g., vascular bed vs. extravascular sites. Due to its size IgM (19S) may be retained to a greater extent within the vascular bed, and trypanosomes in the blood would be more rapidly exposed to high concentrations of IgM antibody. The rate of antigenic variation and the selection of variants by Ab could therefore vary in different anatomical sites. It has been found that antigenic variants appear in the lymph before they are observed in the blood (M. Tanner and L. Jenni, personal communication).

It appears that the antibody specific to Ags on the cell surface can (1) alter the arrangement of the VSA within the surface coat at low concentrations, and (2) at high concentrations produce cell death and lysis. Antibody therefore acts as a selective agent removing the major antigenic types from the population. The temporary rearrangement of the membrane-associated VSA, as discussed previously, may precede a more permanent expression of a new serotype. Once a change in the variant specific Ag has occurred in one or more cells, the selective action by variant specific Ab allows the new minor antigenic variant to proliferate, eventually producing major peaks of parasitemia.

In conclusion, host Igs are suggested to prevent immune recognition of trypanosome Ags in acute infection, as well as to act as selective and/or variation-inducing agents in chronic African trypanosomiasis. The topographic relationships envisioned between the VSA, the host Ig, and the cell membrane are shown in figure 2. The interaction between the VSA and the variant specific host Igs in the trypanosome glycocalyx appears to be complex and dynamic, involving capping of the VSA; metabolic stimulation when the VSA is combined with Ab; possibly the induction of phenotypic change at low Ab concentrations; and the selection of minor variants at higher Ab concentrations. In addition, in hosts with acute infections the trypanosomes appear to acquire substantial amounts of heterospecific Igs on their surface. This interaction between the cell membrane and the heterospecific Igs also appears to be complex. It has been hypothesized that they effectively mask the VSA and prevent their recognition by the host immune system. The trypanosomes therefore appear to have evolved at least two distinct mechanisms for escaping the host's immune response, both possibly involving molecular interactions of parasite antigens with host Igs at the cell surface.

ACKNOWLEDGMENTS

The authors would like to thank Dr. Thomas M. Seed, the Argonne National Laboratory, Argonne, Illinois, for his help with the electron microscopy, and Mr. John Sechelski for his excellent technical assistance.

REFERENCES

Barbet, A. F., and T. C. McGuire. Crossreacting determinants in variant-specific surface antigens of African tryanosomes. Proc. Natl. Acad. Sci. 75:1989–93.

Barry, J. D., and K. Vickerman. 1977. Further studies on the effect of antibodies on the surface of *Trypanosoma brucei*. Parasitology 75:XXX.

Beale, G. H. 1974. Genetics of antigenic variation in *Paramecium aurelia*. *In* Parasites in the immunized host: mechanisms of survival, pp. 21–27. Ciba Foundation Symposium no. 25. Associated Scientific Publishers, Amsterdam.

Bogucki, M. S. 1978. Noncytotoxic host immunoglobulins in experimental trypanosomiasis and malignant neoplasia. Ph.D. dissertation, Texas A&M University.

Bogucki, M. S., and J. R. Seed. 1977. Host antigenic determinants of *Trypanosoma brucei gambiense* and Ehrlich ascites tumor cells. *In* W. M. Kemp (ed.), Experimental parasitology of host-parasite interfaces, pp. 15–27. Texas J. Sci. (Special Publication no. 2).

Bogucki, M. S., and J. R. Seed. 1978. Parasite-bound heterospecific antibody in experimental African trypanosomiasis. J. Reticuloendothel. Soc. 23:89–101.

Cross, G. A. M. 1975. Identification, purification, and properties of clone-specific glyco-protein antigens constituting the surface coat of *Tryanosoma brucei*. Parasitology 71:393–417.

Cross, G. A. M., 1978. Antigenic variation in trypanosomes. Proc. R. Soc. Lond., Ser. B. 202:55–72.

Cross, G. A. M., and Johnson, J. G. 1976. Structure and organization of the variant-specific surface antigens of *Trypanosoma brucei*. *In* H. van den Bossche (ed.), Biochemistry of parasites and host-parasite relations, pp. 413–20. Elsevier/North Holland Biomedical Press, Amsterdam.

deRaadt, P., and J. R. Seed. 1977. Trypanosomes causing disease in man in Africa. In J. P. Kreier (ed.), Parasitic Protozoa, pp. 175–237. Academic Press, New York.

Diffley, P., and B. M. Honigberg. 1977. Flourescent antibody analysis of host plasma components on bloodstream forms of African pathogenic trypanosomes. I. Host specificity and time of accretion in *Trypanosoma congolense*. J. Parasit. 63:599–606.

Diffley, P., and B. M. Honigberg. 1978. Immunologic analysis of host plasma proteins on bloodstream forms of African pathogenic trypanosomes. II. Identification and quantita-tion of surface-bound albumin, non-specific IgG, and complement on *Trypanosoma congolense*. J. Parasit. 64:674–81.

Doyle, J. J. 1977. Antibody induced movement of variant specific surface antigens of *Trypanosoma brucei in vitro*. Abstract Fifth Internatl. Congress of Protozool. No. 245.

Flemming, B., and C. Diggs. 1978. Antibody-dependent cytotoxity against *Trypanosoma rhodesiense* mediated through an alternative complement pathway. Infect. Immun. 19:928–33.

Fruit, J., D. Afchain, A. Petitprez, N. van Meirvenne, D. Le Ray, D. Bout, and A. Capron, 1977. Antigenic analysis of a variant-specific component of *Trypanosoma brucei brucei*: localization on the surface coat with labelled specific antibodies. Parasitology 74:185–90.

Humphrey, J. H., and R. G. While. 1970. Immunology for students of medicine, 3d ed. p. 169. F.A. Davis Co., Philadelphia, Pa.

Inoki, S., M. Osaki, and T. Nakabayasli. 1956. Studies on the immunological variation in *Trypanosoma gambiense*. II. Verification of the new variation system by Ehrlich's and *in vitro* methods. Med.J. Osaka Univ. 7:165–73.

Lamana, C. 1957. Adhesion of foreign particles to particulate antigens in the presence of antibody and complement (serological adhesion). Bac. Rev. 21:30–45.

Le Ray, D., J. D. Barry, and K. Vickerman. 1978. Antigenic heterogeneity of metacyclic forms of *Trypanosoma brucei*. Nature 273:300–302.

Linder, J. D., and L. A. Staehelin. 1977. Plasma membrane specialization in a trypanosoma flagellate. J. Ultrastructure Res. 60:246–62.

Merritt, S. C., and J. R. Seed. 1977. Capping in *Trypanosoma brucei gambiense*. *In* Immune mechanisms in African trypanosomiasis. Workshop at the 26th Annual Meeting of the American Society of Tropical Medicine and Hygiene.

Molday, R. S., W. J. Dreyer, A. Rembaum, and S. P. S. Yen. 1975. New Immunolatex spheres: visual markers of antigens on lymphocytes for scanning electron microscopy. J. Cell. Biol. 64:75–88.

Musoke, A. J., and A. F. Barbet. 1977. Activation of complement by variant-specific surface antigen of *Trypanosoma brucei*. Nature 270:438–40.

Seed, J. R. 1964. Factors responsible for destruction of the culture form of *Trypanosoma rhodesiense in vivo*. J. Inf. Dis. 114:119–24.

Seed, J. R. 1974. Antigens and antigenic variability of the African trypanosomes. J. Protozool. 21:639–46.

Seed, J. R. 1977. The role of immunoglobulins in immunity to *Trypanosoma brucei gambiense*. Int. J. Parasit. 7:55–60.

Seed, J. R., W. M. Kemp, and R. A. Brown. 1977. Antigenic variation in the African trypanosomes: number and sequence. *In* W. M. Kemp (ed.), Experimental parasitology of host-parasite interfaces, pp. 3–13. Texas J. Sci. (Special Publication no. 2).

Seed, T. M., J. R. Seed, and D. Brindley. 1976. Surface properties of bloodstream trypanosomes (*Trypanosoma brucei brucei*) Tropenmed. Parasit. 27:202–12.

Stevens, D. R., and Moulton, J. E. 1978. Ultrastructural and immunological aspects of the phagocytosis of *Trypanosoma brucei* by mouse peritoneal macrophages. Infect. Immun. 19:972–82.

Takayanagi, T., and G. L. Enriquez. 1973. Effects of the IgG and IgM immunoglobulins in *Trypanosoma gambiense* infections in mice. J. Parasitol. 59:644–47.

Takayanagi, T., Y. Nakatake, and G. L. Enriquez. 1974. Attachment and ingestion of *Trypanosoma gambiense* to the rat macrophage by specific antiserum. J. Parasit. 60:336–39.

Vickerman, K. 1978. Antigenic variation in trypanosomes. Nature 273:613–17.

Vickerman, K., and Luckins. 1969. Localization of variable antigens in the surface coat on *Trypanosoma brucei* using ferritin conjugated antibody. Nature 224:1125–26.

Williams, R. O., K. B. Marcu, J. R. Young, L. Rovis, and S. C. Williams. 1978. Characterization of messenger RNA activities and their sequence complexities in *Trypanosoma brucei*-partial purification and properties of VSSA. Nucleic Acid Res. 5:3171–82.

PETER W. PAPPAS

Enzyme Interactions at the Host-Parasite Interface

7

INTRODUCTION

The host-parasite relationship can be described using a number of different criteria, including morphological, pathological, immunological, and biochemical characteristics. In many host-parasite relationships, the ultimate success or failure of the relationship, from the parasite's standpoint, will depend on the properties of the surface or surface membranes of the parasite (Lumsden, 1975a); the properties of this external surface determine, to a great extent, how the parasite and host will interact.

Parasitic organisms have evolved a number of amazing adaptations to their life styles, some of which are readily apparent (e.g., the loss of a digestive tract in the Cestoda and Acanthocephala). Certain groups of parasitic helminths and protozoa have evolved more subtle, but by no means less important, adaptations involving their surface membranes, adaptations that influence host-parasite interactions. An excellent example of this is the ability of schistosomes (blood flukes) to adsorb host blood proteins to their outer surface and "mimic" immunologically the host. In this case the host's immune response does not recognize the antigenic properties of the parasite's surface, and the parasite is unharmed even though circulating antibodies are present (Smithers et al., 1969; Clegg et al., 1970; Damian et al., 1973). An additional example is "modulation of antigenic determinants" in the trypanosomes (see Seed et al., 1980).

The adaptive significance of the above examples is apparent. However, the surfaces of parasitic helminths, in many instances, must not only protect the parasite from a potential host response, but must also serve as a digestive-absorptive surface as well. For example, cestodes and acanthocephalans lack any remnant of a functional digestive tract, and absorption of nutrients (and digestion, when it occurs) must take place across (or at) their external surfaces. (The fact that cestodes lack a digestive tract has made them a favorite "model" for the study of membrane-associated phenomena in parasitic helminths.) Most trematodes possess a functional digestive tract that almost certainly carries out these functions. Despite the presence of a functional digestive tract, the external surfaces of many trematodes also appear to carry out digestive-absorptive functions, as will be discussed below.

As alluded to above, two major functions of the external surfaces of parasitic helminths are protection and nutrition acquisition; the external surfaces of cestodes, trematodes, and acanthocephalans appear ideally suited, morphologically, for an absorptive-digestive role. Ultrastructurally, these surfaces (or teguments) appear as syncytial epithelial, with the distal portion of the syncytium modified in various ways. These modifications include multiple ridges and/or folds in some trematodes, microtriches (structurally similar to microvilli of the vertebrate small intestine) in adult cestodes, and crypts, pits, or channels in adult acanthocephalans.. All trematodes, cestodes, and acanthocephalans studied to date also possess a glycocalyx (fuzzy coat) that overlies the distal portion of the tegument. For excellent reviews of the ultrastructure of the helminth tegument, readers are referred to Lumsden (1975b) and Lyons (1977).

The teguments of trematodes, cestodes, and acanthocephalans do carry out absorptive functions; previous studies have characterized numerous mediated transport systems in these helminth parasites (see review of Pappas and Read, 1975); these mediated transport systems are important when considering the significance of membrane-bound enzymes in helminth parasites. In addition to having intrinsic enzymes (i.e., enzymes of parasite origin), the teguments of these parasitic organisms may also interact with enzymes of the host. Intrinsic tegumentary enzymes, as well as transport systems, must play an important role in the absorptive-digestive capabilities of the helminth tegument. Interactions of host's enzymes with the helminth tegument may also be an important component of these capabilities, in addition to protecting gastrointestinal parasites from the potentially destructive action of these same enzymes.

ENZYMES OF PARASITE ORIGIN

Phosphohydrolase Activity

Of the "surface" (tegumentary, membrane-bound) enzymes associated with helminth surfaces, phosphohydrolases have received the most attention. Although phosphohydrolases have been demonstrated in the surface of numerous cestode species (table 1), these membrane-bound enzymes have been studied in detail in only one species, *Hymenolepis diminuta*.

Evidence of a "functional" surface phosphohydrolase in *H. diminuta* was first reported by Phifer (1960), when he demonstrated that *p*-nitrophenylphosphate (PNØP) is hydrolyzed in the presence of intact *H. diminuta*. Subsequent studies, discussed below, have provided both direct and indirect evidence that numerous phosphorylated monosaccharides and nucleotides are hydrolyzed at the surface of *H. diminuta*, and that this phosphohydrolase activity (involving several enzymes) is localized at the external tegumentary membranes. Rothman (1966), using ultrastructural techniques, demonstrated localized phosphohydrolase activity in the surface membrane of *Hymenolepis citelli*, and Lumsden and colleagues (1968) and Dike and Read (1971a) reported similar localization of phosphohydrolase activity in *H. diminuta* using PNØP, hexose phosphates, and nucleotides as substrates. Additional evidence for the membrane-bound nature of the enzymatic activity was provided by Arme and Read (1970). These authors reasoned that if the phosphohydrolase activity was associated with the tegument of *H. diminuta*, then fructose derived from fructose-1, 6-diphosphate (FDP) hydrolysis should accumulate in the incubation medium, since the cestode is impermeable to this ketohexose. Arme and Read (1970) found significant hydrolysis of FDP in the presence of *H. diminuta* (i.e., fructose from FDP hydrolysis did accumulate in the incubation medium), and their techniques were later utilized by Dike and Read (1971a) to study hydrolysis of various fructose esters by *H. diminuta*. More recently, Oaks and colleagues (1977) have isolated a tegumentary membrane fraction from *H. diminuta* and found significant phosphohydrolase activity associated with this external membrane fraction.

Numerous phosphate esters are hydrolyzed at the surface of *H. diminuta*, and results of available kinetic and inhibitor studies suggest that more than one surface phosphohydrolase is present. Before discussing these data, however, the experimental techniques utilized by various investigators will be described; this is appropriate here, since similar techniques have been used to study different surface enzymes in many species of parasitic helminths. One approach has been to incubate *H. diminuta* with a substrate and then analyze the incubation medium for hydrolysis products. It has been noted above that this technique relies on the impermeability of *H. diminuta* to hydrolysis products (e.g., fructose). This technique has also been used to study hydrolysis of various *p*-

TABLE 1

A SUMMARY OF HELMINTH PARASITES IN WHICH PHOSPHOHYDROLASE
(PHOSPHATASE) ACTIVITY HAS BEEN DEMONSTRATED IN THE TEGUMENT

Parasite	Reference(s)
Trematoda	
Aspidogaster conchicola	Trimble et al., 1971
Diclidophora merlangi	Halton, 1967
Octodactylus palmata	Halton, 1967
Eurytrema coelomaticum	Yamao, 1954
Eurytrema pancreaticum	Yamao, 1954
Dicrocoelium lancetum	Yamao, 1954
Clonorchis sinensis	Yamao, 1954
Fasciola hepatica	Halton, 1967; Barry et al., 1968; Threadgold, 1968*; Thorpe, 1968
Fasciola gigantica	Probert et al., 1972
Plagioporus lepomis cercariae	Porter and Hall, 1972
Posthodiplostomum minimum	Bogitsh, 1966
Schistosoma mansoni	Dusanic, 1959; Lewert and Dusanic, 1961; Halton, 1967; Robinson, 1961; Wheater and Wilson, 1976.
Cestoda	
Anoplocephala perfoliata	Yamao, 1952a
Anoplocephala magna	Yamao, 1952a
Cysticercus bovis	Yamao, 1952b
Cysticercus fasciolaris	Yamao, 1952b
Cysticercus tenuicollis	Erasmus, 1957b
Diphyllobothrium dendriticum (adult and plerocercoid)	Öhman-James, 1968
Dipylidium caninum	Tarazona Vilas, 1958; Waitz and Schardein, 1964
Echinococcus granulosus	Rogers, 1947; Kilejian et al., 1961
Echinococcus hydatid cyst	Yamao, 1952b
Hymenolepis citelli	Rothman, 1966*
Hymenolepis diminuta	Waitz and Schardein, 1964; Lumsden et al., 1968*; Dike and Read, 1971a*
Hymenolepis nana	Waitz and Schardein, 1964
Ligula intestinalis (adult and plerocercoid)	Arme, 1966
Moniezia expansa	Erasmus, 1957b; Tarazona Vilas, 1958
Schistocephalus solidus plerocercoid	Morris and Finnegan, 1968
Spirometra erinacei plerocercoid	Kwa, 1972b
Taenia pisiformis	Erasmus, 1957a
Taenia (-Hydatigera) taeniaeformis adult	Yamao, 1952a; Tarazona Vilas, 1958; Waitz, 1963; Waitz and Schardein, 1964
Taenia taeniaeformis strobilocercus	Waitz, 1963

TABLE 1—*Continued*

Parasite	Reference(s)
Acanthocephala	
Cornynosoma wegneri	Bullock, 1958
Echinorhynchus coregoni	Bullock, 1949
Echinorhynchus gadi	Bullock, 1958
Fessisentis vancleavi	Bullock, 1958
Illisentis furcatus	Bullock, 1958
Macracanthorhynchus hirudinaceus	Bullock, 1958
Macracanthorhynchus ingens	Bullock, 1958
Moniliformis dubius	Bullock, 1958; Rothman, 1967*; Byram and Fisher, 1975*
Neoechinorhynchus cylindratus	Bullock, 1949
Neoechinorhynchus emydis	Bullock, 1949
Polymorphus minutus	Crompton, 1963
Polymorphus sp.	Bullock, 1958
Pomporhynchus bulbocoli	Bullock, 1949
Telosentis tenuicornis	Bullock, 1958

* Denotes those studies that have utilized ultrastructural techniques to localize phospho-hydrolase activity.

nitrophenol derivatives; following incubation of worms with these derivatives, released *p*-nitrophenol can be determined spectrophotometrically. *H. diminuta* is, however, readily permeable to *p*-nitrophenol (the worms take on a distinct yellow color, even during a 2 min incubation), and care must be taken to extract any absorbed *p*-nitrophenol from worms, and then add this value to that obtained from the incubation medium.

Another approach for demonstrating surface phosphohydrolase activity takes advantage of the fact that mediated uptake of specific substrates can be inhibited by similar compounds (i.e., the process shows stereospecificity). Thus, if a compound is hydrolyzed at the worm's surface and a hydrolysis product interacts with a specific transport locus, this interaction can be detected by measuring the inhibition of mediated uptake through that same transport locus. For example, Pappas and Read (1974) demonstrated adenosine triphosphate (ATP) hydrolysis by measuring the inhibitory effect of adenosine (from ATP hydrolysis) on mediated uridine uptake by *H. diminuta*, and Uglem and colleagues (1974) demonstrated glycerophosphate hydrolysis by measuring the effect of glycerol (from glycerophosphate hydrolysis) on mediated glycerol uptake in *H. diminuta*. Both studies also demonstrated that the inhibitory effect of the phosphate esters on mediated uptake is relieved (or reversed) when phosphohydrolase inhibitors are added to the incubation medium, thereby demonstrating that

(1) phosphate esters do not interact directly with the transport locus and (2) hydrolysis of the esters is necessary to produce inhibition of mediated uptake. Such an experimental design necessitates certain requirements of the system being studied. Two obvious requirements are that the organism (*H. diminuta* in this case) possess (1) surface phosphohydrolase activity and (2) a mediated transport system for hydrolysis products. Also, the enzyme(s) and transport locus must be in close proximity such that hydrolysis products do not diffuse away from the worm, but rather are rapidly absorbed. Such a close spatial relationship results in what has been termed a "kinetic advantage" for the absorption of hydrolysis products. There is convincing evidence for such an "advantage" in *H. diminuta*; Dike and Read (1971b) demonstrated that the "apparent" uptake of glucose-6-phosphate (G6P) was inhibited by several nucleotides and hexose phosphates, or by deleting the Na^+ from the medium; glucose transport in this species is inhibited 96%+ in Na^+-free media (Pappas et al., 1974; Read et al., 1974). In order to demonstrate that glucose liberated from glucose phosphate hydrolysis did not diffuse away from the worms, Dike and Read (1971b) incubated worms in G6P and glucose oxidase. Glucose oxidase in the incubation medium would oxidize any glucose diffusing away from the worms to glucuronic acid, and worms were shown experimentally to be impermeable to this glucose derivative. Since glucose oxidase in the incubation medium had no effect on the "apparent" uptake of G6P, Dike and Read (1971b) concluded that none of the glucose liberated during G6P hydrolysis diffused away from the worms.

Hydrolysis of several phosphate esters by *H. diminuta* (and the effects of phosphohydrolase inhibitors on hydrolysis) has been studied. Fructose phosphate hydrolysis has been determined by incubating worms in fructose phosphates (and potential inhibitors) and determining chemically fructose liberated into the surrounding medium (Arme and Read, 1970; Dike and Read, 1971a). Liberation of *p*-nitrophenol was determined spectrophotometrically to measure hydrolysis of various *p*-nitrophenol derivatives (Phifer, 1960; Dike and Read, 1971b; Pappas and Read, 1974; Lumsden and Berger, 1974). Hydrolysis of glucose phosphates (Dike and Read, 1971b; Pappas and Read, 1974), glycerophosphate (Pittman and Fisher, 1972; Uglem et al., 1974) and nucleotides (Lumsden et al., 1968; Pappas and Read, 1974) has been studied by measuring the effects of various phosphate esters on mediated uptake of unphosphorylated substrates (e.g., the effect of G6P on glucose uptake), or by measuring the effects of potential phosphohydrolase inhibitors on the "apparent" uptake of phosphate esters (as discussed above). Despite the numerous inhibitor studies (table 2), kinetic data describing the hydrolysis of only three phosphate esters have been published: V_{max} for PNØP hydrolysis is 20.7 μmoles hydrolyzed/g ethanol extracted dry wt of worm tissue/2 min (Pappas and Read, 1974); the K_m has been reported to be 0.69 mM (Dike and Read, 1971a) and 0.415 mM (Pappas and Read, 1974). Hydrolysis is inhibited in the absence of Zn^{++} (Lumsden and Berger, 1974), and shows a broad pH optimum of 7.4–9 (Dike and Read, 1971a). The K_m for FDP hydrolysis has been reported to be 1.36 mM (Dike and Read, 1971a), but may be as low as 0.7 mM (calculated from the data of Arme and Read, 1970). FDP hydrolysis shows substrate inhibition above 2 mM FDP (Arme and Read, 1970), and a pH optimum of 7–7.4 (Arme and Read, 1970; Dike and Read, 1971a). The K_m and pH optimum for fructose-1-phosphate hydrolysis are 11.78 mM and 7–7.4, respectively (Dike and Read, 1971a).

TABLE 2

A SUMMARY OF INHIBITOR STUDIES DEMONSTRATING THE
SPECIFICITY OF SURFACE PHOSPHOHYDROLASE ACTIVITY IN HYMENOLEPIS DIMINUTA

(A+ indicates inhibition of substrate hydrolysis by the inhibitor, and a– indicates no inhibition; blank spaces indicate no data available. The numbers in parentheses refer to the appropriate reference.)

SUBSTRATE	INHIBITOR												
	FIP	F6P	FDP	G1P	G6P	GP	PNØP	AMP	ATP	NaF	Molybdate	Phlorizin	p-CMB
F1P							+(4)	+(4)	+(4)		+(4)	+(4)	+(4)
FDP				+(4)	+(2) +(4)	+(2) +(4)		+(4)	+(4)	-(4)		+(4)	
G1P							+(6)	+(6)	-(6)				
G6P	+(5)		+(5)	+(5)			+(5) +(6)	+(6)	-(6)		+(5)		
GP							+(7)	+(7)			+(7)		
PNØP	-(4) +(8)	-(4)	-(4)	-(4) +(8)	-(4)	-(4) +(8)		+(4) +(6)	-(4) +(6) +(8)	-(4)	+(3) +(4)	-(4)	-(4) +(8)
AMP				-(6)			-(6)			-(6)	-(6)		
ATP				+(6)			+(6)			+(1) +(6)	+(6)		

Abbreviations: F1P = fructose-1-phosphate; FDP = fructose-1, 6-diphosphate; F6P = fructose-6-phosphate; G1P = glucose-1-phosphate; G6P = glucose-6-phosphate; GP = glycerophosphate; PNØP = p-nitrophenylphosphate; AMP = adenosine monophosphate; ATP = adenosine triphosphate; p-CMP = p-chloromercuribenzoate.

References: (1) = Lumsden et al., 1968; (2) = Arme and Read, 1970; (3) = Phifer, 1960; (4) = Dike and Read, 1971a; (5) = Dike and Read, 1971b; (6) = Pappas and Read, 1974; (7) = Uglem et al., 1974; (8) = Lumsden and Berger, 1974.

A basic objective of the inhibitor studies summarized in table 2 has been to determine the number of surface phosphohydrolases in (or on) *H. diminuta*. From their data Dike and Read (1971a) suggested two distinct enzymes in *H. diminuta*. Lumsden and Berger (1974) characterized these enzymes as "a *p*-CMB (*p*-chloromercuribenzoate)-insensitive activity with a high affinity for PNØP and a low affinity for other phosphate esters," and "a *p*-CMB-sensitive activity with a high affinity for sugar phosphates and nucleotides as well as PNØP." Dike and Read (1971a) postulated a third enzyme based on fluoride sensitivity of surface ATP hydrolysis noted by Lumsden and colleagues (1968). The data of Pappas and Read (1974) supported the existence of at least two surface enzymes, one of which is responsible for ATP hydrolysis (and fluoride-sensitive), and a second that hydrolyzes glucose phosphates and is fluoride-insensitive. Lumsden and Berger (1974) noted that phosphohydrolase activity (against PNØP) is partially inhibited after treating the worms with 1,10-phenanthroline (a chelating agent that "would be expected to maximally affect a zinc-dependent enzyme") indicating that two enzymes may be responsible for PNØP hydrolysis; whether these are the *p*-CMB-sensitive and insensitive enzymes mentioned above is unclear.

The majority of inhibitor studies dealing with phosphohydrolases in *H. diminuta* are summarized in table 2. Careful examination of these data suggest the presence of a minimum of three (and possibly four) distinct enzymes. ATP hydrolysis by *H. diminuta* is fluoride-sensitive, and hydrolysis of PNØP and triose and hexose phosphates is not; this suggests the presence of a surface enzyme specific for ATP hydrolysis. Lumsden and colleagues (1968) reported three pH optima for ATP hydrolysis by pelleted material derived from *H. diminuta* homogenates, suggesting three ATPases in *H. diminuta*, but the cellular distribution of these enzymes was not determined. Hydrolysis of all phosphate esters examined, with the exception of ATP, is insensitive to fluoride but sensitive to molybdate; in addition, the pH optima for PNØP and fructose phosphate hydrolysis are quite different. Thus, at least two additional surface phosphohydrolases may be present, one with a narrow pH optimum (7–7.4) and one with a broad pH optimum (7.4–9); both are fluroide-insensitive but molybdate-sensitive. Whether these two enzymes are the *p*-CMB-sensitive and insensitive enzymes, or the Zn^{++}-dependent and independent enzymes postulated by Lumsden and Berger (1974) is unclear, but these data make one wonder if additional surface phosphohydrolases are present in *H. diminuta*.

Pappas and Read (1974) found that adenosine monophosphate (AMP) inhibited uptake of labeled uridine by *H. diminuta*, and also noted that this effect was not reversed in the presence of possible competitors of AMP hydrolysis (i.e., FDP, PNØP, fluoride, and molybdate). AMP does, however, inhibit hydrolysis of hexose and triose phosphates, demonstrating that AMP interacts with at least one surface phosphohydrolase. Two explanations are possible for these data. First, AMP may interact directly with the uridine transport locus of *H. diminuta*, and may also interact with a surface phosphohydrolase without being hydrolyzed (thus, competitors of hydrolysis would have no effect). Second, AMP may be hydrolyzed at the worm's surface, although this has not been demonstrated experimentally. If hydrolysis of AMP does occur, then a fourth enzyme must be involved. If it exists, this enzyme must be insensitive to fluoride *and* molybdate, and have high specificity for AMP (since FDP and PNØP do not apparently effect AMP hydrolysis). The existence of this specific AMPase might seem to contradict the observation that AMP inhibits hydrolysis of fructose, glucose, glycerol, and *p*-nitrophenol phosphate esters, but AMP may simply bind "non-productively" (i.e., bind but not be hydrolyzed) to these other surface phosphohydrolases. The presence of such a surface AMPase is certainly amenable to experimental study, and such studies should be conducted.

Phosphohydrolase activity has been detected in surface membranes of many trematode species (table 1), yet data relating to ultrastructural localization and "functional" significance of these enzymes in trematodes are sparse. Utilizing ultrastructural techniques, Threadgold (1968) localized acid phosphatase activity in the surface membranes of *Fasciola hepatica*, but numerous studies have failed to detect alkaline phosphatase activity in the tegument of this parasite (Halton, 1967; Barry et al., 1968; Thorpe, 1968; Threadgold, 1968; Probert et al., 1972). Tegumentary phosphohydrolases of *Schistosoma mansoni* are active against several phosphate esters (Levy and Read, 1975a); when *S. mansoni* is incubated in AMP, adenosine and inosine accumulate in the incubation medium. When *S. mansoni* is incubated in uridine monophosphate (UMP), uridine accumulates in the incubation medium. Indirect evidence for inhibition of this apparent phosphohydrolase activity by PNØP and molybdate was presented by Levy and Read (1975a), and these authors also noted the lack of inhibition of this enzyme activity by G6P, glucose-1-phosphate, and fluoride.

Attempts to wash off the phosphohydrolase activity associated with *H. diminuta* have been unsuccessful, suggesting that the enzymes are intrinsic (i.e., of worm origin). The phosphohydrolase activity of *S. mansoni* can, however, be partially washed off and subsequently demonstrated in the incubation medium (Levy and Read, 1975a). However, the loss of enzyme activity from the worm's surface cannot be accounted for by the enzyme activity recovered in the incubation medium, and Levy and Read (1975a) suggested that "either a loss of a coenzyme(s) and/or loss of an adsorbed enzyme(s) which exhibits decreased activity when free from the parasite" had occurred. An alternative hypothesis, however, is that enzyme(s) are liberated into the incubation medium, possibly as a consequence of tegument "turnover," along with an enzyme inhibitor. This is supported by the observation that membrane proteins are elaborated from *S. mansoni* in culture (Kusel et al., 1975), and the observation that the glycocalyx of *S. mansoni* turns over quite rapidly (Kusel and Mackenzie, 1975), perhaps every four to six hours (Wilson and Barnes, 1977). This latter possibility awaits experimental verification.

Levy and Read (1975a) demonstrated that uptake of adenosine by *S. mansoni* is inhibited by AMP and ATP, and that this inhibition of uptake is relieved (reversed) by PNØP and molybdate, but not by glucose phosphates or fluoride. Thus, products of nucleotide hydrolysis appear to be rapidly translocated across the tegument of *S. mansoni*, rather than diffusing into the medium; apparently the spatial relationship of enzyme(s) responsible for AMP and ATP hydrolysis and the adenosine transport system is such that a "kinetic advantage" for absorption of hydrolysis products results (Levy and Read, 1975a). On the other hand, UMP in the incubation medium (which is actively hydrolyzed) does not inhibit the mediated uptake of uridine by *S. mansoni* (Levy and Read, 1975a), suggesting no "kinetic advantage" or close spatial relationship between enzyme(s) and transport locus. However, the "apparent" uptake of AMP (actually adenosine from AMP hydrolysis) is inhibited by UMP and ATP (Levy and Read, 1975a), suggesting that AMP, ATP, and UMP are hydrolyzed, in part, by the same enzyme(s). Thus, the observations that AMP inhibits adenosine uptake, but UMP does not inhibit uridine uptake, and that adenosine and uridine are absorbed, in part, through a common mediated transport system (Levy and Read, 1975b), indicate a complex interaction of phosphohydrolase(s), nucleotide hydrolysis, and nucleoside transport in *S. mansoni*.

Phosphohydrolase activity has been detected in the surface membranes of many acanthocephalans (table 1), and on occasion this activity has been localized to membranes lining the "crypts" of the tegument (Rothman, 1967; Byram and Fisher, 1974). Starling and Fisher (1975) presented data suggesting the action of surface phosphohydrolases of *Moniliformis dubius* on phosphate esters in the surrounding medium. These authors demonstrated that mediated glucose uptake by *M. dubius* is inhibited by G6P in the incubation medium, and suggested that this inhibition is due to the liberated glucose interacting with the glucose transport locus; although this hypothesis is consistent with the data relating to cestodes and trematodes, it has not been examined experimentally.

Aminopeptidase Activity

The aminopeptidases are peptide hydrolases that act upon simple peptides. Crompton (1963) demonstrated localized aminopeptidase (APase) activity in the tegument of *Polymorphus minutus*, and Douch (1978) demonstrated hydrolysis of L-leucyl-β-naphthylamide in the distal tegument of *Moniezia expansa*, but is unknown if either of these enzymes hydrolyze peptides in the incubation medium. Likewise, it is not known if the APase activity detected by Uglem and Beck (1972) in extracts of several species of Acanthocephala is localized to the tegument. Uglem and colleagues (1973) reported that *M. dubius* hydrolyzes L-leucyl-β-naphthylamide, and these latter authors noted that the end product of hydrolysis (β-naphthylamide) is rapidly absorbed by the worms, and that this substrate was therefore not acceptable for kinetic studies of surface APase activity. Uglem and colleagues (1973) were able to measure the hydrolysis of several peptides, however, by incubating *M. dubius* with the peptide and a high concentration of methionine; methionine inhibited absorption of the peptide hydrolysis products (amino acids), so these products accumulated in the incubation medium and could be demonstrated subsequently.

Utilizing this experimental design, Uglem and colleagues (1973) demonstrated hydrolysis of leucylleucine, leycylglycylglycine (to leucine and glycylglycine), and tri- and tetra-alanine (both hydrolyzed to alanine and alanylalanine). Although hydrolysis of leucylleucine and alanylalanine appear to involve a single APase, glycylglycine is not hydrolyzed by *M. dubius*, suggesting some substrate specificity of the enzyme(s) involved. Rather than an adsorbed host enzyme, the APase of *M. dubius* appears to be of worm origin. Repeated washing of worms fails to remove any APase activity, and "activated larvae" (cystacanths treated with taurocholate, lipase, Triton X-100, or Tween 80) demonstrate significant APase activity against leucylleucine, even though these larvae (which grow in cockroaches) have not been in the rat small intestine.

In the absence of excess methionine in the incubation medium, only 7% of the leucine from leucylleucine hydrolysis is recovered in the incubation medium, and Uglem and colleagues (1973) postulated that the remainder of the leucine is absorbed by *M. dubius*. That, in fact, such a "kinetic advantage" does exist is shown by the observation that uptake of radioactive leucine by *M. dubius* is inhibited by the presence of leucylleucine. When Pb^{++} is added to the incubation medium (as an inhibitor of APase activity), the inhibitory effect of leucylleucine on leucine uptake is relieved. This observation supported the hypothesis of Uglem and colleagues (1973) that leucine derived from leucylleucine hydrolysis is rapidly absorbed by the worms (as shown by the inhibition of labeled amino acid uptake), and that inhibition of APase activity blocks dipeptide hydrolysis. These workers also pointed out that although the Pb^{++} data suggest the APase of *M. dubius* is a metalloenzyme, incubation of worms in EDTA has no effect on APase activity; in addition, incubation of worms in other divalent cations (Co^{++}, Cu^{++}, Mn^{++}, and Zn^{++}), which stimulate APase activity in vertebrate tissues, has no effect on *M. dubius* APase activity.

A comparison of the data for *M. dubius* cystacanths and adults provides an excellent example of the relationship of surface (tegumentary) architecture and the phenomenon of "kinetic advantage." As noted above, leucine produced by dipeptide hydrolysis does not diffuse away from adult worms, but is rapidly absorbed via a mediated amino acid transport system. As noted by Uglem and colleagues (1973), this requires the APase activity and amino acid transport locus to be in close proximity, or at least in a region (an "unstirred" region) of which the contents do not exchange rapidly with the external medium. Two observations implicate the "crypts" of the tegument as candidates for this unstirred region. Hydrolysis of leucylleucine by adult worms inhibits uptake of radioactive leucine, whereas hydrolysis of the same peptide by activated cystacanth larvae has no effect on leucine uptake by larvae (Uglem et al., 1973). Uglem and his coworkers attributed this to the fact that whereas surface "crypts" in adult worms are well developed, "crypts" in larvae are poorly developed and very shallow. Hydrolytic enzymes (phosphohydrolases) do occur in the membranes lining these "crypts" (Rothman, 1967; Byram and Fisher, 1974), suggesting that these "crypts" are important sites of extracellular digestion. If extracellular digestion is a function of these "crypts," then localization of APase activity and mediated transport loci in the membranes of these "crypts" would be advantageous to the worms.

Trematodes have not been examined for the presence of surface APase activity, but extracts of *Haematoloechus longiplexus* and *Echinostomum coalitum* contain APase activity (Uglem and Beck, 1972). Although extracts of *Proteocephala ptychocheilus* (Cestoda) contain APase activity (Uglem and Beck, 1972), intact specimens of only two cestode species have been examined for APase activity. Read and colleagues (1963) reported that glycylglycine, glycylglycylglycine, and glycylleucine have no effect on uptake of radioactive methionine by *H. diminuta*. Since both glycine and methionine inhibit methionine uptake in this same species (Read et al., 1963), these data suggest a lack of intrinsic APase activity. These observations were supported by the date of Uglem and colleagues (1973) that showed that leucylleucine is not hydrolyzed in the presence of *H. diminuta*. Alternatively, one might hypothesize that APase activity does occur on the surface of *H. diminuta*, but that this enzyme shows a high degree of substrate specificity. However, this hypothesis seems highly unlikely since Uglem (unpublished) has been unable to detect hydrolysis of valylvaline, phenylalanylphenylalanine, leucylleucine, alanylalanine, methionylmethionine, tri-leucine, tetra-alanine, and tri-serine in the presence of *H. diminuta*.

As mentioned in the first paragraph of this section, Douch (1978) demonstrated histochemically the hydrolysis of L-leucyl-β-naphthylamide in the distal tegument of *M. expansa*, but it is not known whether this enzyme is active against substrates in the surrounding medium.

Disaccharidase Activity

Presently, there is only circumstantial evidence for the existence of membrane-bound disaccharidases in helminth parasites. Read and Rothman (1958) showed an increase in acid production by *Cittotaenia* sp. when this cestode is incubated in sucrose or maltose, while many other cestodes appear unable to utilize disaccharides for acid production (see Read, 1959, for a complete list of cestode species examined). Utilization of disaccharides by *Cittotaenia* would require a disaccharidase, but this enzyme has not been shown to be localized in the tegument.

Maltose and cellobiose have been reported to inhibit uptake of radioactive glucose by *Calliobothrium verticillatum* (a cestode), and it was suggested that these disaccharides are hydrolyzed at the worm's surface, with the resulting monosaccharides inhibiting glucose uptake (Fisher and Read, 1971). However, later experiments were unable to corroborate these studies. These later experiments (Pappas and Read, unpublished) used disaccharides (maltose and cellobiose) that were tested for chemical purity and found to be free of monosaccharide contamination. When *C. verticillatum* was incubated in 0.1 mM radioactive glucose and 10 mM maltose or cellobiose, no inhibition of glucose uptake was noted. Apparently, the disaccharides used by Fisher and Read (1971) were contaminated with monosaccharides, and the later data of Pappas and Read (unpublished) indicate that *C. verticillatum* does not possess intrinsic disaccharidase activity. This would be expected, since *C. verticillatum* cannot metabolize disaccharides to any extent (Read, 1957).

Recent experiments in our laboratory have confirmed the absence of disaccharidases in two additional cestode species, namely *H. diminuta* and *Hymenolepis microstoma*. In these experiments worms were removed from their respective hosts, rinsed thoroughly in buffered saline, and randomized into groups of four worms. Individual groups of worms were incubated in the following solutions (in buffer): (1) 5 mM disaccharide (lactose, maltose, or sucrose); (2) 5 mM disaccharide + 0.5 mM phlorizin; or (3) 5 mM disaccharide + 5 mM galactose. Following the incubation, worms were removed and the media analyzed for the presence of glucose by the glucose oxidase method. In these experiments we reasoned that intrinsic disaccharidase activity would liberate glucose from the disaccharides. However, since glucose is rapidly absorbed by both *H. diminuta* (Pappas et al., 1974; Read et al., 1974) and *H. microstoma* (Pappas and Freeman, 1975), additional incubations included phlorizin or galatose to inhibit glucose uptake. Under these experimental conditions no glucose was detected in the incubation media following a 30 min incubation with worms, and it was concluded (Gamble and Pappas, unpublished) that these cestode species lack disaccharidase activity.

Laurie (1957) reported that *M. dubius* can metabolize maltose, and Starling and Fisher (1975) reported that mediated glucose uptake by *M. dubius* is inhibited by maltose. However, the kinetics of the inhibition of glucose uptake by maltose are quite different from the kinetics displayed by competitive inhibitors such as mannose, glucosamine, fructose, and other monosaccharides. Starling and Fisher (1975) postulated that a surface disaccharidase first hydrolyzes maltose, and that liberated glucose inhibits glucose uptake; these data agree with the metabolism studies of Laurie (1957), but no biochemical data are available to substantiate the existence of this enzyme.

Ribonuclease Activity

Intact *H. diminuta* possesses ribonuclease (RNase) activity that is active against both degraded yeast and native rat liver RNA (particularly the 28s and 18s fractions of the latter) (Pappas et al., 1973; Pappas and Gamble, 1978). Pappas and colleagues (1973) demonstrated that the RNase activity could not be washed off the worms, even after 90 min of washing, and that RNase activity could not be demonstrated in worm washings. In addition, these authors demonstrated that *H. diminuta* RNase and rat pancreatic RNase display distinctly different kinetic characteristics. Rat pancreatic RNase displays substrate inhibition, a pH optimum of 8.0, inhibition by Cu^{++} and Zn^{++}, and stimulation by Ca^{++} (at pH 6.5); *H. diminuta* RNase displays no substrate inhibition, a pH optimum above 8.4, and inhibition by Ca^{++}, Mg^{++}, Cu^{++}, Mn^{++}, and Zn^{++} (at pH 6.5). The fact that the RNase activity could not be washed off worms, and the distinct kinetic differences, lead to the hypothesis that the RNase of *H. diminuta* is an enzyme of worm origin (Pappas et al., 1973). An alternative hypothesis is that *H. diminuta* RNase is a rat pancreatic RNase that is very tightly adsorbed to the worm's surface, and that the different kinetic parameters result from a conformational change in the RNase during the adsorption process. Neither hypothesis has received additional experimental attention recently.

Lipase Activity

As discussed below, *H. diminuta* interacts with extrinsic lipase in the incubation medium. In addition, *H. diminuta* possesses some intrinsic lipase activity that is apparently associated with the tegument. Bailey and Fairbairn (1968) noted a rapid hydrolysis of monoolein in the presence of *H. diminuta*, and also demonstrated that this lipase activity could not be washed off the worms. Similar findings were reported by Ruff and Read (1973) using N-methyl-indoxyl-myristate as a substrate to detect lipase activity. Further characterization of this intrinsic lipase activity does not appear in the literature.

INTERACTIONS OF HOST'S ENZYMES AND PARASITES

Interactions with Amylase

Although the data are equivocal, previous studies indicate that *most* intact helminth parasites lack intrinsic, membrane-bound amylolytic activity. Read (1968, 1973) reported that starch is not digested in the presence of *H. diminuta*, and similar findings were reported by Arkind and Raeva (1971) for *M. expansa*, *M. benedeni*, and *Thysaniezia ovilla*. Pleorcercoids of *Ligula intestinalis* and adult *M. dubius* possess significant amylase activity when removed from their hosts (Davydov and Kosenko, 1972; Ruff et al., 1973; respectively), but this activity is readily removed from the worms by washing. (As discussed below, the amylolytic activity associated with *M. dubuis* has been shown to be adsorbed rat amylase.) Taylor and Thomas (1968) reported that *M. expansa*, which had been maintained in saline for up to 12 hr, demonstrated some amylolytic activity; and although these authors also tested *H. diminuta* and *H. microstoma* for this enzyme activity, they did not report these latter results.

Thus, with the possible exception of *M. expansa*, the data published to date indicate that intact helminth parasites do not contain intrinsic, membrane-bound amylolytic activity.

Taylor and Thomas (1968) published the first report dealing with the *in vitro* interactions of cestodes with digestive enzymes. In their study Taylor and Thomas demonstrated that amylase activity is stimulated in the presence of three cestode species, namely, *M. expansa*, *H. diminuta*, and *H. microstoma*. They demonstrated that the increase in enzymatic activity is proportional to the amount of tapeworm tissue present, and that the increase in activity is a function of the entire strobila and not just isolated areas. Although these authors suggested that increased amylolytic activity is due to an adsorption of enzyme to the cestode's surface, similar to that postulated to occur in the vertebrate small intestine (Ugolev, 1965), they presented no data to substantiate this hypothesis. Reichenbach-Klinke and Reichenbach-Klinke (1970) reported no stimulation of amylase activity in the presence of intact *Proteocephalus longicollis*, but were able to demonstrate a potent amylase inhibitor in *extracts* of this cestode.

Read (1973) conducted additional experiments in order to characterize the interaction of *H. diminuta* with amylase. In addition to corroborating the previous experiments of Taylor and Thomas (1968), Read (1973) showed that interaction of amylase with the cestode tegument, in *in vitro* incubations, causes an increase in the maximal velocity (V_{max}) of the amylase reaction. Preincubation of worms in polyanions, such as poly-L-glutamic acid or heparin, partially reverses the stimulation of enzyme activity; but preincubation of worms in polycations, such as poly-L-lysine, does not alter the stimulatory effect of worms. Thus, partially blocking some positively charged groups of the cestode surface partially reverses the stimulation of amylase activity. Read (1973) reported additionally that the stimulatory effect of worms on amylase activity is reversed by briefly washing the worms. These data support the original hypothesis of Taylor and Thomas (1968) that amylase is adsorbed to the cestode tegument, and that during adsorption the enzyme is altered in some manner such that enzyme activity is increased.

As noted above, *M. dubius* possesses significant amylase activity when removed from the host's (rat) small intestine. However, the data of Ruff and colleagues (1973) demonstrated conclusively that this activity is adsorbed rat amylase, rather than intrinsic worm enzyme. This conclusion was based on the following observations: (1) amylolytic activity associated with *M. dubius* was readily washed off the worms; (2) amylolytic activity removed from worms was inhibited by rabbit anti-rat amylase antiserum; (3) "worm amylase" and authentic rat amylase displayed identical pH optima; and (4) incubation of *M. dubius* in rat intestinal contents partially restored the amylolytic activity to worms. Despite the rather large amounts of amylase adsorbed to the surface of *M. dubius*, this association of enzyme and tegument does not alter the kinetic parameters of the enzyme. Ruff and colleagues (1973) demonstrated that the effects of worms (with adsorbed enzyme) and extrinsic amylase on starch hydrolysis were almost exactly additive, rather than enzyme activity being stimulated in the presence of worms.

Interactions with Lipase

Reference has already been made to the observations of Bailey and Fairbairn (1968) and Ruff and Read (1973) that *H. diminuta* possesses intrinsic lipase activity. In addition, *H. diminuta* interacts with extrinsic lipase, resulting in a reversible inhibition of enzyme activity. This reversible inhibition during *in vitro* incubations was postulated by Ruff and Read (1973) on the basis of the following observations: (1) lipase activity in the presence of worms is inhibited, and inhibition occurs rapidly after adding worms to the incubation medium; (2) increasing the time period worms are exposed to the enzyme does not increase the amount of inhibition; (3) inhibition is readily reversible when worms are removed from the incubation medium; and (4) inhibition is a function of worm surface area rather than total worm weight. The presence of polyions, such as poly-L-glutamic acid, poly-L-aspartic acid, RNA, or heparin, has no effect on the inhibitory action of worms on lipase, suggesting inhibition is independent of charged groups on the worm's surface. Inhibition is pH dependent, with inhibition disappearing in media with a pH greater than 8.3. Thus, Ruff and Read (1973) suggested that lipase inhibition is brought about by a "loose, transitory attachment (adsorption of the enzyme) to the surface of the worm by weak bonding involving Van der Waals forces," resulting in either "(1) temporary stabilization of the enzyme in a configuration not favorable for catalytic activity, or (2) blockage of the 'active' site of the enzyme." This hypothesis has not been tested experimentally.

Interactions with Proteases

Modifications of amylase and lipase activity brought about by interactions of these enzymes with the tegument of *H. diminuta* are reversible, as noted in the previous two sections. However, interactions of some proteolytic enzymes with the cestode tegument during *in vitro* incubations causes an irreversible change in enzyme activity. This was first suggested by the data of Reichenbach-Klinke and Reichenbach-Klinke

(1970), who reported trypsin inhibition in the presence of the fish tapeworm, *P. longicollis*. However, these authors did not determine whether worms excreted some type of enzyme inhibitor into the surrounding medium, or whether the inhibition was due to an interaction of enzyme and the cestode's tegument. Pappas and Read (1972a,b) subsequently demonstrated that the interaction of *H. diminuta* with proteolytic enzymes (trypsin and α- and β-chymotrypsins) causes an irreversible inhibition, or inactivation, of these enzymes. Unlike the adsorption mechanism postulated for interactions of lipase (Ruff and Read, 1973) and amylase (Taylor and Thomas, 1968; Read, 1973) with the cestode tegument, the data of Pappas and Read (1972a,b) suggested a different type of interaction with proteolytic enzymes.

When *H. diminuta* was incubated *in vitro* in the presence of these proteolytic enzymes, and removed prior to the addition of substrate, activity was found to be reduced significantly; control experiments consisting of placing worms in incubation media and removing them prior to the addition of enzyme and substrate indicated that worms did not excrete significant amounts of protease inhibitors, and incubation of worms with substrate (azoalbumin in these experiments) failed to demonstrate any intrinsic proteolytic activity. Since loss of proteolytic activity was demonstrable after removing the worms from the incubation medium, and since loss of activity was apparently irreversible, this was termed irreversible inhibition, or inactivation, by Pappas and Read (1972a), as opposed to reversible inhibition. Although these initial data could be explained by trypsin or chymotrypsin adsorption to the tapeworm's surface, an adsorption mechanism is not consistent with the other experimental data. Inactivation of all three enzymes increases with the time period worms are exposed to the enzyme (for up to 45 min), and inactivation is proportional to the weight of the worms exposed to the enzyme, rather than available tegumental surface area. Previous exposure of worms to enzymes does not alter the ability of worms to inactivate fresh enzyme when exposed to it, and various polyions have no effect on inactivation (Pappas and Read, 1972a,b). In attempting to explain these peculiar characteristics of trypsin inactivation by *H. diminuta*, Pappas and Read (1972a) assumed that trypsin is not absorbed by *H. diminuta*, and that any interaction between parasite and enzyme must take place at the surface (tegument) of the worm. Preliminary evidence that *H. diminuta* does not, in fact, absorb trypsin has been obtained in our laboratory (Schroeder, unpublished). Since Oaks and Lumsden(1971) had shown previously that the glycocalyx of *H. diminuta* has a turnover rate of six hours or less, Pappas and Read (1972a) suggested some type of interaction might occur between a labile tegumentary component and the enzymes. We are currently testing this hypothesis in our laboratory.

Trypsin inactivation can be demonstrated using several high molecular weight substrates, including azoalbumin, casein, and hemoglobin (Pappas and Read, 1972a); however, inactivation of these same enzyme preparations cannot be demonstrated using low molecular weight synthetic substrates, such as benzoyl-arginine-*p*-nitroanilide (BAPA), and tosyl-arginine-methyl ester (TAME) (Schroeder, unpublished). These data suggest that the active site of trypsin is not altered during inactivation, but that inactivation involves a change in the enzyme structure such that activity toward high molecular weight substrates is altered. Whatever the mechanism of inactivation, it appears to be quite specific for trypsin and chymotrypsins, since Schroeder (unpublished) has been unable to demonstrate similar inactivation of subtilisin, elastase, papain, or pepsin.

DISCUSSION

The term "host-parasite interface," as used by Read and colleagues (1963), implies that the interaction between host and parasite occurs at a simple surface. "Interface" is probably a poor term to use, since recent studies have shown that the tegument is not a simple surface but a complex membrane with an accompanying glycocalyx. The membrane is invested with intrinsic enzymes, and enzymes of host origin may also interact with the tegument. Numerous mediated transport systems have been identified in various parasitic helminths (Pappas and Read, 1975), and the glycocalyx contains numerous charged groups (Lumsden, 1975a,b). This latter component of the tegument is synthesized by the parasite, and is known to turn over quite rapidly (Oaks and Lumsden, 1971; Kusel et al., 1975; Kusel and Mackenzie, 1975; Trimble and Lumsden, 1975; Wilson and Barnes, 1977). Thus, the "interface" is a dynamic region composed of tegumentary membranes, the glycocalyx, the "unstirred" region within the glycocalyx, and adsorbed components; events such as extracellular digestion probably occur distal to the tegumentary membranes, and possibly distal to even the glycocalyx in the cases of adsorbed host enzymes.

The intimacy of contact between host and parasite will certainly affect the nature of the interface, as will the period of contact between host and parasite; these factors will, therefore, influence the enzymological interactions of host and parasite. In addition, a single host-parasite relationship may involve multiple interfaces. For example, *H. diminuta* migrates within the rat small intestine (Read and Kilejian, 1968; Chappell et al., 1970), and the interface, at least at the biochemical level, would not be expected to remain unchanged during this circadian migration. Also, parasites may be firmly attached to the host's small intestine, creating one interface, and expose additional surface area to the intestinal lumen, creating a second interface. In these latter instances it is difficult to define the morphological and biochemical characteristics of the host-parasite interface and then determine the relative contribution of each interface to the over-all host-parasite relationship; thus, the results of *in vitro* experiments must be interpreted with caution.

This review has dealt with only two aspects of the host-parasite interface, namely, intrinsic enzymes and adsorbed host enzymes. Because of the difficulties encountered in studying these enzyme interactions (see Smyth, 1973, for a discussion of some problems), only a few species of helminth parasites and a small number of enzymes have been studied. Yet, even these examples demonstrate that enzymological interactions at the host-parasite interface are important in determining the manner in which host and parasite may interact.

As discussed above, the enzymes associated with the tegument of many helminth parasites appear to be intrinsic (i.e., of parasite origin), with the exception of amylase activity associated with *M. dubius*. These intrinsic enzymes are digestive in nature, and one would assume they function in extracellular digestion. Considering the limited permeability of the helminth tegument, and the specificity of mediated transport systems known to occur in various helminth species, the action of these intrinsic enzymes in providing monosaccharides, nucleosides, amino acids, and fatty acids for absorption is apparent. The importance of these intrinsic enzymes is more apparent when one considers that in most instances these enzymes and specific transport loci are so located in (or on) the tegument as to confer a "kinetic advantage" for absorption of hydrolysis products. One can assume also that this spatial relationship of membrane-bound enzymes and transport loci allows the parasites to compete more effectively with the host for nutrients.

In addition to intrinsic enzymes, some helminth parasites interact with enzymes of host origin. In the presence of intact tapeworms, enzymatic activity of amylase is increased significantly. (Taylor and Thomas, 1968; Arkind and Raeva, 1971; Davydov and Kosenko, 1972; Read, 1973), and available data (Read, 1973) suggest that amylase is adsorbed to the parasites' surfaces. *H. diminuta* utilizes glucose almost exclusively as an energy source (Read, 1967), and glucose usually is supplied to the host in the form of starch. Thus, an increased amylolytic activity in the immediate vicinity of worms might be expected to increase the glucose concentration at the tapeworm's surface. This, of course, would be a distinct advantage to the cestode in terms of competing with a host for available carbohydrates. However, this postulated series of interactions has several faults. The action of pancreatic amylase on starch produces mainly maltose, and *H. diminuta* is not only impermeable to this disaccharide but also has no membrane-bound disaccharidase to hydrolyze maltose to glucose. Thus, maltose is useless to *H. diminuta* (and many other helminth parasites) as an energy source. However, the rat small intestine does contain a membrane-bound maltase (Crane, 1975) that could act upon maltose, thus producing glucose; given the intimate contact of the cestode tegument and the intestinal brush border, this glucose could be absorbed by the worms. However, the rates of carbohydrate digestion and absorption in control rats and rats infected with *H. diminuta* do not differ significantly (Mead and Roberts, 1972), indicating that amylase activity is not stimulated *in vivo*. However, the results of Mead and Roberts (1972) could also be explained if one assumes that *in vivo* the amount of amylase activated in the presence of *H. diminuta* is only a small proportion of the total amylase present in the host's small intestine; thus, the total amylase activity would mask activation of relatively small quantities of enzyme. This latter explanation seems likely.

The experiments conducted by Read (1973) used an amylase concentration of 0.1 μg/ml of incubation medium, and although Read did not report the activity of his enzyme preparation, it will be assumed (at least for this discussion) that the activity was about 500 units/mg protein; this is the activity of currently available commercial amylase preparations. Thus, in Read's (1973) experiments the incubation medium contained only 0.05 units of enzyme/ml. McGeachin and Ford (1959) reported the amylase activity of the rat small intestine to be on the order of 64,000 units/g tissue, for a total activity of approximately 250,000 units/small intestine (the small intestine of a 100 g rat weighs about 4 g), so the amount of enzyme in Read's (1973) experiments represents only 0.0002% of the total amylase content of the rat small intestine. Since the interaction of amylase with the cestode tegument appears to involve an adsorption phenomenon, there are probably a limited number of "adsorption sites" on the cestode tegument. Therefore, it would seem physically impossible for the cestode tegument to adsorb enough amylase to alter significantly the rate at which carbohydrate digestion occurs in the rat small intestine, although carbohydrate digestion in the worm's immediate vicinity (its "microhabitat") may be altered significantly.

The possible significance of the reversible inhibition of pancreatic lipase by *H. diminuta* remains unknown, and such inhibition has not been demonstrated to occur *in vivo*. Since *H. diminuta* lacks the ability to synthesize fatty acids *de novo* (Ginger and Fairbairn, 1966; Jacobsen and Fairbairn, 1967), the inhibition of pancreatic lipase *in vivo* would limit the availability of essential fatty acids to the tapeworm. Any such decrease in lipase activity *might* be offset by the intrinsic lipase activity associated with the tegument of *H. diminuta*. Clearly, the interactions of extrinsic lipase and the tapeworm tegument need further study.

When *homogenates* of different cestode species are assayed for proteolytic activity, significant amounts of activity can be demonstrated (Kwa, 1972a; Dubovskaya, 1973; Shishova-Kasatochkina and Dubovskaya, 1976; Matskasi and Juhasz, 1977), yet no cestode species has been shown to possess membrane-bound protease activity (Read, 1968; Kwa, 1972a; Pappas and Read, 1972a,b). However, *H. diminuta* does interact with proteolytic enzymes of host origin (in *in vitro* incubations) causing an irreversible inhibition or inactivation of enzyme activity. Many investigators have reported that intact, living gastrointestinal parasites are resistant to proteolytic enzymes, and also have noted that when these parasites are killed, they are then susceptible to digestion. Possibly the ability of *H. diminuta* to inactivate proteases at its surface is an important aspect of the resistance to digestion. Since gastrointestinal parasites are resistant to digestion, some authorities have suggested the presence of "antienzymes" that are produced and excreted by worms; but much of the evidence supporting the existence of such antienzymes comes from biochemical studies utilizing *extracts* of *Ascaris* (Green, 1957; Rhodes et al., 1963; Rola and Pudles, 1966; Pudles et al., 1967; Fraeful and Acher, 1968). Antienzymes have been demonstrated in extracts of nematodes other than *Ascaris* (Ansari et al., 1976; Willadsen, 1977), and in at least one species of cestode (Matskasi and Juhasz, 1977). Excretion of these antienzymes into the surrounding environment has not been demonstrated under highly controlled experimental conditions, so the physiological and biochemical significance of them is uncertain. But the fact that *H. diminuta* does inactivate proteolytic enzymes necessitates a further study of these antienzymes and, more importantly, the mechanisms by which they may function.

In an attempt to determine whether inactivation of proteases by *H. diminuta* occurs *in vivo*, Pappas (1978) measured the tryptic and protease activities of the normal rat small intestine and the rat small intestine infected with *H. diminuta*. Since *H. diminuta* irreversibly inactivates enzyme activity *in vitro*, a loss of enzyme activity should be detected *in vivo* if such inactivation occurs. No loss of enzyme activity in the infected rat small intestine could be detected, however, indicating that inactivation does not occur *in vivo*. However, when one considers that *H. diminuta* inactivates only microgram quantities of enzyme (Pappas and Read, 1972a,b) and the rat small intestine contains milligram quantities of these same enzymes (Pappas, 1978), it is apparent that inactivation of small amounts of enzyme could be masked by the large amounts of enzyme normally present in the rat small intestine (Pappas, 1978). Thus, the question of whether protease inactivation occurs *in vivo* remains unanswered.

The importance of the helminth tegument in the host-parasite relationship has been stressed in previous review articles (Smyth, 1973; Pappas and Read, 1975; Lumsden, 1975a,b, this volume; Starling, 1975; and others), and this article has attempted to stress the enzymological interactions at this interface. Further studies should include experiments designed to demonstrate the presence or absence of additional hydrolytic enzymes in the helminth tegument, and to characterize more fully those enzymes known to be present. These studies should be facilitated greatly by recently developed techniques allowing for the isolation of the helminth tegument (Oaks et al., 1977). Further studies are also needed to determine the significance of the interactions of host enzymes with parasite surfaces, since even small alterations in enzyme activity in the immediate vicinity of the worm (i.e., its microhabitat) could have profound effects on the availability of nutrients, and also on the effects of these various enzymes on the living worm itself.

ACKNOWLEDGMENTS

Dr. Gary Uglem, Department of Biological Sciences, University of Kentucky, was kind enough to provide unpublished data, and to also read and comment on this manuscript. Some experiments reported in this paper were supported in part by a research grant from the National Science Foundation (PCM78-10478).

REFERENCES

Ansari, A. A., M. A. Khan, and S. Ghatak. 1976. *Ascaridia galli*: trypsin and chymotrpysin inhibitors. Exp. Parasit. 39:74–83.

Arkind, M. V., and I. I. Raeva. 1971. Membrane (parietal) digestion in cestodes. J. Evol. Biochem. Physiol. 7:316–20.

Arme, C. 1966. Histochemical and biochemical studies on some enzymes of *Ligula intestinalis* (Cestoda: Pseudophyllidea). J. Parasit. 52:63–68.

Arme, C., and C. P. Read. 1970. A surface enzyme in *Hymenolepis diminuta* (Cestoda), J. Parasit. 56:514–16.

Bailey, H. H., and D. Fairbairn. 1968. Lipid metabolism in helminth parasites. V. Absorption of fatty acids and monoglycerides from micellar solution by *Hymenolepis diminuta* (Cestoda). Comp. Biochem. Physiol. 26:819–36.

Barry, D. H., L. E. Mawdesley-Thomas, and J. C. Malone. 1968. Enzyme histochemistry of the adult liver fluke *Fasciola hepatica*. Exp. Parasit. 23: 355–60.

Bogitsh, B. J. 1966. Histochemical observations on *Posthodiplostomum minimum*. III. Alkaline phosphatase activity in metacercariae and adults. Exp. Parasit. 19:339–47.

Bullock, W. L. 1949. Histochemical studies on the Acanthocephala. I. The distribution of lipase and phosphatase. J. Morph. 84:185–99.

Bullock, W. L. 1958. Histochemical studies on the Acanthocephala. III. Comparative histochemistry of alkaline glycerophosphatase. Exp. Parasit. 7:51–68.

Byram, J., and F. M. Fisher, Jr. 1974. The absorptive surface of *Moniliformis dubius* (Acanthocephala). II. Functional aspects. Tissue Cell 6:21–42.

Chappell, L. H., H. P. Arai, S. C. Dike, and C. P. Read. 1970. Circadian migration of *Hymenolepis* (Cestoda) in the host intestine. I. Observations on *H. diminuta* in the rat. Comp. Biochem. Physiol. 34:31–46.

Clegg, J. A., J. R. Smithers, and R. J. Terry. 1970. "Host" antigens associated with schistosomes: observations on their attachment and nature. Parasitology 61:87–94.

Crane, R. K. 1975. A digestive-absorptive surface as illustrated by the intestinal brush border. Trans. Amer. Microsc. Soc. 94:929–44.

Crompton, D. W. T. 1963. Morphological and histochemical observations on *Polymorphus minutus* (Goeze, 1782), with special reference to the body wall. Parasitology 53:663–85.

Damian, R. T., N. D. Greene, and W. J. Hubbard. 1973. Occurrence of mouse α_2-macroglobulin antigenic determinants on *Schistosoma mansoni* adults, with evidence of their nature. J. Parasit. 59:64–73.

Davydov, O. N., and L. Ya. Kosenko. 1972. (Membrane digestion in *Ligula intestinalis* plerocercoids.) Parazitologiya 6:269–73. (In Russian; English summary.)

Dike, S. C., and C. P. Read. 1971a. Tegumentary phosphohydrolases of *Hymenolepis diminuta*. J. Parasit. 57:81–87.

Dike, S. C., and C. P. Read. 1971b. Relation of tegumentary phosphohydrolases and sugar transport in *Hymenolepis diminuta*. J. Parasit. 57:1251–55.

Douch, P. G. C. 1978. L-leucyl-β-naphthylamidases of the cestode, *Moniezia expansa*, and the nematode, *Ascaris suum*. Comp. Biochem. Physiol. 60B:63–66.

Dubovskaya, A. Ya. 1973. (Study of the proteolytic activity of some species of cestodes.) Parazitologiya 7:154–59. (In Russian; English summary.)

Dusanic, D. G. 1959. Histochemical observations of alkaline phosphatase in *Schistosoma mansoni*. J. Inf. Dis. 105:1–8.

Erasmus, D. A. 1957a. Studies on phosphatase systems in cestodes. I. Studies of *Taenia pisiformis*. Parasitology 47:70–80.

Erasmus, D. A. 1957b. Studies on phosphatase systems of cestodes. II. Studies on *Cysticercus tenuicollis* and *Moniezia expansa*. Parasitology 47:81–91.

Fisher, F. M., Jr., and C. P. Read. 1971. Transport of sugars in the tapeworm *Calliobothrium verticillatum*. Biol. Bull. 140:46–62.

Fraeful, W., and R. Acher. 1968. The amino acid sequence of a trypsin inhibitor isolated from ascaris (*Ascaris lumbricoides* var. *summ*). Biochim. Biophys. Acta 154:615–17.

Ginger, C. D., and D. Fairbairn. 1966. Lipid metabolism in helminth parasites. II. The major origins of the lipds of *Hymenolepis diminuta* (Cestoda). J. Parasit. 52:1097–1107.

Green, N. M. 1957. Protease inhibitors from Ascaris. Biochem. J. 66:416–19.

Halton, D. W. 1967. Studies on phosphatase activity in Trematoda. J. Parasit. 53:46–54.

Jacobsen, N. S., and D. Fairbairn. 1967. Lipid metabolism in helminth parasites. III. Biosynthesis and interconversion of fatty acids by *Hymenolepis diminuta* (Cestoda). J. Parasit. 53:355–61.

Kilejian, A., L. A. Schinazi, and C. W. Schwabe. 1961. Host-parasite relationships of echinococcosis. I. Histochemical observations on *Echinococcus granulosus*. J. Parasit. 47:181–88.

Kusel, J. R., and P. E. Mackenzie. 1975. The measurement of relative turnover rates of the proteins of the surface membranes and other fractions of *Schistosoma mansoni* in culture. Parasitology 71:261–73.

Kusel, J. R., P. E. Mackenzie, and D. J. McLaren. 1975. The release of membrane antigens into cultures by adult *Schistosoma mansoni*. Parasitology 71:247–59.

Kwa, B. H. 1972a. Studies on the sparganum of *Spirometra erinacei*. II. Proteolytic enzyme(s) in the scolex. Int. J. Parasit. 2:29–33.

Kwa, B. H. 1972b. Studies on the sparganum of *Spirometra erinacei*. I. The histology and histochemistry of the scolex. Int. J. Parasit. 2:23–28.

Laurie, J. 1957. The *in vitro* fermentation of carbohydrates by two species of cestodes and one species of Acanthocephala. Exp. Parasit. 6:245–60.

Levy, M. G., and C. P. Read. 1975a. Relation of tegumentary phosphohydrolases to purine and pyrimidine transport in *Schistosoma mansoni*. J. Parasit. 61:648–56.

Levy, M. G., and C. P. Read. 1975b. Purine and pyrimidine transport in *Schistosoma mansoni*. J. Parasit. 61:627–32.

Lewert, R. M., and D. G. Dusanic. 1961. Effects of a symmetrical diaminodibenzylalkane on alkaline phosphatase of *Schistosoma mansoni*. J. Inf. Dis. 109:85–89.

Lumsden, R. D. 1975a. The tapeworm tegument: a model system for studies of membrane structure and function in host-parasite relationships. Trans. Amer. Microsc. Soc. 94:501–7.

Lumsden, R. D. 1975b. Surface ultrastructure and cytochemistry of parasitic helminths. Exp. Parasit. 37:267–339.

Lumsden, R. D., and B. Berger. 1974. Cytological studies on the absorptive surfaces of cestodes. VIII. Phosphohydrolase activity and cation adsorption in the tegument brush border of *Hymenolepis diminuta*. J. Parasit. 60:744–51.

Lumsden, R. D., G. Gonzalez, R. R. Mills, and J. M. Viles. 1968. Cytological studies on the absorptive surfaces of cestodes. III. Hydrolysis of phosphate esters. J. Parasit. 54:524–35.

Lyons, K. M. 1977. Epidermal adaptations of parasitic platyhelminths. *In* R. I. C. Speakman (ed.), Comparative biology of skin, pp. 97–144. Academic Press, New York. (Also cited as follows: Symp. Zool. Soc. Lond., No. 39:97–144.)

Matskasi, I., and S. Juhasz. 1977. *Ligula intestinalis* (L., 1758): investigation on plerocercoids and adults for protease and protease inhibitor activities. Hung. Parasit. 10:51–60.

McGeachin, R. L., and N. K. Ford, Jr. 1959. Distribution of amylase in the gastrointestinal tract of the rat. Am. J. Physiol. 196:972–74.

Mead, R. W., and L. S. Roberts. 1972. Intestinal digestion and absorption of starch in the intact rat: effects of cestode (*Hymenolepis diminuta*) infection. Comp. Biochem. Physiol. 41A:749–60.

Morris, G. P., and C. V. Finnegan. 1968. Studies on the differentiating plerocercoid cuticle of *Schistocephalus solidus*. I. The histochemical analysis of cuticle development. Can. J. Zool. 46:115–21.

Oaks, J. A., W. J. Knowles, and G. D. Cain. 1977. A simple method of obtaining an enriched fraction of tegumental brush border from *Hymenolepis diminuta*. J. Parasit. 63:476-85.

Oaks, J. A., and R. D. Lumsden. 1971. Cytological studies on the absorptive surfaces of cestodes. V. Incorporation of carbohydrate containing macromolecules into tegument membranes. J. Parasit. 57:1256-58.

Öhman-James, C. 1968. Histochemical studies on the cestode *Diphyllobothrium dendriticum* Nitzsch, 1824. Z. Parasitenk. 30:40-56.

Pappas, P. W. 1978. Tryptic and protease activities in the normal and *Hymenolepis diminuta*-infected rat small intestine. J. Parasit. 64:562-64.

Pappas, P. W., and B. A.. Freeman. 1975. Sodium dependent glucose transport in the mouse bile duct tapeworm, *Hymenolepis microstoma*. J. Parasit. 61:434-39.

Pappas, P. W., and H. R. Gamble. 1978. *Hymenolepis diminuta*: action of worm RNase against RNA. Exp. Parasit. 46:256-61.

Pappas, P. W., and C. P. Read. 1972a. Trypsin inactivation by intact *Hymenolepis diminuta*. J. Parasit. 58:864-71.

Pappas, P. W., and C. P. Read. 1972b. Inactivation of α- and β-chymotrypsin by intact *Hymenolepis diminuta* (Cestoda). Biol. Bull. 143:605-16.

Pappas, P. W., and C. P. Read. 1974. Relation of nucleoside transport and phosphohydrolase activity in *Hymenolepis diminuta*. J. Parasit. 60:447-52.

Pappas, P. W., and C. P. Read. 1975. Membrane transport in helminth parasites: a review. Exp. Parasit. 37:469-530.

Pappas, P. W., G. L. Uglem, and C. P. Read. 1973. Ribonuclease activity associated with intact *Hymenolepis diminuta*. J. Parasit. 59:824-28.

Pappas, P. W., G. L. Uglem, and C. P. Read. 1974. Anion and cation requirements for glucose and methionine accumulation by *Hymenolepis diminuta* (Cestoda). Biol. Bull. 146:56-66.

Phifer, K. 1960. Permeation and membrane transport in animal parasites: on the mechanism of glucose uptake by *Hymenolepis diminuta*. J. Parasit. 46:145-53.

Pittman, R. G., and F. M. Fisher, Jr. 1972. The membrane transport of glycerol by *Hymenolepis diminuta*. J. Parasit. 58:742-49.

Porter, C. E., and J. E. Hall. 1970. Histochemistry of a cotylocercus cercaria. II. Hydrolytic and oxidative enzymes in *Plagioporus lepomis*. Exp. Parasit. 27:378-87.

Probert, A. J., M. Goil, and R. K. Sharma. 1972. Biochemical and histochemical studies on the non-specific phosphomonoesterases of *Fasciola gigantica* Cobbold 1855. Parasitology 64:347-53.

Pudles, J., F. H. Rola, and A. K. Matida. 1967. Studies on the proteolytic inhibitors from *Ascaris lumbricoides* var. *suum*. II. Purification, properties, and chemical modification of the trypsin inhibitor. Arch. Biochem. Biophys. 120:594-601.

Read, C. P. 1957. The role of carbohydrates in the biology of cestodes. III. Studies on two species from dogfish. Exp. Parasit. 6:288-93.

Read, C. P. 1959. The role of carbohydrates in the biology of cestodes. VIII. Some conclusions and hypotheses. Exp. Parasit. 8:365-82.

Read, C. P. 1967. Carbohydrate metabolism in *Hymenolepis* (Cestoda). J. Parasit. 53:1023-29.

Read, C. P. 1968. Some aspects of nutrition in parasites. Amer. Zool. 8:139-49.

Read, C. P. 1973. Contact digestion in tapeworms. J. Parasit. 59:672-77.

Read, C. P., and A. Z. Kilejian. 1969. Circadian migratory behavior of a cestode symbiote in the rat host. J. Parasit. 55:574–78.

Read, C. P., and A. H. Rothman. 1958. The role of carbohydrates in the biology of cestodes. VI. The carbohydrates metabolized *in vitro* by some cyclophyllidean cestodes. Exp. Parasit. 7:217–23.

Read, C. P., A. H. Rothman, and J. E. Simmons, Jr. 1963. Studies on membrane transport, with special reference to parasite-host integration. Ann. N.Y. Acad. Sci. 113:154–205.

Read, C. P., G. L. Stewart, and P. W. Pappas. 1974. Glucose and sodium fluxes across the brush border of *Hymenolepis diminuta* (Cestoda). Biol. Bull. 147:146–62.

Reichenbach-Klinke, H.-H., and K.-E. Reichenbach-Klinke. 1970. Enzymuntersuchungen an Fischen II. Trypsin- und α-amylase-Inhibitoren. Arch. Fischereiwiss.. 21:67–72.

Rhodes, M. D., C. L. Marsh, and G. W. Kelley, 1963. Trypsin and chymotrpysin inhibitors from *Ascaris*. Exp. Parasit. 13:266–72.

Robinson, D. L. H. 1961. Phosphatases of *Schistosoma mansoni*. Nature 191:473–74.

Rogers, W. P. 1947. Histological distribution of alkaline phosphatase in helminth parasites. Nature 159:374–75.

Rola, F. H., and J. Pudles. 1966. Studies on the chymotryptic inhibitor from *Ascaris lumbricoides* var. *suum*. Purification and properties. Arch. Biochem. Biophys. 113:134–42.

Rothman, A. H. 1966. Ultrastructural studies of enzyme activity in the cestode cuticle. Exp. Parasit. 19:332–38.

Rothman, A. H. 1967. Ultrastructural enzyme localization in the surface of *Moniliformis dubius* (Acanthocephala). Exp. Parasit. 21:42–46.

Ruff, M. D.,and C. P. Read. 1973. Inhibition of pancreatic lipase by *Hymenolepis diminuta*. J. Parasit. 59:105–11.

Ruff, M. D., G. L. Uglem, and C. P. Read. 1973. Interactions of *Moniliformis dubius* with pancreatic enzymes. J. Parasit. 59:839–43.

Seed, J. R., M. S. Bogucki, and S. C. Merritt. 1979. Interactions between immunoglobulins and the trypanosome cell surface. *In* this volume.

Shishova-Kasatochkina, O. A., and A. Ya. Dubovskaya. 1975. Proteinase activity in certain cestode species parasitizing vertebrates of different classes. Acta Parasit. Pol. 23:389–93.

Smithers, S. R., R. J. Terry, and D. J. Hockley. 1969. Host antigens in schistosomiasis. Trans. Roy. Soc. B. 171:483–94.

Smyth, J. D. 1973. Some interface phenomena in parasitic protozoa and platyhelminths. Can. J. Zool. 51:367–77.

Starling, J. A. 1975. Tegumental carbohydrate transport in intestinal helminths: correlation between mechanisms of membrane transport and the biochemical environment of absorptive surfaces. Trans. Am. Microsc. Soc. 94:508–23.

Starling, J. A., and F. M. Fisher, Jr. 1975. Carbohydrate transport in *Moniliformis dubius* (Acanthocephala). I. The kinetics and specificity of hexose transport. J. Parasit. 61:977–90.

Tarazona Vilas, J. M. 1958. Contribución al estudio de las gliceromonofosfatasas en los platemintos parásitos. Rev. Iber. Parasit. 18:233–42.

Taylor, E. W., and J. N. Thomas. 1968. Membrane (contact) digestion in the three species of tapeworm *Hymenolepis diminuta*, *Hymenolepis microstoma*, and *Moniezia expansa*. Parasitology 58:535–46.

Thorpe, E. 1968. Comparative enzyme histochemistry of immature and mature stages of *Fasciola hepatica*. Exp. Parasit. 22:150–59.

Threadgold, L. T. 1968. Electron microscope studies of *Fasciola hepatica*. IV. The ultrastructural localization of phosphatase. Exp. Parasit. 23:264–76.

Timble, J. J., III, H. H. Bailey, and E. N. Nelson. 1971. *Aspidogaster conchicola* (Trematoda: Aspidobothrea): histochemical localization of acid and alkaline phosphatase. Exp. Parasit. 29:457–62.

Trimble, J. J., III, and R. D. Lumsden. 1975. Cytochemical characterization of tegument membrane-associated carbohydrates in *Taenia crassiceps* larvae. J. Parasit. 61:665–76.

Uglem, G. L., and S. Beck. 1972. Habitat specificity and correlated aminopeptidase activity in the acanthocephalans *Neoechinorhynchus cristatus* and *N. crassus*. J. Parasit. 58:911–20.

Uglem, G. L., P. W. Pappas, and C. P. Read. 1973. Surface aminopeptidase in *Moniliformis dubius* and its relation to amino acid uptake. Parasitology 67:185–95.

Uglem, G. L., P. W. Pappas, and C. P. Read. 1974. Na$^+$-dependent and Na$^+$-independent glycerol fluxes in *Hymenolepis diminuta* (Cestoda). J. Comp. Physiol. 93:157–71.

Ugolev, A. M. 1965. Membrane (contact) digestion. Physiol. Rev. 45:555–95.

Waitz, J. A. 1963. Histochemical studies on the cestode *Hydatigera taeniaeformis* Batsch, 1786. J. Parasit. 49:73–81.

Waitz, J. A., and J. L. Schardein. 1964. Histochemical studies of four cyclophyllidean cestodes. J. Parasit. 50:271–77.

Wheater, P. R., and R. A. Wilson. 1976. The tegument of *Schistosoma mansoni*: a histochemical investigation. Parasitology 72:99–109.

Willadsen, P. 1977. A chymotrypsin inhibitor from the parasitic nematode, *Oesophagostomum radiatum*. Aust. J. Biol. Sci. 30:411–19.

Wilson, R. A., and P. E. Barnes. 1977. The formation and turnover of the membranocalyx on the tegument of *Schistosoma mansoni*. Parasitology 74:61–71.

Yamao, Y. 1952a. (Histochemical studies on endoparasites. VII. Distribution of the glycerol-mono-phosphatases in the tissues of the cestodes, *Anoplocephala perfoliata*, *A. magna*, *Moniezia expansa*, and *Taenia taeniaeformis*.) Zoo. Mag. (Tokyo) 61:254–60. (In Japanese.)

Yamao, Y. 1952b. (Histochemical studies on endoparasites. VIII. Distribution of the glycerol-mono-phosphatases in various tissues of larvae of cestodes, *Cysticercus bovis*, *Echinococcus cysticus fertilis*, and *Cysticercus fasciolaris*.) Zoo. Mag. (Tokyo) 61:290–94. (In Japanese.)

Yamao, Y. 1954. (Histochemical studies on endoparasites. V. Distribution of glycerol-mono-phosphatases in various tissues of flukes, *Eurytrema coelomaticum*, *E. pancreaticum*, *Dicroceolium lanceatum*, and *Clonorchis sinensis*.) J. Coll. Arts Sci., Chiba Univ. 1:9–13. (In Japanese.)

L. H. RHODES AND J. W. GERDEMANN

Nutrient Translocation in Vesicular-Arbuscular Mycorrhizae

8

INTRODUCTION

During the past twenty-five years, much progress has been made toward a better understanding of the effects of vesicular-arbuscular (VA) mycorrhiza, a mutualistic symbiosis between roots of higher plants and certain fungi. The most frequently reported observation is that VA mycorrhizae enhance plant growth by increasing inorganic nutrient uptake from the soil. This growth response can be quite dramatic, with mycorrhizal plants growing some 25 to 30 times larger that their non-mycorrhizal counterparts (e.g., Kleinschmidt and Gerdemann, 1972). Because of the potential for increased plant growth and yield, and since most crop plant species form VA mycorrhizae, increasing attention is being given to this symbiosis by agricultural and biological scientists. It should be emphasized, however, that the mycorrhizal condition is the normal condition of roots of the vast majority of species of vascular plants. Absence of mycorrhiza usually occurs only when the symbiotic fungi are eliminated by man.

In this paper we will briefly describe the processes that account for the VA mycorrhizal growth response. To understand these processes, it is important that attention be given not only to the biology of the mycorrhizal relationship but also to the physical and chemical factors that govern soil nutrient supply to roots. In this way it becomes possible to make predictions about the significance of VA mycorrhizae for particular plant nutrients under specific conditions. Consequently, some space has been devoted to these aspects of VA mycorrhizal studies. Initially, we will present a brief review of the morphology of VA mycorrhizae and the cytological changes that occur during the infection process. Secondly, we will concentrate on the means by which nutrients are supplied to the plant by the fungal partner and the secondary interactions that result. Finally, the current status of our knowledge of nutrient movement from host to fungus will be discussed.

MORPHOLOGY OF VA MYCORRHIZAE

The widespread occurrence and the nature of VA mycorrhizal infections appears to have been first reported by Schlicht (1888; 1889). He described and illustrated the morphology of the fungal structures in plant roots, noted the absence of damage to root tissue, and suggested a nonpathogenic role of the fungi involved. The excellent work of Janse (1896), Stahl (1900), Shibata (1902), and Gallaud (1905) greatly added to an understanding of the distribution and morphology of VA mycorrhizae. Much of the information amassed by these early investigators has been confirmed only recently by electron microscopy and autoradiography.

Currently, the fungi that form VA mycorrhizae are placed in four genera in the family Endogonaceae of the Mucorales, namely, *Glomus*, *Sclerocystis*, *Gigaspora*, and *Acaulospora* (Gerdemann and Trappe, 1974). All species in these genera that have been thoroughly investigated have been found to form VA mycorrhizae. All appear to be obligate symbionts, and have not been grown in axenic culture. In contrast to most obligate parasites, they have extremely wide host and geographic ranges. For example, *Glomus mosseae* has been found in North America (Gerdemann, 1961), Europe (Mosse, 1953), Australia (Mosse and Bowen, 1968), Africa (El-Giahmi et al., 1976), and Asia (Khan, 1971), and readily forms mycorrhiza on such vastly different hosts as bald-cypress (*Taxodium distichum*), tomato (*Lycopersicum esculentum*), tulip tree (*Liriodendron tulipifera*), and wheat (*Triticum aestivum*).

Vesicular-arbuscular mycorrhizal fungi form resting spores, which are borne either singly in soil or in hypogeous (or rarely epigeous) sporocarps. The spores range from less than 15 μm in diameter (*Glomus tenuis*) to over 500 μm in diameter (*Gigaspora gigantea*). Under favorable conditions spore germination takes place. Germ tubes grow through soil and may eventually come in contact with roots of a suitable host. Here appressoria form and hyphal penetration into or between epidermal cells of the root takes place. Hyphae grow through the epidermis and into the cortex, where ramification of hyphae is both inter- and intracellular. Tissues medial to the cortex are not invaded, nor are epidermal cells after initial penetration. Longitudinal spread of infection appears to be limited to approximately 5 mm around a single penetration point (Cox and Sanders, 1974).

A hypha growing into an individual cortical cell may undergo repeated dichotomous branching, with the ultimate branches being approximately 1 μm in diamter. These highly branched, haustorium-like structures are called "arbuscules" (Gallaud, 1905), a term meaning "little trees," which the arbuscules resemble. Individual arbuscular branches are surrounded by the intact plasmalemma of the root cortical cell (Janse, 1896; Cox and Sanders, 1974; Kinden and Brown, 1975; Scannerini et al., 1975; Kaspari, 1973, 1975).

One result of arbuscule formation is a substantial increase in host cell cytoplasm (Kaspari, 1973). For onion roots infected with *Glomus mosseae*, Cox and Tinker (1976) calculated a 15-fold increase in the cytoplasm of arbuscule-containing cells over that of uninfected cells. Increases in cell organelles, including mitochondria, ribosomes, endoplasmic reticula, and dictyosomes, also occur (Kaspari, 1973; Cox and Sanders, 1974, Kinden and Brown, 1975), indicating that the increased volume occupied by host cytoplasm represents synthesis of cytoplasm rather than expansion of existing cytoplasm into a larger volume. Vacuoles also are smaller and more numerous (Kinden and Brown, 1976). Disappearance of starch reserves is also associated with the synthesis of cytoplasm (Endrigkeit, 1937; Gallaud, 1905; Janse, 1896; Lihnell, 1939, Kinden and Brown, 1975). In addition, nuclei of infected cells become greatly enlarged (Demeter, 1923, Gallaud, 1905; Lihnell, 1939; McLuckie and Burges, 1932; Ali, 1969; Mosse, 1963).

Arbuscules remain intact for only a brief period. Averages of 4 days (Cox and Tinker, 1976) or 4–15 days (Bevege and Bowen, 1975) have been calculated for the processes that include arbuscule formation to complete arbuscular disintegration. It is not known whether the breakdown of arbuscules is strictly an autolytic process or whether digestion by the host is involved. The process begins with deterioration of cytoplasm in arbuscular hyphae followed by the collapse of hyphal walls of the tips of individual arbuscular branches (Cox and Sanders, 1974; Kinden and Brown, 1976). This may occur in only one region of the arbuscule at first, but proceeds until the entire arbuscule, with the exception of the main trunk, has disintegrated. Remnants of the arbuscule are seen as somewhat globular clumps of fungal material within cells. Electron microscopy reveals that these remnants are usually encapsulated in an amorphous matrix, presumably of host origin, surrounded by host cell membrane (Kaspari, 1973; Cox and Sanders, 1974; Kinden and Brown, 1976). After deterioration of the arbuscule, the nucleus of the host cell returns to its normal size and starch may reappear (Gallaud, 1905).

Vesicles (Janse, 1896) are globose to ellipsoid (sometimes irregularly lobed) hyphal swellings that are either inter- or intracellular. They are usually borne terminally, but may be intercalary. Vesicles contain oil droplets similarly to spores borne outside the root, and it is usually assumed that these organs serve a storage function for the fungal endophyte. If vesicles become thick-walled, they resemble spores in morphology and function. Vesicles within roots are nearly always lacking in VA mycorrhizae formed by *Gigaspora* species. In this genus vesicles are borne outside the root (Nicolson and Gerdemann, 1968; Gerdemann and Trappe, 1974). Such vesicles are probably not homologous to the internal vesicles produced by *Glomus, Sclerocystis*, and *Acaulospora* spp., although their function may be similar.

The morphology of infection is influenced by the host as well as the fungus. For example, infection of corn (*Zea mays*) by *Glomus fasciculatus* follows Gallaud's "*Arum*-type infection" with hyphae and vesicles primarily intercellular and arbuscules produced terminally on hyphae within cortical cells. In tulip tree *G. fasciculatus* produces a "*Paris*-type infection", the hyphae and vesicles being intracellular, with arbuscules formed on lateral branches from coiled hyphae and confined to the inner cortical cells (Gerdemann, 1965).

Hyphae that extend into the soil surrounding the root apparently originate only as branches of penetration hyphae. Points of egress from internal fungal structures in living roots have not been observed (Mosse, 1959; Nicolson, 1959). External hyphae do, however, grow along the root surface and repeatedly penetrate the root. Such hyphae have been referred to as "runner" hyphae (Nicolson, 1959). The number of penetration points is important, since it represents the maximum possible number of connections between mycelium in the root with that in the soil, through which all nutrient translocation must take place. Counts of entry points range from approximately 1 to 20 per mm root length (table 1). Hyphae grow through soil pore spaces, forming a rather loose network, and may grow from the root surface into the soil for considerable distances. The extent of external mycelium development is usually not evident since the mycelium breaks off readily when roots are removed from soil. However, by allowing mycorrhizae to develop in root observation chambers, external hyphae can be easily observed. Six weeks after onion roots were inoculated with *Glomus fasciculatus*, external hyphae had grown up to 8 cm from roots (Rhodes and Gerdemann, 1975). In later stages of infection, older hyphae of some VA fungi, e.g., *Glomus fasciculatus*, may be up to 30 μm diameter near the root surface, and have very thick hyphal walls. Such hyphae appear to be better adapted for nutrient translocation than for absorption.

TABLE 1

NUMBER OF PENETRATIONS PER UNIT OF ROOT BY VA MYCORRHIZAL ENDOPHYTES

Plant Species	No. of entry points per unit of root	Reference
Eriostemon crowei	2.5–4.1 per mm^2 root surface	McLuckie & Burges, 1932
Juniperus communis	4–16 per mm root length	Lihnell, 1939
Various cereal species	10–20 per cm root length	Winter, 1953
Various grass species	1 per mm root length	Nicolson, 1959
Fragaria sp.	2.6–21.1 per mm root length	Mosse, 1959
Malus sp.	4.6–10.7 per mm root length	Mosse, 1959
Allium cepa	6 per cm root length	Sanders & Tinker, 1973

NUTRIENT UPTAKE BY VA MYCORRHIZAE

Mosse (1957) appears to have been the first to compare mineral compositions of VA mycorrhizal versus nonmycorrhizal plants. Certain nutrients were found to be in higher concentrations in tissues of mycorrhizal apple seedlings, and other nutrients were in significantly lower concentrations, suggesting that the growth response reflected an improvement in plant nutrition brought about by mycorrhizal infection. Since that time, a number of similar studies have been done, (e.g. Baylis 1959; Ross and Harper, 1970; Kleinschmidt and Gerdemann 1972). It has been noted that phosphorus (P) is by far the most important nutrient involved in the VA mycorrhizal growth response, and is consistently found in higher concentrations in mycorrhizal plants as compared to nonmycorrhizal controls (see reviews by Mosse, 1973a; Gerdemann, 1975; Tinker, 1975). However, increases or decreases in the accumulated concentrations of host plant nutrients other than P also result from VA mycorrhizal infection, either as direct effects of the mycorrhiza or as secondary effects of changes in P concentration in the host plant.

Several hypotheses for mechanisms of increased P uptake by VA mycorrhizae have been proposed. These have been summarized by Tinker (1975), and include: (1) that infection by a mycorrhizal endophyte may alter plant morphology in some way, e.g., by inducing root hypertrophy, which would result in an enlarged root system and thus greater surface area for P absorption; (2) that mycorrhizae have access to sources of P that are not available to nonmycorrhizal roots, including relatively insoluble, inorganic P such as apatite, tri-calcium P, and iron and aluminum P compounds; (3) that mycorrhizal infection alters host metabolism so that P absorption or utilization by infected roots is enhanced, i.e., an increase in "absorbing power" (Nye, 1966) of individual roots; (4) that hyphae in soil absorb P and translocate it to infected roots, where it is tranferred to the mycorrhizal host, thus in effect increasing the volume of soil utilized by the nutrient-absorbing system of the plant; or (5) that infected regions of a root remain active in nutrient absorption for a longer period than comparable regions that are not infected.

The hypothesis that mycorrhizae are able to utilize slowly soluble forms of P, normally not available to nonmycorrhizal roots (hypothesis 2) has received considerable attention. Murdoch and colleagues (1967) grew mycorrhizal and nonmycorrhizal corn in pots, using sources of P fertilizer of varying availability. In all cases mycorrhizal plants grew better and took up more P than nonmycorrhizal plants from the soil that received relatively unavailable forms of added P (rock P and tri-calcium P). However, there were no significant differences between growth and P uptake of mycorrhizal and nonmycorrhizal corn when the P source was readily available (superphosphate or monocalcium P). Similar results were obtained by Daft and Nicolson (1966), Hall (1975), Ross and Gilliam (1973), and Powell and Daniel (1978). Although mycorrhizae were more efficient in removing P from these soils, all current evidence points to the fact that mycorrhizal and nonmycorrhizal plants absorb from similar pools of soil P. Sanders and Tinker (1971), Hayman and Mosse (1972), and Powell (1975) conducted experiments in which ^{32}P was allowed to equilibrate with the labile pool of soil P. In all cases the specific activity of P in plant tissue (^{32}P/^{31}P) was similar for mycorrhizal and nonmycorrhizal plants, indicating that both drew from the same labile pool of soil P. The better utilization of relatively insoluble forms of P by mycorrhizae in the previously mentioned experiments can therefore be attributed to greater uptake by mycorrhizae of P in solution and those forms that are in rapid equilibrium with it, rather than an increase in P solubilization by mycorrhizae.

The hypothesis that root hypertrophy is the primary effect of mycorrhizal infection meets with the objection that mycorrhizal plants usually have lower root-to-shoot ratios than comparable nonmycorrhizal control plants. Mycorrhizal root systems may, in fact, be smaller than uninfected root systems yet support significantly greater top growth.

Sanders and Tinker (1971, 1973) dismissed the possibility of enhanced absorbing power by mycorrhizal roots (hypothesis 3) as being the important mechanism in increased P uptake. On the basis of their calculations, nonmycorrhizal roots were already absorbing P from solution at the maximum rate possible. Infected roots would thus have received no benefit from increased efficiency of P uptake since it had been depleted in the immediate vicinity of the root. This does not, however eliminate the possibility of an increase in P uptake efficiency of mycorrhizal roots, and its importance under some conditions.

Since a mycorrhizal growth response may be evident 4 or 5 weeks after inoculation it seems unlikely that hypothesis 5, i.e., longer duration of root functioning, could account for the enhancement of P uptake. This seems particularly true for many herbaceous plants where roots may function for the entire life of the plant.

Direct evidence, which is presented below, indicates that translocation of P and other nutrients does occur, and thus hypothesis 4 is assumed to be of primary importance in increasing nutrient uptake.

INORGANIC NUTRIENT TRANSFER FROM FUNGUS TO HOST

The process by which inorganic nutrients are supplied to the higher plant by the fungal symbiont can best be divided into three phases. The first phase involves absorption of nutrient ions from soil by external hyphae. The second phase involves translocation of nutrients from external to internal mycelia. In the third phase the nutrients are released from fungal structures to root cells.

Nutrient Uptake by External Hyphae

Relatively little is known about ion absorption by external hyphae. The fact that VA mycorrhizal fungi are obligate symbionts has made nutrient uptake studies with axenic fungal cultures impossible. It is assumed that active ion transport across the fungal plasmalemma occurs in external hyphae much the same as it does in numerous other microorganisms and higher plants. Bowen and colleagues (1975) found that uptake of P by germ tubes of VA mycorrhizal fungi was inhibited at 10°C, indicating an active uptake mechanism. Since the concentration of P in the cytoplasm of external hyphae is approximately 0.3% (Pearson and Tinker, 1975), which is on the order of 3×10^3 times greater than P concentration in soil solution (Reisenauer, 1966), it can be assumed that P is accumulated by an energy-requiring process.

Cress and colleagues (1979) have applied enzyme kinetics analysis to data for ion uptake from solution by mycorrhizal and nonmycorrhizal tomato plants. Their results indicate that for solution concentrations of P closely approximating those in soil, mycorrhizae had lower K_m values than did nonmycorrhizal roots; however, V_{max} values were similar. One interpretation of this data was that within this concentration range affinity for P at absorption sites was greater in external hyphae than in uninfected roots. No differences between mycorrhizae and nonmycorrhizal roots for either K_m or V_{max} values were found for manganese (Mn) or zinc (Zn) uptake.

Translocation from External to Internal Fungal Structures

Once absorbed by external hyphae, nutrient ions are translocated to internal fungal structures in roots. The rate at which this process occurs in hyphae relative to the rate at which ions may move through soil by mass flow or diffusion is one of the factors determining the importance of mycorrhizae for that particular nutrient. For example, P is found in extremely low concentrations in soil solutions, usually less than 1 ppm (Reisenauer, 1966). Also, P in solution is readily absorbed by soil colloids. Movement of P to plant roots by mass flow or by diffusion in soil is, thus, very slight, and P is said to be highly immobile in soil. Rapid accumulation of P by an individual plant root results in a P depletion zone extending approximately 1-2 mm around the root (Bhat and Nye, 1974; Lewis and Quirk, 1967). Further uptake of P is then limited by the relatively slow rate of P movement to the root surface. Sanders and Tinker (1971) measured accumulation of P at regular intervals in tissues of mycorrhizal and nonmycorrhizal onion plants and determined that the rates of P inflow to mycorrhizal plants could not be accounted for by mass flow or diffusion of P through soil. They concluded that the rapid rate of P movement to roots could only be accounted for by assuming the occurrence of P translocation through hyphae.

The use of chambers that confine roots to one region of a growth medium while allowing external hyphae to grow into an adjacent region has helped elucidate some of the facts about nutrient translocation in VA mycorrhizae. In such systems a test plant is grown in a chamber provided with a barrier through which roots do not pass, but through which external hyphae can grow. Radioactively labeled plant nutrients may then be applied to external mycelia, and the movement of these tracers can be followed. The exact design of such chambers has varied among different investigators, and they have been termed "soil chambers," "split plates," or "nutrient translocation chambers." Similar systems have been used to study translocation phenomena in ectomycorrhizae (Melin, 1958; Skinner and Bowen, 1974) orchid mycorrhizae (Smith, 1967; Hadley and Purves, 1974), and ericoid mycorrhizae (Pearson and Read, 1973).

The ability of mycorrhizal fungi to translocate P was demonstrated by Hattingh and colleagues (1973). The mycorrhizal fungi *Endogone mosseae* (= *Glomus mosseae*) and *E. fasciculata* (= *G. fasciculatus*) formed mycorrhizae on onion, and hyphae grew from confined roots into adjacent soil. When ^{32}P was injected into soil in which the external hyphae were present, it was detected in host roots after 2 days. When ^{32}P was injected into soil in which nonmycorrhizal plants were grown, it was not detected in roots. Also, when external hyphae were severed from roots prior to tracer injection, ^{32}P transport was not detected. Using a similar soil chamber system and injecting ^{32}P into soil at 1 cm intervals from confined roots, Rhodes and Gerdemann (1975) showed that hyphal transport of P may occur from as much as 7 cm from the root surface. Nonmycorrhizal plants did not obtain ^{32}P even when it was injected one cm from root surfaces. Thus, mycorrhizal roots have access to a much greater volume of soil from which they can obtain P. Pearson and Tinker (1975) measured the steady state flux of P through external hyphae to clover plants. A calculated flux of approximately 1×10^{-9} moles/cm^2/sec was obtained. This was in reasonable agreement with a flux of 3.8×10^{-8} moles/cm^2/sec calculated for the mycorrhizal system of onion in pot cultures (Sanders and Tinker, 1973). The fact that P is readily translocated to roots through external hyphae has thus been repeatedly confirmed.

There is increasing evidence that P is translocated in the form of polyphosphate. Cox and colleagues (1975) and Ling-Lee and colleagues (1975) observed bodies in hyphae of VA fungi that showed a metachromatic reaction when stained with toluidine blue, a characteristic of polyphosphate. On the basis of this evidence and other histochemical tests, they suggested that these bodies were granules of polyphosphate. Cox and colleagues (1975) found one granule of polyphosphate per fungal vacuole, although faint staining of vacuolar contents indicated polyphosphate also in the vacuolar solution. Vacuoles could be seen to move rapidly with the cytoplasmic flow in the fungal lumen. Consequently, it was suggested that polyphosphate could serve as an efficient means of P translocation by external hyphae (Cox et al., 1975; Tinker, 1975).

Using phenol-detergent extraction and polyacrylamide gel electrophoresis, Callow and colleagues (1978) were able to separate and identify polyphosphate from mycorrhizal onion roots. Polyphosphate was not detected in extracts from nonmycorrhizal roots. Staining of mycorrhizae with toluidine blue showed polyphosphate restricted to hyphae. Their calculations indicated that the concentration of polyphosphate in hyphae (40% of total P) was sufficient to account for previously determined P fluxes in mycorrhizal systems.

The efficiency of P translocation in VA fungi is borne out by comparison of its rate of movement into plant tissue with that of calcium (Ca). Rhodes and Gerdemann (1978a) injected ^{32}P and ^{45}Ca tracers into soil 4.5 cm from confined onion roots. Each tracer was applied at a rate designed to label the available pool of each nutrient at equivalent specific activities, i.e., accumulation of the nutrients in amounts proportional to plant requirements would result in equal radioactivity in plant tissue. Radioactivity resulting from ^{32}P was some 40 times greater in roots and 30 times greater in shoots than was ^{45}Ca, indicating that P was more readily translocated than was Ca. Neither radionuclide was present in non-mycorrhizal plants, indicating that all radioactive atoms in plants were translocated to them through hyphae. Thus, a situation of Ca immobility relative to P appears to exist in the hyphal lumen.

The mechanisms that act to translocate nutrients over considerable distances in hyphae of VA mycorrhizal fungi are as yet imperfectly understood. By gently extracting external hyphae from soil and examining them under the light microscope, it is easy to observe rapid bidirectional cytoplasmic streaming. This streaming bears little resemblance to the flow of cytoplasm in Oomycetes such as *Pythium* and *Phytophthora*, or even to that of other members of the Mucorales such as *Rhizopus*, where cytoplasmic flow occurs principally in one direction at a given time (Arthur, 1897). Instead, cytoplasmic streaming in *Glomus* spp. is characterized by the simultaneous, rapid bidirectional movement of cytoplasm containing vacuoles whose diameters are usually less than one third the diameter of the hyphal lumen. Bodies moving in opposite directions often collide, indicating that there are not separate channels within the lumen for bodies moving in opposite directions (Rhodes, unpublished). Streaming eventually slows under the intense heat of a microscope lamp, and finally comes to a complete halt. The stimulus responsible for cytoplasmic streaming is unknown. Similar cytoplasmic flow occurs in germ tubes of VA fungi, indicating that it is an inherent property of the fungus and not a response induced by the host following mycorrhizal infection.

Mycorrhizal infection decreases with increasing levels of P in the soil or growth medium (Daft and Nicolson, 1969, 1972; Mosse and Phillips, 1971; Mosse, 1973b; Hall, 1975). Consequently, at very high soil P levels, only a small fraction of the root system may be infected. It is the concentration of P in plant tissue, rather than P in soil, that regulates the degree of mycorrhizal infection. Sanders (1975) found that foliar application of P to onion plants resulted in decreased mycorrhizal infection. A similar result was obtained by Menge and colleagues (1978) who used a split root system technique. They found that when one half of a sudangrass (*Sorghum sudanense*) root system grew in a high P soil, mycorrhizal infection was reduced in the other half of the root system even though it grew in a separate container of relatively low P soil.

The biochemical mechanism whereby P acts to inhibit mycorrhizal infection is as yet unknown. Recent evidence, however, indicates a mycorrhiza-specific alkaline phosphatase (MSAP) that has optimal activity closely coinciding with the most active period of mycorrhizal infection (Gianinazzi-Pearson and Gianinazzi, 1976, 1978). It was suggested that MSAP may be involved in the establishment of mycorrhizal infection. Since MSAP is repressed in the presence of P, a possible mechanism exists by which mycorrhizal infection may be suppressed by an increase in P concentration in root tissue.

Very little attention has been given to the effect of P on external mycelium development, although a close correlation exists between external mycelium and development of the fungus in the root cortex (Nicolson, 1959; 1967; Sanders and Tinker, 1973; Sanders et al., 1977). Thus, P levels that reduce internal mycelium probably also cause a corresponding reduction of external mycelium. Such a reduction of hyphae in the soil is probably of little consequence in terms of reducing P obtained by the plant, since it is, in fact, the high level of P in plant tissue that causes infection to be suppressed. However, if external hyphae are functioning in the transfer of nutrients other than P, reduction of infection brought about by P fertilization could be of importance in reducing uptake of these essential ions.

It has often been reported that certain micronutrient deficiencies, particularly of Zn and copper (Cu), result from high P fertilization (Olsen, 1972). Reduction of mycorrhizal infection at high P levels could account for at least a portion of these P-induced micronutrient deficiencies. For example, application of $196 g P/m^2$ to nursery beds resulted in reduction of mycorrhizal infection and uptake of Cu by sour orange (*Citrus aurantium*) (Timmer and Leyden, 1978). In further experiments seedlings given high rates of P eventually developed Cu deficiency symptoms. Both Gilmore (1971) and LaRue and colleagues (1975) related Zn deficiency in peach (*Prunus persica*) to absence of mycorrhizae following methyl bromide fumigation of nursery beds. In the latter study mycorrhizal inoculation increased Zn uptake and improved growth more than fertilization with Zn and P, indicating that mycorrhizae have a direct effect on Zn uptake. Mycorrhizae of *Araucaria cunninghamii* absorbed Zn at greater rates from solution in short-term uptake experiments than did nonmycorrhizal roots (Bowen et al., 1974). Cooper and Tinker (1978) demonstrated the ability of the mycorrhizal fungus *Glomus mosseae* to translocate Zn to host roots. A mean Zn flux of 10^{-12} mole/cm^2/sec was calculated for uptake from $2 \times 10^{-6}M$ Zn over a 10-day period. To test the hypothesis that P nutrition could influence micronutrient uptake by affecting infection levels, Rhodes and colleagues (1978) grew mycorrhizal and nonmycorrhizal onions in soil chambers at two soil P levels. Inoculated onions grown in soil having a moderate P level (45 μg P/g soil) had mycorrhizal infection greater than 70% and were able to obtain ^{32}P or ^{65}Zn injected 4.5 cm from roots. Nonmycorrhizal control onions did not have access to injected tracers, indicating that hyphal translocation was supplying the nutrients to mycorrhizal roots. Inoculated plants grown in high P soil (97 μg P/g soil) had less than 8% infection and were also not able to obtain ^{32}P or ^{65}Zn. Thus, suppression of mycorrhizal infection and consequent reduction of external mycelium may reduce or eliminate hyphal translocation of Zn and may be a contributing factor in the phenomenon of P-induced Zn deficiency.

Sulfur (S) uptake can also be greatly increased if plants are mycorrhizal (Gray and Gerdemann, 1973), and S can also be translocated to roots through external hyphae (Cooper and Tinker, 1978; Rhodes and Gerdemann, 1978b). Sulfur is taken up by plants primarily in the form of sulfate (SO_4^{-2}). Since SO_4^{-2} is present in much higher concentrations in soil solutions than is P (often 25 ppm or greater [Reisenauer, 1966]), and is less readily absorbed to soil colloids (Kamprath et al., 1956), it is much more mobile than P in soil. Therefore, it would be expected that the importance of hyphal translocation for increased SO_4^{-2} uptake would be considerably less than for P uptake, since SO_4^{-2} may move to roots through soil as rapidly as through hyphae.

In soil chamber experiments it was found that movement of ^{35}S to roots, as determined by ^{35}S content of soil cores, was similar for treatments in which external hyphae were intact, severed, or absent (Rhodes and Gerdemann, 1978b). However, mycorrhizal onions from both severed and intact hyphae treatments took up more ^{35}S than did nonmycorrhizal plants, indicating an increase in "absorbing power" (Nye, 1966) of roots of mycorrhizal plants. Bowen and colleagues (1975) suggested that increases in absorbing power could result from improved assimilate supply to roots of mycorrhizal plants, and that this could subsequently affect the uptake of mobile nutrient ions. There is evidence that assimilate supply can affect absorbing power (Rovira and Bowen, 1973). This hypothesis was tested for SO_4^{-2} by growing mycorrhizal and nonmycorrhizal onions at 2 soil P concentrations and measuring uptake of applied ^{35}S after a 3-day period (Rhodes and Gerdemann, 1978c). Concentrations of ^{35}S in roots of mycorrhizal and nonmycorrhizal plants grown at high P levels were significantly greater than those in nonmycorrhizal plants grown in low P soil. In a second experiment detached nonmycorrhizal onion roots from plants given a nutrient solution containing P for 26 days before short-term uptake experiments absorbed ^{35}S from solution at greater rates than roots given a complete minus P nutrient solution. This occurred at all three concentrations of S tested, 1 mM, 10 μM, and 0.1 μM. Thus, at least a portion of increased S absorption by roots of mycorrhizal plants appears to be a secondary effect brought about by improved P nutrition, and a portion can be attributed to hyphal translocation of S.

Decreased resistance to water transport in soybean roots also occurs as a result of mycorrhizal infection (Safir et al., 1971). This apparently also results from an improvement in plant nutrition brought about by mycorrhizal infection (Safir et al., 1972). However, hyphal translocation of water may also take place and may be important under certain conditions, as in the case with SO_4^{-2}.

Mycorrhizae are involved in nitrogen (N) nutrition of legumes through an interaction with the root nodule symbiosis formed by *Rhizobium* spp. Nitrate (NO_3^-), which is the primary form of N absorbed by plant roots, is similar to P in that it is required in relatively large amounts. However, it differs radically from P in its characteristics in soil. NO_3^- is highly mobile and moves readily with the soil solution. Thus, it would not be expected that movement to the root surface would be the rate-limiting step for N uptake, and hyphal translocation of N would probably be of little significance for plant N nutrition.

Ross and Harper (1970) obtained increased growth and yield following inoculation of soybeans with mycorrhizal fungi. Mycorrhizal plants had higher N concentrations in leaf tissue, suggesting an enhancement of N uptake brought about by mycorrhizae. Asai (1948) suggested that mycorrhizae were indispensable for normal growth of many legumes and noted an increase in nodulation of plants that had mycorrhiza. Crush (1974) and Daft and El-Giahmi (1974) confirmed the observation of increased nodulation of legumes and showed that this resulted from improved P nutrition of mycorrhizal plants. Mycorrhiza or P fertilization had similar effects in increasing nodule numbers. Additionally, Daft and El-Giahmi (1974) noted that nitrogenase activity, as measured by acetylene reduction, was increased either by P fertilization or by mycorrhizal infection. This result has been confirmed for nitrogenase and nitrogen reductase activity (Carling et al., 1978). The facts that mycorrhizal infection usually does not increase N concentration of nonlegumes, and that mycorrhizae did not increase growth of non-nodulating isolines of soybean under N-deficiency conditions (Shenck and Hinson, 1973; Carling et al., 1978) argue against a direct effect of mycorrhizae in uptake of N. The importance of P in nodulation and N-fixation by *Rhizobium* spp. has been well documented (Van Schreven, 1958). Thus, improved N nutrition of legumes does not appear to be a direct effect of hyphal translocation of N, but rather a secondary effect of increased N fixation by *Rhizobium* resulting from greater P supply to roots.

The concept of the rate-limiting step in nutrient supply and utilization as set forward by Fried and Shapiro (1961) is useful in summarizing the importance of hyphal translocation and of mycorrhizae in general for plant nutrition. For ions such as P, and in some cases Zn and Cu, the rate-limiting step for uptake by plants is likely to be movement to the root surface. The rate of P uptake is thus substantially increased if P ions are translocated to roots through external hyphae rather than diffusing slowly through soil. In contrast, SO_4^{-2} may move to roots through soil by mass flow with little additional advantage being conferred by translocation through hyphae, the rate-limiting step now occurring at root surfaces. Mycorrhizal roots, having an improved P supply, are able to absorb S more rapidly. The rate-limiting step for N supply in soils is nearly always conversion to the nitrate or ammonium forms. Again, the effect of mycorrhiza is secondary, such as through the action of enhanced N fixation by *Rhizobium* in legume root nodules.

A further aspect of nutrient translocation in VA mycorrhizal fungi is the possibility of nutrient transfer between different plants through a mutually shared mycorrhizal fungus. Woods and Brock (1964) injected ^{32}P and ^{45}Ca into a red maple (*Acer rubrum*) stump and found the radioisotope present in 19 broadleaf tree species in the area. Since interspecific root grafting was unlikely, they hypothesized that the tracers may have been translocated through mycorrhizal hyphae from one tree to another. Hirrel and Gerdemann (1979) tested this hypothesis using a "donor" and "recipient" onion plant grown in soil chambers. The donor plant was exposed to $^{14}CO_2$ and the nonexposed recipient plant was subsequently assayed for ^{14}C activity. The fact that the highest amount of ^{14}C occurred in recipient plants that were mycorrhizal indicated that ^{14}C transfer through external hyphae was probably occurring. The fungitoxicant pentachloronitrobenzene (PCNB) applied to mycorrhizal plants reduced but did not eliminate ^{14}C activity in recipient plants, suggesting that a portion of the transferred ^{14}C was moving by means of a "wick" effect on the outside of hyphal walls (Read and Stribley, 1975). However, it was not possible to determine if

PCNB had completely inhibited translocation in the mycorrhizal fungus. The possibility of an interconnecting system of mycorrhizal hyphae in plant communities deserves further study.

Release of Nutrients from Fungus to Host

The site of release of nutrients to the mycorrhizal host has usually been assumed to be the arbuscule. There were two reasons for this assumption. First, arbuscular branches were believed to greatly increase the area of surface contact between host and fungus, providing more surface area for nutrient exchange. Second, arbuscules were observed to be broken down within host cells, indicating either digestion of the arbuscule by the host or autolysis, either process resulting in a release of hyphal contents to host cells.

Current evidence indicates that breakdown of arbuscules cannot account for sufficient transfer of nutrients from endophyte to host. Cox and Tinker (1976) estimated that less than 1% of P inflow to the host could be attributed to arbuscular breakdown. They suggested instead that transfer of P across the membranes of living fungus and host was a more likely mechanism.

Techniques using radioisotopes have shed some light on the role of the arbuscule and other internal fungal structures in the release of nutrients from fungus to host. After application of a solution containing ^{32}P to onion pot cultures, Gray and Gerdemann (1969) found greatest labeling in infected root segments as compared with uninfected segments or root tips from mycorrhizal plants. In autoradiographs Ali (1969) found ^{32}P labeling restricted to hyphae after mycorrhizal roots were incubated in a solution of ^{32}P. However, no comparison in amount of labeling was made between hyphae, vesicles, and arbuscules. These two studies indicate that at least initially there is an accumulation of P in internal fungal structures prior to its release to host cells. Using X-ray microanalysis, Schoknecht and Hattingh (1976) found that cytoplasm of cells containing arbuscules had higher P concentrations than adjacent cortical cells lacking arbuscules. Also, arbuscular branches had higher P concentrations than the host cytoplasm surrounding them. They found no appreciable differences in the concentrations of other nutrients between infected and uninfected cells.

Electron micrographs showed an absence of polyphosphate granules in the tips of arbuscular branches, although these granules were present throughout the remainder of the mycelium (Cox et al., 1975). This indicated a possible unloading of polyphosphate from the arbuscular branches. Silver grain counts from autoradiographs of mycorrhizal roots following incubation in a solution of ^{32}P showed some 4 times the activity in cortical cells with arbuscules as compared to uninfected cells (Bowen et al., 1975).

There can be little doubt as to the importance of the arbuscule in nutrient transfer. However, it remains to be demonstrated that the arbuscule is more efficient for nutrient release than undifferentiated hyphae or vesicles. For an entire onion root system, surface contact between fungus and host plasmalemma was increased only 1.5 times by arbuscules (Cox and Tinker, 1976). It is likely that an increase of this magnitude or greater would be conferred by intracellular vesicles or hyphae. An autoradiograph by Bowen and colleagues (1975) shows heavy labeling in vesicles that is equal to, or greater than, that in an adjacent arbuscule. Further work is needed to determine the rate at which vesicular P is transferred to host cells.

NUTRIENT TRANSFER FROM HOST TO FUNGUS

Since VA mycorrhizal endophytes are obligate symbionts, it has been generally assumed that they obtain essential organic nutrients from their hosts, and that transfer of carbon compounds from host to fungus must, therefore, occur. However, few of the details of this process are known. Ho and Trappe (1973) exposed aerial portions of mycorrhizal fescue (*Festuca occidentalis*) to $^{14}CO_2$. They found ^{14}C in spores of the mycorrhizal endophyte *Endogone (Glomus) mosseae* 2 months after the exposure period, indicating that a portion of the carbon fixed by the host was eventually utilized by the mycorrhizal fungus. Because of the long duration of the experiment, a portion of the ^{14}C accumulating in spores may have resulted from root exudation of ^{14}C and absorption of exuded compounds by external hyphae. However, Cox and colleagues (1975) found ^{14}C in both internal and external fungal structures 27 hr after exposure of onion tops to $^{14}CO_2$. Since 27 hr was the earliest harvest, ^{14}C may have actually been present in the mycorrhizal fungus somewhat sooner.

Apparently ^{14}C is preferentially accumulated by the mycorrhizal fungus as compared to root cells. The amount of ^{14}C per unit weight of hyphae was nearly 4 times higher than that of associated clover roots (Bevege et al., 1975). Autoradiographs of Cox and colleagues (1975) indicate accumulation of ^{14}C in endophyte structures. It is interesting that the fungal symbiont appears to have a greater affinity for host assimilate than the host itself.

The form in which C is transferred from host to endophyte is not known. In ectomycorrhizae, sucrose appears to be the principal carbohydrate transferred from host to fungus. Sucrose is rapidly converted to trehalose and mannitol, and eventually to glycogen, which is stored by the fungus (Lewis and Harley, 1965; Bevege et al., 1975). Neither trehalose nor mannitol has been detected in VA mycorrhizae (Hayman, 1974; Bevege et al., 1975). Although glycogen in hyphae of a mycorrhizal fungus contained approximately 10% of the total ^{14}C label extracted from fungal material after exposure of tops to ^{14}C, a larger proportion (56.2%) was found in the metal-precipitated fraction, suggesting storage as lipids or lipo-proteins (Bevege et al., 1975). This result agrees with histochemical results (Cox et al., 1975; Cooper and Lösel, 1978) that show a large proportion of the fungal volume occupied by lipid. Cooper and Lösel (1978) found *G. mosseae* mycelium to contain 43.8% lipid. Triglyceride, diglyceride, and free fatty acids predominated the neutral lipid fraction, and phosphatidyl ethanolamine was the principal phospholipid present.

Further work on nutrient transfer from host to fungus may elucidate some of the facts about the nutrient requirements and thus the obligate nature of VA mycorrhizal fungi. Multiply-labeled carbohydrates fed to hosts, followed by sequential extraction and fractionation of carbon compounds in external mycelium, should give a better indication as to what compounds are transferred and their pathways for synthesis of growth and storage compounds in the fungus.

CONCLUSION

Vesicular-arbuscular mycorrhiza plays an extremely important role in plant nutrition. Hyphae of VA fungi grow from the root into the soil and absorb and translocate nutrients to the host root; therefore, they are most important in supplying the host with those essential ions that are highly immobile in soil. Vesicular-arbuscular mycorrhizae are particularly important in assuring adequate P nutrition of the host. Phosphate is absorbed and rapidly translocated to the host, probably as polyphosphate granules, and it is released to the host within the root cortex.

There is a unique self regulating mechanism with regard to P that assures the host of obtaining near optimal levels of P. At low soil P levels the plant is highly mycorrhizal, and much of the P used by the host is supplied by fungal hyphae. As levels of P in soil are increased and the concentration of P within the host rises, the level of mycorrhizal infection drops. At high P levels mycorrhizal infection may be totally inhibited, and the much less efficient nonmycorrhizal root supplies the plant with adequate amounts of P. The system would seem to provide adequate P to the plant over a wide range of P levels in soil and tend to avoid luxury consumption of P by the host. However, by restricting mycorrhizal infection, high levels of P in soil may reduce uptake and cause deficiencies of other essential nutrients such as Zn or Cu, which are also highly immobile in soil.

From measurements of the fungal component of clover mycorrhizae, Bevege and colleagues (1975) estimated that external hyphae accounted for only 1% of the total plant weight, yet resulted in a 150% increase in plant growth. This is indeed, as they stated, "a handsome return on the investment." In terms of economy of plant photosynthate, formation of VA mycorrhizae may well be more efficient than synthesis of additional root tissue for uptake of immobile nutrients.

LITERATURE CITED

Ali, B. 1969. Cytochemical and autoradiographic studies of mycorrhizal roots of *Nardus*. Arch. Mikrobiol. 689:236–45.

Asai, T. 1948. Über die Mykorrhizenbildung der Leguminosen-Pflanzen. Jpn. J. Bot. 13:463–85.

Arthur, J. C. 1897. The movement of protoplasm in coenocytic hyphae. Ann. Bot. 11:491–507.

Baylis, G. T. S. 1959. Effect of vesicular-arbuscular mycorrhizae on growth of *Griselinia littoralis* (Cornaceae). New Phytol. 58:274–80.

Bevege, D. T., and G. D. Bowen. 1975. Endogone strain and host plant differences in development of vesicular-arbuscular mycorrhizas. *In* F. E. Sanders, B. Mosse, and P. B. Tinker (eds.), Endomycorrhizas, pp. 77–86. Academic Press, London.

Bevege, D. I., G. D. Bowen, and M. F. Skinner. 1975. Comparative carbohydrate physiology of ecto- and endomycorrhizas. *In* F. E. Sanders, B. Mosse, and P. B. Tinker (eds.), Endomycorrhizas, pp. 149–74.

Bhat, K. K. S., and P. H. Nye. 1974. Diffusion of phosphate to plant roots in soil. II. Uptake along the roots at different times and the effect of different levels of phosphorus. Plant Soil 41:365–82.

Bowen, G. D., D.I. Bevege, and B. Mosse. 1975. Phosphate physiology of vesicular-arbuscular mycorrhizas. *In* F. E. Sanders, B. Mosse, and P. B. Tinker (eds.), Endomycorrhizas, pp. 241–60. Academic Press, London.

Bowen, G. D., M. F. Skinner, and D. I. Bevege. 1974. Zinc uptake by mycorrhizal and uninfected roots of *Pinus radiata* and *Araucaria cunninghamii*. Soil Biol. Biochem. 6:141–44.

Callow, J. A., L. C. M. Capaccio, G. Parish, and P. B. Tinker. 1978. Detection and estimation of polyphosphate in vesicular-arbuscular mycorrhizas. New Phytol. 80:125–34.

Carling, D. E., W. G. Riehle, M. F. Brown, and D. R. Johnson. 1978. Effects of a vesicular-arbuscular mycorrhizal fungus on nitrate reductase and nitrogenase activities in nodulating and non-nodulating soybeans. Phytopathology 68:1590–96.

Cooper, K. M., and P. B. Tinker. 1978. Translocation and transfer of nutrients in vesicular-arbuscular mycorrhizas. II. Uptake and translocation of phosphorus, zinc, and sulphur. New Phytol. 81:43–52.

Cooper, K. M., and D. M. Lösel. 1978. Lipid physiology of vesicular-arbuscular mycorrhiza. I. Composition of lipids in roots of onion, clover, and ryegrass infected with *Glomus mosseae*. New Phytol. 80:143–51.

Cox, G., and F. E. Sanders. 1974. Ultrastructure of the host-fungus interface in a vesicular-arbuscular mycorrhiza. New Phytol. 73:901–12.

Cox, G., F. E. Sanders, P. B. Tinker, and J. A. Wild. 1975. Ultrastructural evidence relating to host-endophyte transfer in a vesicular-arbuscular mycorrhiza. *In* F. E. Sanders, B. Mosse, and P. B. Tinker (eds.), Endomycorrhizas, pp. 298–312. Academic Press, London.

Cox, G., and P. B. Tinker. 1976. Translocation and transfer of nutrients in vesicular-arbuscular mycorrhizas. I. The arbuscule and phosphorus transfer: a quantitative ultrastructural study. New Phytol. 77:371–78.

Cress, W. A., G. O. Throneberry, and D. L. Lindsey. 1979. Comparative kinetics of phosphate, zinc, and manganese absorption by mycorrhizal tomatoes. Plant Physiology 63:(in press).

Crush, J. R. 1974. Plant growth responses to vesicular-arbuscular mycorrhiza. VII. Growth and nodulation of some herbage legumes. New Phytol. 73:743–49.

Daft, M. J., and A. A. El-Giahmi. 1974. Effect of *Endogone* mycorrhiza on plant growth. VII. Influence of infection on the growth and nodulation in French bean (*Phaseolus vulgaris*). New Phytol. 73:1139–47.

Daft, M. J., and T. H. Nicolson. 1966. Effect of *Endogone* mycorrhiza on plant growth. New Phytol. 65:343–50.

Daft, M. J., and T. H. Nicolson. 1969. Effect of *Endogone* mycorrhiza on plant growth. II. Influence of soluble phosphate on endophyte and host in maize. New Phytol. 68:945–52.

Daft, M. J., and T. H. Nicolson. 1972. Effect of *Endogone* mycorrhiza on plant growth. IV. Quantitative relationships between the growth of the host and the development of the endophyte in tomato and maize. New Phytol. 71:287–95.

Demeter, K. 1923. Über "Plasmoptysen"-Mykorrhiza. Flora (Jena) 116:405–56.

El-Giahmi, A. A., T. H. Nicolson, and M. J. Daft. 1976. Endomycorrhizal fungi from Libyan soils. Trans. Br. Mycol. Soc. 67:164–69.

Endrigkeit, A. E. 1937. Beitrage zum ernahrungsphysiologischen Problem der Mykorrhiza unter besonderer Berucksichtigung des Baues und den Funktion der wurzel- und Pilzmembrane. Botan. Arch. 39:1–87.

Fried, M., and R. E. Shapiro. 1961. Soil-plant relationships in ion uptake. Ann. Rev. Plant Physiol. 12:91–112.

Gallaud, I. 1905. Etudes sur les mycorrhizes endotrophes. Rev. Gén. Bot. 17:5–48, 123–36, 223–39, 313–25, 423–33, 479–500.

Gerdemann, J. W. 1961. A species of *Endogone* from corn causing vesicular-arbuscular mycorrhiza. Mycologia 53:254–61.

Gerdemann, J. W. 1965. Vesicular-arbuscular mycorrhizae formed on maize and tuliptree by *Endogone fasiculata*. Mycologia 57:562–75.

Gerdemann, J. W. 1975. Vesicular-arbuscular mycorrhizae. *In* J. G. Torrey and D. T. Clarkson (eds.), The development and function of roots, pp. 575–91. Academic Press, New York.

Gerdemann, J. W., and J. M. Trappe. 1974. The Endogonaceae in the Pacific Northwest. Mycol. Mem. No. 5, New York Botanical Garden and The Mycological Society of America, New York. 76 pp.

Gianinazzi-Pearson, V., and S. Gianinazzi. 1976. Enzymatic studies on the metabolism of vesicular-arbuscular mycorrhiza. I. Effect of mycorrhiza formation and phosphorus nutrition on soluble phosphatase activity in onion roots. Physiol. Veg. 14:833–41.

Gianinazzi-Pearson, V. and S. Gianinazzi. 1978. Enzymatic studies on the metabolism of vesicular-arbuscular mycorrhiza. II. Soluble phosphatase specific to mycorrhizal infection in onion roots. Physiol. Plant Pathol. 12:45–53.

Gilmore, A. E. 1971. The influence of endotrophic mycorrhizae on the growth of peach seedlings. J. Amer. Soc. Hortic. Sci. 96:35–38.

Gray, L. E., and J. W. Gerdemann. 1969. Uptake of phosphorus-32 by vesicular-arbuscular mycorrhizae. Plant Soil 30:415–22.

Gray, L. E., and J. W. Gerdemann. 1973. Uptake of sulphur-35 by vesicular-arbuscular mycorrhizae. Plant Soil 39:687–89.

Hadley, G., and S. Purves. 1974. Movement of ^{14}Carbon from host to fungus in orchid mycorrhiza. New Phytol. 73:475–82.

Hall, I. R. 1975. Endomycorrhizas of *Metrosideros umbellata* and *Weinmannia racemosa*. N. Z. J. Bot. 13:463–72.

Hattingh, M. J., L. E. Gray, and J. W. Gerdemann. 1973. Uptake and translocation of ^{32}P-labeled phosphate to onion roots by endomycorrhizal fungi. Soil Sci. 116:383–87.

Hayman, D. S. 1974. Plant growth responses to vesicular-arbuscular mycorrhiza. VI. Effect of light and temperature. New Phytol. 73:71–80.

Hayman, D. S., and B. Mosse. 1972. Plant growth responses to vesicular-arbuscular mycorrhiza. III. Increased uptake of labile P from soil. New Phytol. 71:41–47.

Hirrel, M. C., and J. W. Gerdemann. 1979. Enhanced carbon transfer between onions infected with a vesicular-arbuscular mycorrhizal fungus. New Phytol. 82:(in press).

Ho, I., and J. M. Trappe. 1973. Translocation of ^{14}C from *Festuca* plants to their endomycorrhizal fungi. Nature New Biol. 244:30–31.

Janse, J. M. 1896. Les endophytes radicaux de quelques plantes javanaises. Ann. de Jard. Bot. de Buitenzorg 14:53–212.

Kamprath, E. J., W. L. Nelson, and J. W. Fitts. 1956. The effect of pH, sulfate, and phosphate concentrations on the adsorption of sulfate by soils. Soil Sci. Soc. Amer. Proc. 20:463–66.

Kaspari, H. 1973. Elektronemmikroskopische Untersuchung zur Feinstruktur der endotrophen Tabakmykorrhiza. Arch. Mikrobiol. 92:201–7.

Kaspari, H. 1975. Fine structure of the host-parasite interface in endotrophic mycorrhiza of tobacco. *In* F. E. Sanders, B. Mosse, and P. B. Tinker (eds.), Endomycorrhizas, pp. 325–34. Academic Press, London.

Khan, A. G. 1971. Occurrence of *Endogone* spores in West Pakistan soils. Trans. Br. Mycol. Soc. 56:217–24.

Kinden, D. A., and M. F. Brown. 1975. Electron microscopy of vesicular-arbuscular mycorrhizae of yellow poplar. III. Host-endophyte interactions during arbuscular development. Can. J. Microbiol. 21:1930–39.

Kinden, D. A., and M. F. Brown. 1976. Electron microscopy of vesicular-arbuscular mycorrhizae of yellow poplar. IV. Host-endophyte interactions during arbuscular deterioration. Can. J. Microbiol. 22:64–75.

Kleinschmidt, G. D., and J. W. Gerdemann. 1972. Stunting of citrus seedlings in fumigated nursery soils related to the absence of endomycorrhizae. Phytopathology 62:1447–53.

LaRue, J. H., W. D. McClellan, and W. L. Peacock. 1975. Mycorrhizal fungi and peach nursery nutrition. Calif. Agric. 29:6–7.

Lewis, D. G., and J. P. Quirk. 1967. Phosphate diffusion in soil and uptake by plants. III. P^{31}-movement and uptake by plants as indicated by P^{32}-autoradiography. Plant Soil 26:445–53.

Lewis, D. H., and J. L. Harley. 1965. Carbohydrate physiology of mycorrhizal roots of beech. III. Movement of sugars between host and fungus. New Phytol. 64:256–69.

Lihnell, D. 1939. Untersuchungen über die Mycorrhizen und die Wurzelpilze von *Juniperus communis*. Symbolae Bot. Upsalienses 3:1–143.

Ling-Lee, M., G. A. Chilvers, and A. E. Ashford. 1975. Polyphosphate granules in three different kinds of tree mycorrhiza. New Phytol. 75:551–54.

McLuckie, J., and A. Burges. 1932. Mycotrophism in Rutaceae. I. The mycorrhiza of *Eriostemon crowei*. Proc. Linn. Soc. N. S. Wales 57:291–312.

Melin, E. 1958. Translocation of nutritive elements through mycorrhizal mycelia to pine seedlings. Botan. Notiser 111:251–56.

Menge, J. A., D. Steirle, D. J. Bagyaraj, E. L. V. Johnson, and R. T. Leonard. 1978. Phosphorus concentrations in plants responsible for inhibition of mycorrhizal infection. New Phytol. 80:575–78.

Mosse, B. 1953. Fructifications associated with mycorrhizal strawberry roots. Nature (London) 171:974.

Mosse, B. 1957. Growth and chemical composition of mycorrhizal and non-mycorrhizal apples. Nature (London) 179:922–24.

Mosse, B. 1959. Observations on the extra-matrical mycelium of a vesicular-arbuscular endophyte. Trans. Br. Mycol. Soc. 42:439–48.

Mosse, B. 1963. Vesicular-arbuscular mycorrhiza: an extreme form of fungal adaptation. Symp. Soc. Gen. Microbiol. 13:146–70.

Mosse, B. 1973a. Advances in the study of vesicular-arbuscular mycorrhiza. Ann. Rev. Phytopathol. 11:171–96.

Mosse, B. 1973b. Plant growth responses to vesicular-arbuscular mycorrhiza. IV. In soil given additional phosphate. New Phytol. 72:127–36.

Mosse, B., and G. D. Bowen. 1968. A key to the recognition of some *Endogone* spore types. Trans. Br. Mycol. Soc. 51:469–83.

Mosse, B., and J. M. Phillips. 1971. The influence of phosphate and other nutrients on the development of vesicular-arbuscular mycorrhiza in culture. J. Gen. Microbiol. 69:157–66.

Murdoch, J. A., J. A. Jackobs, and J. W. Gerdemann. 1967. Utilization of phosphorus sources of different availability by mycorrhizal and non-mycorrhizal maize. Plant Soil 27:329–34.

Nicolson, T. H. 1959. Mycorrhiza in the Gramineae. I. Vesicular-arbuscular endophytes, with special reference to the external phase. Trans. Br. Mycol. Soc. 42:421–38.

Nicolson, T. H. 1967. Vesicular-arbuscular mycorrhiza—a universal plant symbiosis. Sci. Prog., Oxf. 55:561–81.

Nicolson, T. H., and J. W. Gerdemann. 1968. Mycorrhizal *Endogone* species. Mycologia 60:313–25.

Nye, P. H. 1966. The effect of the nutrient intensity and buffering power of a soil; and the absorbing power, size and root hairs of a root, on nutrient absorption by diffusion. Plant Soil 25:81–105.

Olsen, S. R. 1972. Micronutrient interactions. *In* J. J. Mortvedt, P. M. Giordano, and W. L. Lindsay (eds.), Micronutrients in agriculture, pp. 243–64. Soil Science Society of America, Madison, Wis.

Pearson, V., and P. B. Tinker. 1975. Measurement of phosphorus fluxes in the external hyphae of endomycorrhizas. *In* F. E. Sanders, B. Mosse, and P. B. Tinker (eds.), Endomycorrhizas, pp. 277–87. Academic Press, London.

Pearson, V., and D. J. Read. 1973. The biology of mycorrhiza in the Ericaceae. II. Transport of carbon and phosphorus by the endophyte and the mycorrhiza. New Phytol. 72:1325–39.

Powell, C. Ll. 1975. Plant growth responses to vesicular-arbuscular mycorrhiza. VII. Uptake of P by onion and clover infected with different *Endogone* spore types in ^{32}P labelled soils. New Phytol. 75:563–66.

Powell, C. Ll., and J. Daniel. 1978. Mycorrhizal fungi stimulate uptake of soluble and insoluble phosphate fertilizer from a phosphate deficient soil. New Phytol. 80:351–58.

Read, D. J., and D. P. Stribley. 1975. Diffusion and translocation in some fungal culture systems. Trans. Br. Mycol. Soc. 64:381–88.

Reisenauer, H. M. 1966. Mineral nutrients in soil solution. *In* P. S. Altman and D. S. Dittmer (eds.), Environmental biology, pp. 507–8. Federation of American Societies for Experimental Biology, Bethesda, Md.

Rhodes, L. H., and J. W. Gerdemann. 1975. Phosphate uptake zones of mycorrhizal and non-mycorrhizal onions. New Phytol. 75:555–61.

Rhodes, L. H., and J. W. Gerdemann. 1978a. Translocation of calcium and phosphate by external hyphae of vesicular-arbuscular mycorrhizae. Soil Sci. 126:125–26.

Rhodes, L. H., and J. W. Gerdemann. 1978b. Hyphal translocation and uptake of sulfur by vesicular-arbuscular mycorrhizae of onion. Soil Biol. Biochem. 10:355–60.

Rhodes, L. H., and J. W. Gerdemann. 1978c. Influence of phosphorus nutrition on sulfur uptake by vesicular-arbuscular mycorrhizae of onion. Soil Biol. Biochem. 10:361–64.

Rhodes, L. H., M. C. Hirrel, and J. W. Gerdemann. 1978. Influence of soil phosphorus on translocation of ^{65}Zn and ^{32}P by external hyphae of vesicular-arbuscular mycorrhizae of onion. Phytopathology News 12(9):197 (Abstract).

Ross, J. P., and J. A. Harper. 1970. Effect of *Endogone* mycorrhiza on soybean yields. Phytopathology 60:1552–56.

Ross, J. P., and J. W. Gilliam. 1973. Effect of *Endogone* mycorrhiza on phosphorus uptake by soybeans from inorganic phosphates. Soil Sci. Soc. Amer. Proc. 37:237–39.

Rovira, A. D., and G. D. Bowen. 1973. The influence of root temperature on ^{14}C assimilate profiles in wheat roots. Planta 114:101–7.

Safir, G. R., J. S. Boyer, and J. W. Gerdemann. 1971. Mycorrhizal enhancement of water tranpsort in soybean. Science 172:581–83.

Safir, G. R., J. S. Boyer, and J. W. Gerdemann. 1972. Nutrient status and mycorrhizal enhancement of water transport in soybean. Pl. Physiol. 49:700–703.

Sanders, F. E. 1975. The effect of foliar-applied phosphate on the mycorrhizal infections of onion roots. *In* F. E. Sanders, B. Mosse, and P. B. Tinker (eds.), Endomycorrhizas, pp. 261–76. Academic Press, London.

Sanders, F. E., and P. B. Tinker. 1971. Mechanism of absoprtion of phosphate from soil by *Endogone* mycorrhizas. Nature (London) 233:278–79.

Sanders, F. E., and P. B. Tinker. 1973. Phosphate flow into mycorrhizal roots. Pestic. Sci. 4:385–95.

Sanders, F. E., P. B. Tinker, R. L. B. Black, and S. M. Palmerly. 1977. The development of endomycorrhizal root systems. I. Spread of infection and growth-promoting effects with four species of vesicular-arbuscular endophyte. New Phytol. 78:257–68.

Scannerini, S., P. F. Bonfante, and A. Fontana. 1975. An ultrastructural model for the host-symbiont interaction in the endotrophic mycorrhizae of *Ornithogalum umbellatum* L. *In* F. E. Sanders, B. Mosse, and P. B. Tinker (eds.), Endomycorrhizas. pp. 313–24. Academic Press, London.

Schenck, N. C., and K. Hinson. 1973. Response of nodulating and nonnodulating soybeans to a species of *Endogone* mycorrhiza. Agron. J. 65:849–50.

Schlicht, A. 1888. Ueber neue Fälle von Symbiose der Pflanzenwurzeln mit Pilzen. Ber. Dt. Bot. Gesell. 6:269–72.

Schlicht, A. 1889. Beitrag zur Kenntniss der Verbreitung und der Bedeutung der Mykorhizen. Landw. Jbr. 18:477–506.

Shoknecht, J. D., and M. J. Hattingh. 1976. X-ray microanalysis of elements in cells of VA mycorrhizal and nonmycorrhizal onions. Mycologia 68:296–303.

Shibata, K. 1902. Cytologische Studien über die endotrophen Mykorrhizen. Pringsh. Jbr. Wiss. Bot. 37:643–84.

Skinner, M. F., and G. D. Bowen. 1974. The uptake and translocation of phosphate by mycelial strands of pine mycorrhizas. Soil Biol. Biochem. 6:53–56.

Smith, S. E. 1967. Carbohydrate translocation in orchid mycorrhizas. New Phytol. 66:371–78.

Stahl, E. 1900. Der Sinn der Mykorhizenbildung. Jbr. Wiss. Botan. 34:539–668.

Timmer, L. W., and R. F. Leyden. 1978. Stunting of citrus seedlings in fumigated soils in Texas and its correction by phosphorus fertilization and inoculation with mycorrhizal fungi. J. Amer. Soc. Hortic. Sci. 103:533–37.

Tinker, P. B. H. 1975. Effects of vesicular-arbuscular mycorrhizas on higher plants. Symp. Soc. Expt. Biol. 29:325–49.

Van Schreven, D. A. 1958. Some factors affecting the uptake of nitrogen by legumes. *In* E. G. Hallsworth (ed.), Nutrition of the legumes, pp. 137–63. Butterworths, London.

Winter, A. G. 1953. Zum Problem der Mykorrhiza bei Landwirtshaftlichen Kulturpflanzen. I. Die Mycorrhizen der Gramineen. Z. PflErnähr. Düng. 60:221–43.

Woods, F. W., and K. Brock. 1964. Interspecific transfer of Ca-45 and P-32 by root systems. Ecology 45:886–89.

D. C. SMITH

Mechanisms of Nutrient Movement between the Lichen Symbionts

9

INTRODUCTION

Biotrophy

One of the commonest types of cellular interaction in symbiosis is the movement of nutrient molecules between the associating organisms. In some cases of parasitic symbiosis, this movement is *necrotrophic* and involves the destruction of part or all of one organism by the other. In many cases of advanced parasitic symbiosis, and in almost all cases of mutualistic symbiosis, the movement is mainly *biotrophic*. Biotrophy involves the movement of nutrient molecules from the living cells of one symbiont into the living cells of the other. The flow of nutrient molecules out of the cells of an organism is usually very much greater when it is in symbiosis than when it is grown in isolation. A major problem in studying biotrophy is therefore the mechanism of this efflux and how it is induced by existence in symbiosis.

Biotrophy is an important and widespread biological phenomenon. The aspect of biotrophy that has been explored in greatest detail at the molecular level is the flow of photosynthate from the alga to the fungus of lichens. These plants are particularly suitable experimental material because they are simple in structure, and it is much easier to separate the symbionts than in the morphologically more complex associations involving higher organisms. Unlike many alga/invertebrate or higher plant/fungus associations, the algal symbiont of lichens is extracellular and not often penetrated by haustoria. The process of nutrient transfer between cells can therefore be studied directly.

The main object of this paper will be to examine how far we understand, at the molecular level, the mechanism of the transfer of photosynthate from the alga to the fungus of lichens. This should provide a background to the general problem of the phenomenon of biotrophy.

Structure and Other Relevant Characteristics of Lichens

The algal symbionts are either unicellular or in short, simple filaments. In about 10% of lichen species, the algae are blue-green (with *Nostoc* the commonest genus), but in the remainder they are green (the commonest genus being *Trebouxia* of the Chlorococcaceae, occurring in approximately 70% of species). Altogether about 27 different genera of algae have been found in lichens.

The bulk of the lichen thallus is composed of fungal hyphae. In the great majority of species, the algal symbionts are restricted to a thin layer just beneath the surface. Above the algae the hyphae form a rather dense but thin cortex; below the algae the hyphae are thick-walled and more loosely arranged into a medulla.

Within the algal layer the hyphae are thin-walled. They become closely appressed to the algal cells, typically with the terminal portion of a hypha curling around the alga. Occasionally in a number of species, and regularly in a very few, short projections from the fungus either indent the algal wall, causing it to invaginate, or appear to rupture the cell wall. In the latter case, however, the plasma membrane normally remains intact except in a very few lichens. These projections are termed "haustoria," but should not necessarily be considered analagous to the haustoria of parasitic fungi. The frequency of haustoria in lichens is controversial and will be discussed later, but they are not considered to be of particular significance in photosynthate transfer that is believed to occur over the whole area of contact between the symbionts, and not restricted to specific regions (Collins and Farrar, 1978; Peveling, 1973a).

In moving from alga to fungus, therefore, photosynthate molecules have to cross the outer membrane of the alga, traverse wall material of alga and fungus, and cross the fungal membrane into the hypha. The distance between the symbiont membranes is of the order of 0.3 to 0.4 μm. The space between the symbiont membranes is accessible to solutions diffusing in from outside the thallus, so that it is possible to inhibit either movement across the algal membrane, or uptake into the fungus. If fungal uptake is blocked, material released from the alga can diffuse out of the thallus.

Two further points should be borne in mind about lichens. First, they are slow-growing and long-lived. Their maximum ages are measured in terms of hundreds of years, and, in extreme cases in the Arctic, in thousands of years (Beschel, 1958). It is very unlikely that individual symbiont cells persist for these lengths of time, although there is no information on rates of turnover of algal cells in a thallus. Nevertheless, the pathway for the transport of photosynthate molecules from an algal cell to its associated fungal hypha probably persists for much longer than in any other kind of symbiosis.

Second, almost all lichens undergo, in nature, frequent cycles of drying out and rewetting. In dry weather the water content may often fall to about 5% of the dry weight. Under these conditions the symbiont membranes are no longer functional. When a lichen is rewetted by immersing it in water, the membranes recover their properties remarkably rapidly. Their ability to act as permeability barriers is restored in about 50-90 seconds; before this recovery period is completed, solutes may be lost from the symbiont cells (Smith and Molesworth, 1973; Farrar and Smith, 1976). As soon as the alga recovers the capacity for photosynthesis (normally a few minutes—the actual time varies with the severity of the previous desiccation) the pathway for photosynthate transport from alga to fungus is fully operative.

The pathway between the symbionts is thus endowed with remarkable stability and persistence.

General Features of Photosynthate Transport between the Symbionts

The general features of photosynthate transport from alga to fungus have already been reviewed in detail (Smith et al., 1969; Smith, 1974, 1975). The principal characteristics can be summarized as follows:

1. Transport is substantial, and a high proportion of all the carbon fixed by the alga passes to the fungus.
2. Fixed carbon moves predominantly as a single type of molecule. Table 1 shows that in lichens with blue-green symbionts the transported molecule is glucose, and in lichens with green symbionts, it is a polyol (ribitol, sorbitol or erythritol).
3. The rapid and massive efflux of carbohydrate ceases quite rapidly as soon as the algae are isolated from the lichen, and may be scarcely detectable two to three hours later. At the same time, the transported molecule becomes much less prominent among both intracellular and extracellular products of the alga.

In order to analyze the cellular interactions between the symbionts, more precise characterization of these general features is required. First,

TABLE 1

This table summarizes investigations into the identity of the mobile carbohydrate in 42 lichen species. The type of carbohydrate depends upon the type of alga. All symbiotic green algae release polyhydric alcohols, and blue-green algae release glucose. Note the identity between the last three carbon atoms in each carbohydrate.

Symbiont	Algal genus	Number of lichens investigated		Mobile carbohydrate	Formula
		Genera	Species		
Green	Trentepohlia	4	6	Erythritol	CH_2OH H—C—OH H—C—OH CH_2OH
Green	Trebouxia	10	13	Ribitol	CH_2OH H—C—OH H—C—OH H—C—OH CH_2OH
	Myrmecia	2	4		
	Coccomyxa	3	3		
Green	Hyalococcus	1	2	Sorbitol	CH_2OH HO—C—H H—C—OH H—C—OH H—C—OH CH_2OH
	Stichococcus .	2	2		
Blue-Green	Nostoc	5	10	Glucose	CHO HO—C—H H—C—OH H—C—OH H—C—OH CH_2OH
	Calothrix	1	1		
	Scytonema	1	1		

however, it is very necessary to consider the experimental methods used to study photosynthate transport.

METHODS IN THE STUDY OF PHOTOSYNTHATE TRANSPORT IN LICHENS

All methods for studying photosynthate transport begin with the intact thallus and not the cultured symbionts. This is because photosynthate efflux ceases soon after the algal symbionts are isolated (fig. 1). Release of

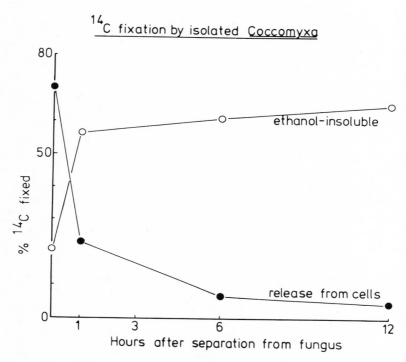

Fig. 1. The changes in release of fixed ^{14}C and incorporation of ^{14}C into ethanol-insoluble material (almost all polysaccharide) in *Coccomyxa* isolated from *Peltigera aphthosa*. The graph is based on the data of Green (1970), and Hill and Smith (1972).

substances by lichen algae in culture is small in extent and similar in nature to that from free-living algae. Both symbionts are also different in morphology when they are grown apart from each other: the fungus in culture shows none of the tissue differentiation in the thallus, and the algae have thicker cell walls. It has not been possible to induce cultured symbionts by artificial means to revert to the behavior and morphology they show in the lichens.

Six general methods have been used to study photosynthate transport. They are all based on the photosynthetic fixation of the radioactive isotope ^{14}C, normally achieved by incubating lichen samples in the light in aqueous media containing $NaH^{14}CO_3$.

Separation of Alga-Free from
Alga-Containing Regions of a Lichen

In those lichens where the algal layer and upper cortex are relatively even and not convoluted, they can be separated from the underlying fungal tissue by careful and delicate manipulation with a sharp scalpel. Experiments are then carried out as follows: appropriate samples of lichen are exposed to ^{14}C in the light, and then dissected into alga-free and alga-containing regions at various time intervals after the start of the experiment.

A progressive accumulation of photosynthetically fixed ^{14}C in the alga-free region is observed, unequivocal evidence that photosynthate moves between the symbionts. The rate at which it accumulates sets a lower limit to the rate of transfer between the symbionts, and analysis of the ^{14}C-labeled compounds in the fungal region shows the compounds into which photosynthate received from the alga is ultimately converted. Furthermore, feeding such excised portions of lichen fungus with various ^{14}C-labeled compounds can provide useful supplementary evidence of pathways of metabolism of photosynthate in the hyphae.

The method was first used by Smith (1961) for the lichen *Peltigera polydactyla*. Photosynthetically fixed ^{14}C was detectable in the medulla within 10 minutes of the exposure of the intact thallus to NaH^{14}CO$_3$, and fixed ^{14}C accumulated in the medulla at approximately 40% of the over-all rate of photosynthesis. The predominant labeled compound accumulating in the medulla was ^{14}C-mannitol, a polyol not found in the algal symbiont, *Nostoc*.

Such experiments show that transport between the symbionts has occurred, but they give little direct information about the precise intracellular interactions between alga and fungus, since these occur within the alga-containing region.

Physical Separation of Symbionts by Homogenization

If lichen thalli are gently homogenized, a more or less pure preparation of algae can be obtained by differential centrifugation of the homogenate (Drew and Smith, 1966, 1967a). Gradient centrifugation can also be used (Millbank and Kershaw, 1969).

Such freshly isolated algae already show much less efflux of fixed ^{14}C (fig. 1), although the mobile carbohydrate released in symbiosis is still detectable among the products in the medium. The changes occurring after isolation are all in the direction of regaining the properties of cultured algae, and show that the effects of symbiosis on algae are reversible and not permanent. The most striking change is that while the proportion of fixed ^{14}C accumulating in polysaccharides in the thallus is small (often about 5%), it increases sharply on isolation, usually reaching about 50% within 6 hours.

This technique is primarily useful in providing direct evidence for the identity of the mobile carbohydrates, but it has three main disadvantages:

1. The very close physical attachment between alga and fungus means that fragments of hyphae still remain attached to many of the algal cells after they have been homogenized. If these algae are excluded by repeated centrifugation, the yield of remaining algae may become very small and it is not known if they are atypical (i.e., those algae not firmly attached to fungal hyphae may not be in the same physiological condition as those which are).

2. In those lichens where the algal symbiont occurs in short filaments, these inevitably become disrupted. This is particularly important for *Nostoc* sp., since the resultant separation of heterocysts from vegetative cells will disrupt the physiological organization of the filament. Photosynthesis may not be markedly affected, but nitrogen fixation ceases altogether.

3. Because homogenization also produces fungal fragments whose mass may be in the same range as the algal cells, it may take repeated centrifugation to obtain a reasonably pure preparation of algae. This may take up to an hour or more, and during this period extensive changes already occur in the algal cells (see fig. 1). Thus, the final preparation is so different in behavior from algae in the thallus that precise and quantitative analysis of photosynthate efflux by this technique is of little value.

Isotope Trapping or "Inhibition Techniques"

In the lichen *P. polydactyla*, the carbohydrate moving between the symbionts is glucose. Drew and Smith (1967b) originally observed that if this lichen was incubated in the light in an aqueous medium containing $NaH^{14}CO_3$ and 1% (w/v) ^{12}C-glucose, more than 50% of the photosynthetically fixed ^{14}C was released to the medium as ^{14}C-glucose.

The original explanation of this phenomenon was that the high concentration of ^{12}C-glucose in the medium successfully competed with ^{14}C-glucose released from the alga at the fungal uptake sites (fig. 2). Hence, ^{14}C-glucose diffused into the medium because the uptake sites were saturated by ^{12}C-glucose. The method was misleadingly termed the "Inhibition Technique" by this author, since fixed ^{14}C was inhibited from entering the fungus. In retrospect, "isotope trapping" would have been a more satisfactory term.

The method was then applied to a range of other lichens (Richardson and Smith, 1968; Richardson et al., 1968; Hill and Smith, 1972). In almost every case incubation of the lichen in $NaH^{14}CO_3$ and a high concentration of the nonradioactive mobile carbohydrate resulted in the radioactive form of the mobile carbohydrate being released to the medium. The effect was

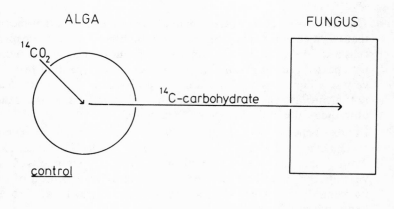

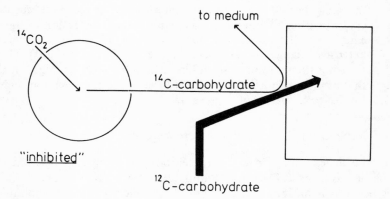

<u>Original explanation of "inhibition"</u>

Fig. 2. The original explanation of the mechanism of the "Inhibition Technique" as given by Smith et al., (1969).

specific because closely related carbohydrates induced no release of fixed ^{14}C from the lichens. A rapid and simple method was then developed for screening a large number of lichens to determine the identity of the mobile carbohydrate. A screening experiment consisted simply of incubating a lichen in solutions of a range of different types of carbohydrates, adding $NaH^{14}CO_3$, and determining in which medium large amounts of fixed ^{14}C were released from the lichens.

At first it was believed that this technique would also prove an excellent method for the precise analysis of intercellular interactions in photosynthate transport. Numerous experiments were carried out in our laboratory to study the effects of various factors on photosynthate transport, the latter being measured as the amount of fixed ^{14}C released to "inhibition" or isotope trapping media.

Unfortunately, it became progressively apparent that ^{14}C release to "inhibition media" was *not* an exact measure of photosynthate transport. Values obtained for the rate and amount of transport were always less— sometimes substantially less—than those obtained by other methods. For example, in the lichens containing *Trebouxia*, such as *Xanthoria aureola*, it is about an hour before significant ^{14}C release can be detected in "inhibition media" (Richardson 1967), but only 5–10 minutes before photosynthetically fixed ^{14}C appears in fungal metabolites in control samples in media lacking exogenous carbohydrate (Bednar and Smith, 1966). After 24 hours 23.6% of the total fixed ^{14}C is released to "inhibition media" containing ribitol, but at this time well over 50% of the fixed ^{14}C is in the fungal polyols in control samples (Richardson and Smith, 1968).

Clearly, the interpretation of "inhibition" phenomena shown in figure 2 is not correct. This began to become evident when it was found that the method could also induce the specific release of fixed ^{14}C from invertebrates such as corals containing *intracellular* symbiotic algae (Lewis and Smith, 1971), and from the mollusc *Elysia viridis* which contains *intracellular* chloroplasts (Trench et al., 1974). In the latter case the specific release could be induced of ^{14}C-galactose, known to be an *animal* product *not* synthesised by the chloroplasts; ^{14}C-galactose is formed in animal cells from labeled products of photosynthesis received from symbiotic chloroplasts. Trench and colleagues (1974) suggested that the release of ^{14}C-galactose occurred by the process of "exchange diffusion" between ^{12}C-galactose in the medium and ^{14}C-galactose in animal cells. The phenomenon of "exchange diffusion" can be simply illustrated by the experiments of Robbie and Wilson (1969) on *Escherichia coli*. They labeled cells by allowing them to take up ^{14}C-thiomethyl-β-galactoside (TMG). The efflux of labeled TMG was then markedly stimulated by adding ^{12}C-TMG to the medium. They explained the phenomenon on the basis that in the transport process, a carrier-substrate complex could move more rapidly across the membrane than the carrier alone. Many other examples are known of this phenomenon of "exchange diffusion."

In the lichen much of the ^{14}C released by "inhibition techniques" may well occur by exchange diffusion, and a more likely explanation of the "inhibition technique" is shown in figure 3. Direct proof of the mechanism suggested there is lacking, but it would explain the sometimes substantial discrepancies between the rate of efflux of ^{14}C in "inhibition" experiments and the rate of photosynthate transport measured by other methods. Thus, the reason why it is almost an hour before detectable ^{14}C release occurs from lichens containing *Trebouxia* in "inhibition experiments" may be because *algal* transport carriers are initially swamped by ^{12}C-carbohydrate, and the process of exchange diffusion occurs only slowly.

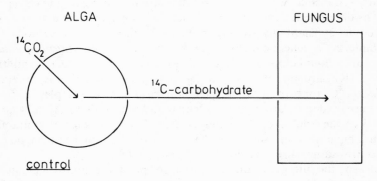

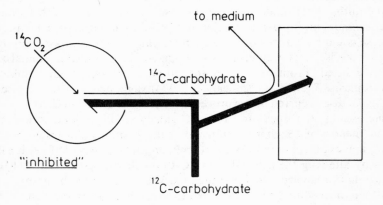

Revised explanation of "inhibition"

Fig. 3. A revised explanation of the "Inhibition Technique" (see text).

Using "inhibition technique" experiments, Richardson and colleagues (1968) developed the concept that the rate of transfer of ^{14}C between the symbionts could vary widely between different lichens, and they categorized transfers as "slow," "fast," "intermediate," and so forth, depending on the rate of release of ^{14}C into "inhibition media." This concept is quite misleading and should be abandoned. All the other methods for studying photosynthate transport show that it is always "fast."

A further misunderstanding needs to be cleared up. When it was believed that the "inhibition technique" operated by the mechanism shown in figure 2, it was thought that the algal symbionts could excrete the mobile carbohydrate against the substantial concentration gradient created by the presence of unlabeled carbohydrate in the medium. This would have implied an appreciable input of energy by the alga into photosynthate transport. However, if the mechanism shown in figure 3 is operative, then it does not follow that there is an active excretion against a concentration gradient by the alga. More recently, Tapper (1979) has raised the question of whether the algal membrane transport system responsible for carbohydrate efflux in symbiosis is necessarily the same as that involved in ^{14}C release in "inhibition" phenomena.

Although "inhibition techniques" have proved an extremely useful aid to the identity of mobile carbohydrates, they are quite misleading for the analysis of cellular interactions, especially since the technique *may* operate by transport mechanisms that differ from those involved in photosynthate transport in symbiosis.

Short-term Photosynthesis Studies in Intact Thalli

If labeled mobile carbohydrate is fed to portions of lichen containing only the fungus, conventional methods of analysis will show the products into which the mobile carbohydrate is converted by the fungus. Such experiments show that most of the mobile carbohydrates are immediately converted to fungal polyols, especially mannitol and arabitol. A small amount of label occurs in other fungal products, including ethanol-insoluble material (presumably polysaccharide), and some is lost as CO_2. Fungal polyols such as mannitol and arabitol have not been detected in lichen algae, whether as freshly isolated or cultured symbionts. Furthermore, mobile carbohydrates (including algal polyols such as ribitol and sorbitol) have never been found to accumulate in detectable amounts in the fungus. In a complete and intact lichen thallus, it is therefore possible to infer which soluble carbohydrates are in the alga and which are in the fungus.

Thus, the progress of fixed carbon from alga to fungus can be followed in simple photosynthesis experiments. Intact lichens are incubated in solutions of $NaH^{14}CO_3$ in the light, and at rapid intervals after the start of the experiment samples are removed, killed in hot ethanol, and the distribution of fixed ^{14}C among algae and fungal carbohydrates analyzed by conventional radio-chromatographic techniques. Typical results of such an experiment are shown in figure 4. The rapid appearance of fixed ^{14}C in the mobile carbohydrate (glucose) at a short but significant time interval before its appearance in the fungal polyol mannitol is good confirmation that the mobile carbohydrate has been correctly identified. Figure 4 also illustrates that after the initial period the amount of fixed ^{14}C appearing in polysaccharides and other metabolites is relatively low compared with the amount in mannitol. Thus, following the path of fixed ^{14}C through the mobile carbohydrates into fungal polyols gives a reasonable, general approximation of the rate and amount of photosynthate transport from alga to fungus. However, the method cannot be used as a precise and exact tool for the study of photosynthate transport because the partitioning of metabolites other than simple carbohydrates between alga and fungus is not known, and because the extent to which mobile carbohydrate is respired to CO_2 is also not known. The method can be used to set reliable *lower* limits to the rate and amount of transport, but cannot give exact data.

Short-term photosynthesis experiments have, however, given useful qualitative information about synthesis of mobile carbohydrates in the alga. In the very early stages of this type of experiment, there is always a temporary sharp rise in the proportion of fixed ^{14}C in polysaccharide (fig. 4) before it declines to a lower but relatively constant level. The significance of this observation is considered later.

Inhibitors of Transport

The space between the symbionts is accessible to externally supplied solutions. This has led to a search for compounds that can specifically block movement between the symbionts. So far, four inhibitors have been found. They have primarily been used to study lichens in which glucose is the mobile carbohydrate and the algal symbiont is *Nostoc*. The value of the inhibitors in lichens with green algae is unfortunately limited.

Sorbose. Chambers and colleagues (1976) found that pretreatment of the lichen *P. polydactyla* with sorbose reduced photosynthate transport (measured as ^{14}C accumulation in the fungal polyol mannitol). Since photosynthesis was not reduced and fixed ^{14}C was not released to the medium, it was assumed that efflux from the alga had been reduced. The most likely explanation was that sorbose can enter the algal cell and then compete with glucose for the algal transport system, so partially reducing efflux. These workers point out that, in other organisms, the glucose

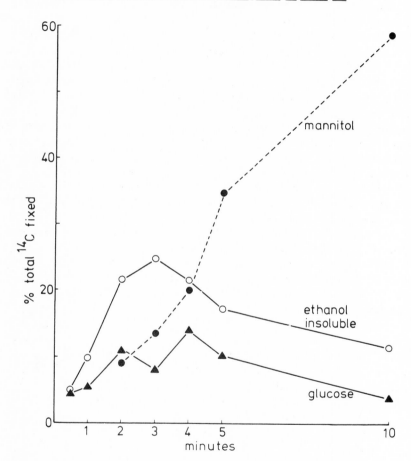

Fig. 4. The distribution of fixed ^{14}C between glucose, mannitol, and ethanol-insoluble compounds (almost all polysaccharide) in the early stages of photosynthesis (based on the data of Drew and Smith, 1967b). For the sake of clarity, incorporation into other compounds has been omitted (e.g., at 30 seconds, almost all the ^{14}C is in PGA, etc.)

transport systems sensitive to sorbose are those that operate by facilitated diffusion.

2-deoxy-glucose. When 2-deoxy-glucose is added to the medium of *P. polydactyla* during ^{14}C-fixation, large amounts of ^{14}C-glucose are released from the lichen. This effect was originally thought to operate by "exchange diffusion," as described for isotope trapping (or "inhibition technique") above (cf. figs 2 and 3). However, Chambers and colleagues (1976) found that thalli *pretreated* in 2-deoxy-glucose, and then incubated in ^{14}C fixation media lacking this compound, still show substantial release of ^{14}C-glucose. Clearly, this release cannot be the result of "exchange diffusion" with the alga, and must result from partial blocking of fungal uptake. It is not known if this results from binding of 2-deoxy-glucose to the fungal glucose uptake sites, or from interference with the intermediate metabolism of glucose after uptake. It is not known if the alga is completely unaffected; if glucose can penetrate algae, then one might expect at least limited penetration by 2-deoxy-glucose, but it clearly does not prevent substantial ^{14}C-glucose efflux. Using this inhibitor, Tapper (1979) has attempted to produce kinetic data on the efflux mechanism. He also found some effect of 2-deoxy-glucose with lichens in which glucose is not the mobile carbohydrate.

Digitonin. Digitonin is a detergent that binds to non-esterified sterol groups. These are universal in eukaryote membranes, but virtually absent from prokaryote membranes. In lichens with prokaryote symbionts such as *Nostoc*, digitonin therefore inactivates the fungus but not the alga. Thus, in *P. polydactyla*, digitonin treatment abolishes the accumulation of fixed ^{14}C in the fungal polyol mannitol, but there is massive release of fixed ^{14}C-glucose to the medium. This result shows that fungal uptake is unnecessary to induce algal efflux. There are two disadvantages to the use of digitonin. First, it cannot be used where the algal symbiont is eukaryotic, and this applies to 85% or more of all lichens. Second, a prolonged period of digitonin treatment, lasting several hours, is necessary to achieve full inactivation of the fungus, and it is not clear that the alga remains completely unaffected during this period.

Uranyl ions. These large ions cannot penetrate membranes, but will specifically bind to sites of sugar transport. Unlike any of the other inhibitors mentioned above, they do not interfere with the internal metabolism of the symbionts, and have therefore been used as sensitive probes to investigate certain aspects of transport mechanisms by Tapper (1979). The principal disadvantage of uranyl ions is that they have to be used at pH 4.0 and below, known to be an unsatisfactory pH for prolonged incubation of lichens.

Apart from these inhibitors that have positive effects on transport, there are a large number of compounds that have no effect (see Smith 1975). Some of these compounds have specific effects such as ouabain, which affects Na^+—K^+ ATPases. Such negative results are helpful in excluding the possibility that certain types of mechanism are operative in transport between the symbionts.

Techniques Based upon Microscopy

Microscopy has been used in two principal ways to get more information about photosynthate transport. First, microautoradiography of sections of lichens has been carried out to locate ^{14}C after fixation (Hessler and Peveling, 1978; Peveling and Hill, 1974). Until very recently, a major difficulty was that preparative techniques involved removal of all water-soluble compounds from the tissue, so that only the small proportion of fixed ^{14}C that accumulates in insoluble material could be located. Nevertheless, such information is useful since no other technique gives such good and direct information about the distribution of insoluble ^{14}C between the symbionts. A substantial advance has now been made by Tapper (1979), who has developed preparative techniques that do not remove soluble ^{14}C compounds. Using this for quantitative or dynamic studies is at present laborious and involves uncomfortably large degrees of error, but offers the potential that future refinement and development may produce a particularly valuable technique for studying the partitioning total fixed ^{14}C between the symbionts.

Second, histochemical techniques can be used to localize both enzymes (Boissière, 1973) and metabolites such as polyglucosides (Boissière, 1972a, b), believed to be connected with photosynthate transport. The value of these techniques will improve substantially when the role of such enzymes and metabolites has been positively confirmed by other methods.

MECHANISM OF PHOTOSYNTHATE EFFLUX IN SYMBIOSIS

The preceding section shows that no single method of studying photosynthate transport is entirely satisfactory on its own. This section attempts to assess the extent to which the mechanism of photosynthate efflux from algae in symbiosis can be understood using various combinations of these methods.

Identity of Mobile Carbohydrates

Much of the information on the identity of mobile carbohydrates in table 1 was obtained primarily by isotope trapping (or "inhibition techniques"). As explained above, this method is now believed to operate mainly by exchange diffusion with algal cells rather than by simple and direct interference with transport between the symbionts. This raises the obvious question of whether the compounds released by "inhibition techniques" are identical to the mobile carbohydrates. Two main lines of evidence suggest that they are indeed identical. First, other methods of identification always give the same result; the clearest illustration of this is the massive release of free glucose from lichens containing blue-green algae during incubation in digitonin. Second, release of labeled carbohydrate by isotope trapping is invariably accompanied by specific inhibition of ^{14}C accumulation in fungal polyols.

The only doubts about the identities shown in table 1 come from the results of some isotope-trapping experiments where glucose is the mobile carbohydrate. In these, labeled glucose polymers as well as labeled glucose are sometimes released (Richardson et al., 1968). This has given rise to suggestions that glucose polymers rather than glucose are released, the former then being broken down to glucose by the fungus (e.g., Hill, 1972). However, release of labeled polymers could be an artifact arising from a combination of the presence of high concentrations of glucose in the medium, and the action of surface transglycosylation.

The fact that mobile carbohydrates can always be released by exchange diffusion could be a helpful clue in the eventual understanding of the mechanism of photosynthate efflux.

Synthesis of Mobile Carbohydrate by the Alga

In all lichen algae the incorporation of fixed ^{14}C into ethanol-insoluble material (largely polysaccharide) is very much less in symbiosis than in isolation. After isolation, incorporation of fixed ^{14}C into polysaccharide increases as that into mobile carbohydrate declines. There is therefore a clear link between polysaccharide formation and mobile carbohydrate synthesis. Short-term photosynthesis experiments reinforce the existence of this link. The incorporation of ^{14}C into polysaccharides begins at a very early stage simultaneous with, or perhaps even earlier than, incorporation into mobile carbohydrate; this is true for lichens with blue-green (fig. 4) and green algae (Farrar, 1973). Using pulse-label techniques, Hill (1972) showed that ^{14}C-polysaccharide formed early in photosynthesis by *P. polydactyla* broke down rapidly with a concomitant rise in ^{14}C-mannitol, as if there was a small pool of polysaccharide that turned over rapidly to produce mobile carbohydrate that then moved to the fungus (fig. 5). However, as Hill points out, his experiments do not show conclusively whether polysaccharide is a precursor, or whether carbon is diverted from polysaccharide formation to glucose synthesis.

Peat (1968), in his study of the ultrastructure of *P. polydactyla*, reported that polyglucoside granules were absent from *Nostoc* in lichens, as if

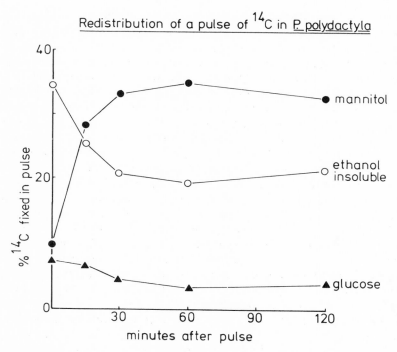

Fig. 5. Redistribution of a short pulse of fixed ^{14}C in *Peltigera polydactyla* between glucose, mannitol, and ethanol-insoluble compounds (after Hill, 1972). The lichen was exposed to ^{14}C for 10 min, rapidly washed, and incubated in a nonradioactive medium. Almost all the ethanol-insoluble material is polysaccharide. For the sake of clarity, other compounds have been omitted.

polysaccharide accumulation had been *completely* inhibited. This would have been contrary to the results of ^{14}C-labeling experiments above. More recently, Boissière (1972 a,b) has demonstrated polyglucoside reserves by histochemical techniques in both alga and fungus of *Peltigera canina*.

In blue-green algae the mobile carbohydrate, glucose, ceases to be detectable soon after isolation. However, free glucose rarely occurs in more than very small quantities in the cells of free-living organisms, and instead it occurs as the mono- or diphosphate. The occurrence of glucose in the nonphosphorylated form in symbiosis is thus a problem whose solution might provide a key to understanding photosynthate transport.

In green algae, by contrast, mobile polyols continue to be detectable after isolation, and are normal constituents of the soluble carbohydrates. The polyols are not unique to lichen algae; for example, free-living

Trentepohlia sp. contains erythritol in amounts equivalent to 1–2% dry weight (Tischer, 1936). The level of incorporation of fixed ^{14}C into mobile polyols is low in isolated algae, suggesting that they do not turn over rapidly. Thus, there seems to be no necessity to invoke novel pathways for the synthesis of mobile polyols when the algae enter symbiosis.

In both green and blue-green symbiosis in lichens, Farrar (1976b, 1978) suggests that some experimental results would be consistent with the existence of two pools of carbohydrate in the alga. This is particularly the case in *Trebouxia*, where changes in the specific activity of ribitol following a pulse of $H^{14}CO_3$ imply that one pool contains ribitol available for release, and the other not (Green, 1970; Farrar, 1976b).

The accumulation of fixed ^{14}C into fungal polyols is the most reliable method for placing a lower limit on the amount of photosynthate released from the alga. In *P. polydactyla*, 62% of the fixed ^{14}C is in mannitol after 45 min fixation (Drew and Smith, 1967b), and 72% after 17 h (Chambers et al., 1976). Using estimates for the amount of glucose a fungus would either respire or convert to other products, Farrar (1978) concludes that as much as 90% of all the carbon fixed in photosynthesis passes to the fungus. Data for lichens containing green algae are less good, but Farrar again estimates a 90% transfer for *Trebouxia*-containing lichens.

Farrar (1976b) shows that rates of photosynthesis for algae in lichens are comparable to free-living forms such as *Chlorella*. Thus, photosynthate efflux from lichen algae is of massive proportions, and involves the path of carbon being almost entirely diverted to the synthesis of a single carbohydrate. Nearly all the fixed carbon used by algae such as *Chlorella* to achieve high growth rates is, in algae such as *Trebouxia*, transferred to the fungus, so that lichen algae consequently have extremely low growth rates.

Rate of movement

Much of the data on the rate of movement of photosynthate from alga to fungus is based on measurements of ^{14}C. The interpretation of these data is complex for three reasons. First, as explained above, the rate of efflux of ^{14}C to the medium in "inhibition experiments" may be a very misleading indicator of ^{14}C movement between the symbionts. Second, the rate of movement of ^{14}C will, in the early stages, depend on the pool size of mobile carbohydrates in the alga. In *Nostoc* the pool size is very small and the turnover rapid, so that ^{14}C moves rapidly to the fungus. In *Trebouxia*, and perhaps especially in *Trentepohlia*, the pool size is much larger, and ^{14}C movement appears much slower. Conversion of a rate of ^{14}C movement into an absolute rate of carbohydrate movement can only be made if both the pool size of mobile carbohydrate and the specific activity of fixed ^{14}C

are known. Third, the possibility has been raised that there is more than one pool of mobile carbohydrate in the alga, so that even if the total amount of mobile carbohydrate was known, accurate calculation of rates of movement might still not be possible.

In *P. polydactyla* ^{14}C is detectable in mannitol about 90 sec after fixation by the alga (Drew and Smith, 1967b; and see fig. 4). This implies that the movement of a glucose molecule out of the alga, across the intersymbiont space, and into the fungal hypha is quite rapid. This should be borne in mind when considering hypotheses, based on electron microscope studies, that substances move between the symbionts in vesicles. There is no reason at all to suppose that the rate of movement of carbohydrate in lichens with green algae is any slower, even though rates of ^{14}C movement are substantially slower. Given that algal growth rates are very slow, and photosynthesis rates high, rapid export of fixed carbon must occur once the storage capacity of the cell is saturated.

Under natural conditions most lichens are exposed to rapid alternations of wetting and drying, so that they are metabolically active for only short periods. It is consequently of obvious advantage for fixed carbon to be transferred to the fungus as rapidly as possible.

Algal Membrane Carriers Involved in Efflux.

Various attempts to characterize the type of carrier systems involved in efflux yield the following fragments of evidence.

1. Glucose efflux from the *Nostoc* symbiont of *P. polydactyla* is probably by a carrier-mediated facilitated diffusion mechanism. Like such mechanisms in other organisms, it is sensitive to sorbose but insensitive to a wide range of other inhibitors (Chambers et al., 1976).

2. Boissière (1973) observed ATPase activity on the internal side of the outer membrane of the *Nostoc* symbiont of *Peltigera canina*. The histochemical reaction was particularly strong in regions of contact with the fungus, but was absent from free-living *Nostoc* (or the *Nostoc* symbiont of the gelatinous lichen *Collema*). She concluded this was evidence of active transport out of the alga. However, this enzyme could also have been concerned with glucose formation, the transport of other material, or with a number of other cellular processes.

3. A detailed kinetic analysis by Tapper (1979) of the very early stages of ^{14}C efflux from the *Nostoc* symbiont of *P. polydactyla*, and the *Trebouxia* symbiont of *Cladonia convoluta* provided no evidence

that the efflux mechanism depends on a novel protein induced by symbiosis. It therefore seems to depend upon an existing carrier protein.

4.   The massive extent of efflux is not due to a very powerful fungal uptake mechanism acting as a "sink." As described above, substantial efflux continues when fungal uptake is blocked by digitonin in *P. polydactyla*, or 2-deoxy-glucose pretreatment in both *P. polydactyla*, and *C. convoluta*.

Summary and Conclusions

When the lichen is metabolically active, there is a rapid and massive efflux of fixed carbon from the alga to the fungus. For *Nostoc* symbionts (and probably other lichen algae) the efflux certainly involves over 70% of the carbon fixed in photosynthesis, and may be as much as 90%. All the evidence suggests that most if not all of the carbon leaves the algal cell as a single, simple carbohydrate; the type of carbohydrate depends on the alga. The fact that efflux is through a single export channel should, in theory, give the alga an opportunity for regulating its carbon loss, but no evidence of this has been obtained so far. The conversion of most of the fixed carbon into a single compound does not appear to involve novel pathways, but simply a diversion of existing ones. The mobile carbohydrate can probably be released from algal cells by exchange diffusion ("inhibition" experiments), but it is not definitely clear that this is the same transport system involved in symbiont efflux. The symbiont efflux is almost certainly carrier-mediated, but probably by a system that requires no energy, such as facilitated diffusion. There is no evidence that a novel transport protein is involved in efflux.

HOW IS MASSIVE EFFLUX OF CARBOHYDRATE INDUCED?

There is no massive efflux of carbohydrate from cultured lichen algae. It is therefore probably induced by some aspect of existence in symbiosis. There are three interconnected aspects to the problem of the induction of efflux. (1) How does algal cell metabolism change so that almost all the photosynthetically fixed carbon becomes channeled into the synthesis of a single carbohydrate? (2) Why does efflux of this carbohydrate occur? (3) What is the external stimulus that causes these events to occur in the algal cell?

Changes in Algal Cell Metabolism Leading to Extensive Synthesis of Mobile Carbohydrate

One possibility that has been considered is that rapid removal by the fungus of any mobile carbohydrate released would lead to continued synthesis of the latter, as it never reaches its limiting-pool size. However, since massive efflux continues when fungal uptake is inhibited by digitonin or 2-deoxy-glucose pretreatment, it is unlikely that this could be more than a very minor factor in the synthesis of mobile carbohydrate. Alternative possibilities must therefore be considered.

It was pointed out in the previous section that synthesis of mobile carbohydrate is accompanied by marked reduction in polysaccharide formation. In this connection it is valuable to consider parallel work on the movement of fixed nitrogen from blue-green symbionts to lichen fungi. Stewart and Rowell (1977), using digitonin as an inhibitor of fungal uptake, showed that fixed nitrogen moves in large amounts, primarily as ammonia. Within the alga the first enzyme in the pathway of ammonia assimilation is glutamine synthetase. The activity of this enzyme in symbiosis is repressed, leading to accumulation of free ammonia available for efflux.

The analogous situation for photosynthate transport would be that enzymes of polysaccharide synthesis become repressed, enabling the mobile carbohydrate to accumulate. However, it may be over-simplifying to regard the parallel as exact. There is no firm evidence that mobile carbohydrates are on the direct route to polysaccharide synthesis. Indeed, there is some evidence from short-term photosynthesis experiments that mobile carbohydrates may be derived from polysaccharides, rather than formed before them. Nevertheless, it is tempting to assume that the same basic mechanism induces the efflux of both fixed carbon and fixed nitrogen from blue-green symbionts. Since, in cephalodiate lichens, the same fungus associates with green as well as blue-green algae, the mechanism could apply to all lichens.

Why Does Efflux of Carbohydrate Occur?

As explained above, continued removal of carbohydrate by the fungus cannot itself explain the massive efflux. Smith (1974) originally suggested that since the vectorial characteristics of some membrane transport systems are partly dependent on the membrane potential, the massive efflux could result from a reduction or even reversal of the algal membrane potential in symbiosis. He noted that published values for the membrane potentials of fungi were substantially higher than for algae such as *Chlorella*, and suggested that the algal membrane potential was affected by an electrical charge carried between the symbiont membranes by the matrix separating them. This hypothesis became untenable when Chambers and colleagues (1976) showed that efflux continued when the fungal membrane was inactivated by digitonin. These workers suggested the alternative possibility that the cell wall of the fungus contained groups with fixed negative charges that affected the alga.

Using various redox and other dyes, Tapper (1979) was unable to obtain any histochemical evidence in support of this hypothesis. Nor was he able to show pH differences between the inside and outside of the cell of the order to be expected on this hypothesis. Furthermore, a change in a general characteristic such as membrane potential would be unlikely to affect only a single specific transport system, and the efflux of other compounds should also be detectable.

As explained in the previous section, the transport mechanism involved in efflux is probably facilitated diffusion. Massive efflux by such a system could occur if a high concentration of mobile carbohydrate occurred at the transport sites at the inner surface of the algal membrane. Farrar (1978) and Collins and Farrar (1978) give detailed calculations based upon observed data to show that simple diffusion could explain the level of efflux observed.

External Stimulus Inducing Synthesis and Efflux of Carbohydrate

So far, it has proved impossible to induce the massive efflux of photosynthate from isolated symbionts. There is therefore no direct clue as to how this is induced in symbiosis. Three main possibilities, not mutually exclusive, offer themselves.

First, the fungus could release one or more specific compounds that enter the alga to achieve the necessary alterations in metabolic pathways. Although a search has been made for these compounds, none have been found. Water-soluble "factors" inducing photosynthate release have been isolated from a number of marine invertebrate hosts of symbiotic algae. So far, such "factors" have been without effect on lichen algae, and analogous factors have not been detected in lichens. Failure to detect does not prove such factors do not exist, but it does compel the consideration of the alternative possibilities.

Second, the fungus could exert a physical rather than a chemical influence on the alga. The fact that efflux only occurs when the fungus is in direct physical contact with the alga supports this view. Precise formulation of an experimentally testable hypothesis is, however, difficult. As discussed above, Smith (1974) and Chamber and colleagues (1976) suggested various effects of the fungus on the membrane potential of the alga, but Tapper (1979) failed to find any supporting evidence. Furthermore, it is not easy to see how a change in membrane potential could cause the changes in internal metabolism. Alternatively, the effect might result from physical restriction of algal growth by the encircling fungal hypha, but there seems no obvious means of testing this hypothesis.

Third, other aspects of the microenvironment of the alga in the lichen could be having an effect—such as the oxygen and carbon dioxide tensions. Millbank (1977) has shown that oxygen tensions within lichen thalli are below atmospheric but it is not known how this affects algal metabolism.

The stimulus inducing carbohydrate synthesis and efflux thus remains a mystery. It is difficult to see how worthwhile experimental investigations can proceed apart from a continuing search for "factors" excreted by the fungus. There is, however, one further experimental approach that might be pursued, and that is the study of the induction of photosynthate release during lichen synthesis. The single paper of Hill and Ahmadjian (1972) gave preliminary evidence that in *Cladonia cristatella*, no photosynthate transport could be detected until 12 weeks after the symbionts initially came into contact. If this can be confirmed, then a detailed study of this slow induction might prove rewarding, though the labor of setting up the large number of artificial syntheses needed to provide experimental material is daunting!

Finally, there are some purely speculative suggestions. Ahmadjian (1966) originally advanced a hypothesis that one mechanism of regulating transport between the symbionts might involve the enzyme urease in the fungus. Under conditions of nitrogen deficiency, urease might break down urea to ammonia and CO_2. The release of CO_2 would increase the rate of photosynthesis. Ammonia would not only be converted to amino acids but some would also pass to the alga. Since Syrett (1962) has shown that ammonia decreases the membrane permeability barriers of *Chlorella*, it is suggested that the ammonia might increase photosynthate efflux. This hypothesis has been extended by Vicente and colleagues (1978), and Vicente, Palasi, and Estavez (1978), who have found that l-usnic acid, a product of an appreciable number (but not all) lichen fungi, will inactivate urease. They suggest that usnic acid formation would increase as photosynthesis increases, and could thus act as a kind of feedback inhibitor in the over-all regulatory process.

Although this theory is attractive, it lacks any experimental evidence to support the central idea that the effects of CO_2 and ammonia will increase carbohydrate release. There are no data on the amount of urea in the thallus. Considerable quantities would presumably be needed to maintain photosynthate efflux over prolonged periods, and to make a significant increase in CO_2 over and above that emanating from fungal respiration. There is no evidence for the production of usnic acid in the region of contact between the symbionts. In general, there is no evidence that fungal hyphae are nitrogen-deficient.

MORPHOLOGICAL ASPECTS OF PHOTOSYNTHATE TRANSPORT

A number of authors considered that there may be special types of morphological contact between the symbionts that facilitate transport of photosynthate or conduction of the inducing stimulus. An extreme form of this suggestion is by Walker (1968), who described electron microscope observations showing channels originating from the algal chloroplast leading to the space between the symbionts in the lichen *Cornicularia normoerica*. Numerous subsequent authors—admittedly studying other lichen species—have failed to observe such channels, so Walker's original observations may have been artifacts.

Physical penetration of the fungus into the alga has been reported by some authors. Most often, this appears to involve an apparent rupture of the outer wall, leaving the algal membrane intact. More rarely, penetration and apparent rupture of the algal membrane have been described, analogous to the haustoria of pathogens of higher plants. It is usually suggested that these penetrations are either absent or only occasionally present in many species, and in some others are restricted only to degenerate cells (i.e., in which photosynthesis may well have ceased). They are regularly present in only a few lichens, and according to Collins and Farrar (1978), there are only two species in which more than 25% of all algal cells are reported to be regularly invaded.

Collins and Farrar (1978) conclude that "none of the published evidence convinces us that haustoria have a role in the biotrophic nutrition of lichen fungi." They show that in the lichen *Xanthoria parietina*, only 20% of the surface area of the algal symbionts is in contact with the fungus, but that this area of contact would be adequate for the observed rate and amount of carbohydrate transfer to occur by simple diffusion. The apparent unimportance of haustoria in lichens should not necessarily be regarded as a "primitive" characteristic. In higher plants a fungal hypha ramifying through a multicellular tisue may have access to substantially less than 20% of the surface area of a host cell, so that haustorial penetration may be essential to achieve an adequate area of contact. Also, in higher plants only a very small proportion of host cells are in contact with the fungus, but in lichens every algal cell is normally in contact.

Peveling (1969, 1973a) has described how the plasmalemma of the fungus may become strongly convoluted in the region of contact with the alga. By analogy with the similar appearance of the plasmalemma in "transfer cells" in higher plants, it seems very likely that this is an adaptation to provide a large surface area for the uptake of the massive flux of carbohydrate from the alga.

Other structures have been seen by electron microscopy that have also been considered to be involved in transfer. In *Lichina pygmaea*, Peveling (1973b) has described numerous small vesicles in the fibrillar sheath of the algal symbiont *Calothrix*, and suggested they may be involved in a mechanism for promoting transport between the symbionts. Boissière (1977) described mesosome-like structures apparently derived from folds in the fungal plasmalemma containing polyglucosides. These appear to be collected by autophagic vacuoles with acid phosphatase activity. Boissière therefore suggests that these may represent a mechanism of absorption and digestion of metabolic substances coming from the phycobiont.

It seems unlikely that any structures involving continuous movement or dissolution of vacuoles would explain the massive and rapid nature of photosynthate transport. On the other hand, they might well be involved in much slower and smaller movements of other compounds.

After passing from alga to fungus, fixed carbon may rapidly move away from the region of contact. The medulla has long been considered a region of carbohydrate storage (Smith, 1961). More recently, using microautoradiography, Tapper (1979) has shown that in *P. polydactyla*, fixed ^{14}C accumulates in the pseudoparenchymatous upper cortex, a tissue formerly considered to be purely protective in function.

MOVEMENT OF COMPOUNDS OTHER THAN
CARBOHYDRATE FROM ALGA TO FUNGUS

All the methods of studying transport from alga to fungus mentioned above involve tracing the movement of ^{14}C recently fixed by photosynthesis. They clearly demonstrate the rapid bulk flow of carbohydrate, but they may well have been inadequate to show whether there are other compounds moving in smaller quantities and possibly at slower speeds, and involving carbon not derived from recent photosynthesis. For example, it is very likely that algal symbionts contain nitrate reductase, and it would be surprising if some products of nitrate reduction did not become available to the fungus, either as ammonia or amino acids. Feige (1976) shows that there is little mass carbohydrate flow from the alga to the fungus in cephalodia of *Peltigera aphthosa*, and situations such as this might repay closer examination for the movement of other compounds.

It is particularly important to remember that, as explained below, the primary role of the bulk carbohydrate flow is probably not nutritive but associated with resistance to extremes of environmental conditions. Since lichens grow very slowly, the movement of essential molecules and ions exclusively involved in nutrition may be one or two orders of magnitude smaller than bulk carbohydrate flow. The probable small nature of such movements should be borne in mind in designing appropriate experiments to determine if they occur. The possibility should not be ignored that structures seen by electron microscopy and described in the preceding section may be involved in such slower transport.

IS THERE MOVEMENT FROM FUNGUS TO ALGA?

Because lichens have been regarded as an outstanding example of a mutualistic symbiosis, it has been virtually automatic to believe that substances move from fungus to alga. Yet, so far, no single item of experimental evidence exists to show that movement occurs. In fact, it may be questioned whether there is need for any substances to move from fungus to alga. Since the alga is extracellular, it can presumably absorb solutes from solutions permeating the thallus. Collins and Farrar (1978) calculated that only 20% of the surface area of the algal symbiont of *Xanthoria parietina* is in contact with the fungus, leaving 80% available for uptake from solutions saturating the inter-symbiont spaces in the algal layer. Experiments on photosynthesis show that it takes 15–20 secs for $H^{14}CO_3^-$ to penetrate from the external solution to the photosynthetic centers of the alga in *P. polydactyla* (Smith, 1973). It seems hardly likely that this process involves uptake and excretion by the fungus, followed by uptake by the alga. It is well known that algal symbionts are more sensitive to SO_2 pollution than fungal, but nowhere is it suggested that the toxic ions are first absorbed by the fungus and then released to the alga.

Since the growth rate of lichens in nature is extremely slow, nutrient requirements of the symbionts are correspondingly low. Farrar (1976c) has calculated that the phosphate content of rainwater and the frequency of wetting by rain showers is adequate to support the observed growth of the lichen *Hypogymnia physodes*. The situation is presumably the same for other minerals.

There is thus no *a priori* need for any movement of minerals from fungus to alga; absorption from solutions passing through the thallus should be adequate to explain observed growth rates of the alga. On the other hand, the fungal hyphae, which constitute the bulk of the thallus, are highly efficient at accumulating substances from dilute solution, and it would be surprising if some mechanism had not evolved in which some of the advantages of this property were not passed on to the alga. Smith (1978) has suggested such a mechanism, based upon the experimental observation that when dry lichens are rapidly rewetted in water, there is a period of up to 60–90 sec before the membrane permeability barriers are reestablished, and during which solutes are lost from the thallus (Smith and Molesworth, 1973). Rapid immersion of dry lichens in water would be extremely rare in nature, but localized effects may well occur when raindrops first hit a dry thallus. In an area of the algal layer that has been rewetted by a rain drop, there would be a nonspecific release of solutes from *both* symbionts, followed by rapid reabsorption. It must be emphasized, strenuously, that no experimental verification of this hypothesis has yet been carried out. Neither is a more conventional transport from fungus to alga, relying on membrane carriers mediating efflux and uptake, excluded.

Studies of the growth requirements of isolated symbionts in culture are notoriously misleading indicators of the requirements in symbiosis. Nevertheless, the fact that the commonest algal symbiont, *Trebouxia*, thrives better in organic than inorganic medium *might* signal some dependence on exogenous organic substances in nature.

ECOLOGICAL ASPECTS OF PHOTOSYNTHATE TRANSPORT

The previous sections have stressed that since the growth of lichens is so slow, movements of substances essential to growth between symbionts also need only be very slow. It is therefore remarkable that photosynthate transport should be so large and rapid. Farrar (1978) has calculated that the supply of photosynthate may be 10–20 times greater than that necessary to supply the growth and respiratory requirements of the fungus. He explains this apparent paradox on the basis that to survive the extreme and varying environmental conditions found in most lichen habitats, a high concentration of polyols is essential to the fungus. This is not simply to maintain a high osmotic pressure because of the water stress in the habitat, but also to combat the carbon losses that occur in phenomena such as "resaturation respiration" and leakage during rewetting of dry thalli. He terms this general phenomenon "physiological buffering." Thus, a key role of algal photosynthesis may be that of maintaining a high internal polyol concentration in the fungus, enabling it to withstand environmental extremes and variables. The consequential massive movement of carbohydrate has fortuitously provided a picture of the mechanisms of biotrophy, a picture that is still tantalizingly blurred.

SUMMARY

One of the commonest types of interaction in symbiosis is the biotrophic movement of nutrient molecules between the associating organisms. One of the simplest movements to analyze experimentally is the massive flow of carbohydrate from the alga to the fungus in lichens. This paper therefore attempts to assess the extent to which this movement can be analyzed at the membrane and molecular level.

The various experimental techniques used to study carbohydrate flow in lichens are reviewed. Each has disadvantages, and none, on its own, can yield data whose precision enables an exact analysis of movement to be made. Although the identity of the mobile carbohydrate in many lichens is known, pathways of synthesis within the algal cell are not. Synthesis of mobile carbohydrate in symbiosis is clearly linked to depression in polysaccharide formation, but the nature of the linkage is unclear. The mechanism by which the mobile carbohydrate leaves the algal cell is probably carrier-mediated facilitated diffusion. This requires no metabolic energy, but requires a high concentration of mobile carbohydrate to build up at transport sites within the cell to give the massive and rapid carbohydrate movement that is observed. Although haustoria are occasionally observed, they are unnecessary to explain the degree and level of movement. The stimulus that, in symbiosis, causes the synthesis and release of mobile carbohydrate, remains completely unknown.

The amount of carbohydrate moving to the fungus greatly exceeds that required for fungal respiration and growth. The primary role of the movement is believed to maintain a high concentration of soluble carbohydrate in the fungus, this being an essential mechanism for withstanding environmental extremes. Since lichens grow very slowly, movement between the symbionts of substances exclusively concerned in nutrition may be very much slower than photosynthate transport, and not detectable by the methods used to study it. Certain mechanisms of movement postulated on ultrastructural observations, such as the possible movement of vesicles, may be involved in these slower movements; it is very unlikely that they could explain the mass carbohydrate flow. So far, no evidence exists of movement of substances from fungus to alga.

LITERATURE CITED

Ahmadjian, V. 1966. Lichens. *In* S. M. Henry (ed.), Symbiosis, pp. 36–97. Academic Press, New York.

Bednar, T. W., and D. C. Smith, 1966. Studies in the physiology of lichens. VI. Preliminary studies of photosynthesis and carbohydrate metabolism of the lichen *Xanthoria aureola*. New Phytol. 65:211–20.

Beschel, R. 1958. Lichenometrical studies in West Greenland. Arctic. 11:254.

Boissière, M-C. 1972a. Cytologie du *Peltigera canina* (L.) Willd. en microscopie électronique. I. Premières observations. Rev. Gén. Bot. 79:167–85.

Boissière, M-C. 1972b. Mise en évidence cytochimique en microscopie électronique de polyglucosides de réserve chez des *Nostoc* libres et lichenisés. C. R. Acad. Sci. Paris. 274:2643–46.

Boissière, M-C. 1973. Activité phosphatasique neutre chez le phycobionte de *Peltigera canina* comparée à celle d'un *Nostoc* libre. C. R. Acad. Sci. Paris. 277:1649–51.

Boissière, M-C. 1977. Un mécanisme possible d'absorption des glucides d'origine cyanophytique par les hyphes de quelques lichens. Rev. Bryol. Lichenol. 43:19–25.

Chambers, S., M. Morris, and D. C. Smith. 1976. Lichen physiology. XV. The effect of digitonin and other treatments on biotrophic transport of glucose from alga to fungus in *Peltigera polydactyla*. New Phytol. 76:485–500.

Collins, C. R., and J. F. Farrar. 1978. Structural resistances to mass transfer in the lichen *Xanthoria parietina*. New Phytol. 81:71–83.

Drew, E. A., and D. C. Smith. 1966. The physiology of the symbiosis in *Peltigera polydactyla* (Neck.) Hoffm. Lichenologist 3:197–201.

Drew, E. A., and D. C. Smith. 1967a. Studies in the physiology of lichens. VII. The physiology of the *Nostoc* symbiont of *Peltigera polydactyla* compared with cultured and free-living forms. New Phytol. 66:379–88.

Drew, E. A., and D. C. Smith. 1967b. Studies in the physiology of lichens. VIII. Movement of glucose from alga to fungus during photosynthesis in the thallus of *Peltigera polydactyla*. New Phytol. 66:389–400.

Farrar, J. F. 1973. Physiological lichen ecology. Ph. D. diss., Oxford University.

Farrar, J. F. 1976a. Ecological physiology of the lichen *Hypogymnia physodes*. II. Effects of wetting and drying cycles and the concept of "physiological buffering." New Phytol. 77:105–13.

Farrar, J. F. 1976b. The lichen as an ecosystem. *In* B. W. Ferry, M. S. Baddeley, and D. L. Hawksworth (eds.), Air pollution and lichens, pp. 238–82. Athlone Press, London.

Farrar, J. F. 1976c. The uptake and metabolism of phosphate by the lichen. *Hypogymnia physodes*. New Phytol. 77:127–34.

Farrar, J. F. 1978. Symbiosis between fungi and algae. *In* M. Reichigl (ed.), CRC Handbook Series in Nutrition and Food. Cleveland CRC Press (in press).

Farrar, J. F., and D. C. Smith 1976. Ecological physiology of the lichen *Hypogymnia physodes*. III. Importance of the rewetting phase. New Phytol 77:115–25.

Feige, G. B. 1976. Untersuchungen zur Physiologie der Cephalodien der Flechte *Peltigera aphthosa* (L.) Willd. II. Das photosynthetische ^{14}C-Markierungsmuster und der kohlenhydrattransfer zwischen Phycobiot und Mycobiot. Z. Pflanzenphysiol. 80:386–94.

Green, T. G. A. 1970. The biology of lichen symbionts. Ph. D. diss., Oxford University.

Hessler, R., and E. Peveling. 1978. Die Lokalisation von ^{14}C-Assimilaten in Flechtenthalli von *Cladonia incrassata* Floerke und *Hypogymnia physodes* (L.) Ach. Z. Pflanzenphysiol. 86:287–302.

Hill, D. J. 1972. The movement of carbohydrate from the alga to the fungus in the lichen *Peltigera polydactyla*. New Phytol. 71:31–39.

Hill D. J., and V. Ahmadjian. 1972. Relationship between carbohydrate movement and the symbiosis in lichens with green algae. Planta 103:267–77.

Hill, D. J., and D. C. Smith. 1972. Lichen physiology. XII. The inhibition technique. New Phytol. 71:15–30.

Lewis, D. H., and D. C. Smith. 1971. The autotrophic nutrition of symbiotic marine coelenterates with special reference to hermatypic corals. I. Movement of photosynthetic products between the symbionts. Proc. R. Soc. Lond. B. 178:111–29.

Millbank, J. W. 1977. The oxygen tension within lichen thalli. New Phytol. 79:649–57.

Millbank, J. W., and K. A. Kershaw. 1969. Nitrogen metabolism in lichens. I. Nitrogen fixation in the cephalodia of *Peltigera aphthosa*. New Phytol. 68:721–9.

Peat, A. 1968. Fine structure of the vegetative thallus of the lichen *Peltigera polydactyla*. Arch. Mikrobiol. 61:212–22.

Peveling, E. 1969. Elektronoptische Untersuchungen an Flechten. III. Cytologische Differenzierungen der Pilzzellen im Zusammenhang mit ihrer symbiontischen Lebensweise. Z. Pflanzenphysiol. 61:151–64.

Peveling, E. 1973a. Fine structure. *In* V. Ahmadjian and M. E. Hale (eds.), The lichens, pp. 147–82. Academic Press, New York and London.

Peveling, E. 1973b. Vesicles in the phycobiont sheath as possible transfer structures between the symbionts in the lichen *Lichina pygmaea*. New Phytol. 72:343–47.

Peveling, E. and D. J. Hill. 1974. The localization of an insoluble intermediate in glucose production in the lichen *Peltigera polydactyla*. New Phytol. 73:767–69.

Richardson, D.H.S. 1967. Studies on the biology and physiology of lichens with special reference to *Xanthoria parietina* (L.) Ph.D. diss., Oxford University.

Richardson, D.H.S., and D. C. Smith. 1968. Lichen physiology. IX. Carbohydrate movement from the *Trebouxia* symbiont of *Xanthoria aureola* to the fungus. New Phytol. 67:61–68.

Richardson, D. H. S., D. J. Hill, and D. C. Smith 1968. Lichen physiology. XI. The role of the alga in determining the pattern of carbohydrate movement between lichen symbionts. New Phytol. 67:469–86.

Robbie, J. P., and T. H. Wilson, 1969. Transmembrane effects of β-galactosides on thiomethyl-β-galactoside transport in *E. coli*. Biochim. Biophys. Acta. 173:234–44.

Smith, D.C. 1961. The physiology of *Peltigera polydactyla* (Neck) Hoffm. Lichenologist 1:209–26.

Smith D. C. 1963. Experimental studies of lichen physiology. Symp. Soc. Gen. Microbiol. 13:31–50.

Smith, D. C. 1973. The lichen symbiosis. Oxford Biology Readers, no. 42. Oxford University Press.

Smith, D. C. 1974. Transport from symbiotic algae and symbiotic chloroplasts to host cells. Symp. Soc. Exp. Biol. 28:437–508.

Smith, D. C. 1975. Symbiosis and the biology of lichenised fungi. Symp. Soc. Exp. Biol. 29:373–405.

Smith, D. C. 1978. What can lichens tell us about REAL fungi? Mycologia 70:915–34.

Smith, D. C., and S. Molesworth. 1973. Lichen physiology. XIII. Effects of rewetting dry lichens. New Phytol. 72:525–34.

Smith, D. C., L. Muscatine, and D. H. Lewis. 1969. Carbohydrate movement from autotrophs to heterotrophs in parasitic and mutualistic symbiosis. Biol. Rev. 44:17–90.

Stewart, W. D. P., and P. Rowell. 1977. Modifications of nitrogen-fixing algae in lichen symbiosis. Nature 265:371–72.

Syrett, P. J. 1962. Nitrogen assimilation. *In* R. A. Lewin (ed.), Physiology and biochemistry of algae, pp. 171–88. Academic Press, New York.

Tapper, R. C. 1979. Studies on carbohydrate efflux from lichen algae. Ph.D. diss., University of Bristol.

Tischer, J. 1936. Über die Carotinoide und die bildung von Jonon in *Trentepohlia nebst* Bemerkungen über den Gehalt dieser alge an Erythrit. Carotinoide der susswasseralgen II. Hoppe-seylers Z. physiol. Chem. 243:103–18.

Trench, R. K., J. E. Boyle, and D. C. Smith. 1974. The association between chloroplasts of *Codium fragile* and the mollusc *Elysia viridis*. III. Movement of photosynthetically fixed ^{14}C in tissues of intact living *E. viridis* and *Tridachia crispata*. Proc. R. Soc. Lond. B. 185:453–64.

Vicente, C., A. Azpiroz, M. P. Estevez, and M. L. Gonzalez. 1978. Quaternary structure changes and kinetics of urease inactivation by L-usnic acid in relation to the regulation of nutrient transfer between lichen symbionts. Plant, cell, and environment 1:29–33.

Vicente, C., M. Palasi, and M. P. Estevez. 1978. Urease regulation mechanism in *Lobaria pulmonaria* (L.) Hoffm. Rev. Bryol. Lichenol. 44:83–87.

Walker, A. T., 1968. Fungus-alga ultrastructure in the lichen *Cornicularia normoerica*. Amer. J. Bot. 55:641–48.

Uptake, Retention, and Release of Dissolved Inorganic Nutrients by Marine Alga-Invertebrate Associations

10

INTRODUCTION

A cardinal feature of many tropical marine communities is the high incidence and relatively great abundance of alga-invertebrate symbiotic associations. Correlative with this, tropical waters are often poor in dissolved inorganic nutrients. Whereas alga-invertebrate associations are able to sustain high gross primary productivity as a result of fixation of carbon by the symbiotic algae and subsequent bilateral movement of reduced carbon and nitrogen, their ability to tolerate environmental nutrient deficiency is not well understood. There is a current trend toward research on nutrient interactions in alga-invertebrate associations (Muscatine and Porter, 1977). The purpose of this paper is to review extant information on the uptake, retention, and release of dissolved inorganic nutrients by selected marine alga-invertebrate associations. Because it reflects the bulk of the current literature, flux of inorganic nitrogen in reef corals will be considered in greatest detail.

Corals, such as colonial cnidarians, have a high ratio of prey capture surface to living tissue biomass (Yonge, 1930), but recent studies show that only a small fraction of their daily caloric needs comes from available zooplankton (Porter, 1974), the rest presumably coming from translocation products of zooxanthellae. In fact, ingested zooplankton are viewed as an important source of nutrients rather than calories (Johannes et al., 1970; Johannes, 1974). Although the contribution of zooplankton to coral daily nitrogen budgets is not yet quantified, the apparent lack of emphasis on zooplanktivory has shifted attention to other potential sources of nitrogen.

In an aquatic carnivore such as a marine cnidarian, nitrogen is acquired by uptake of organic nitrogen by holozoic feeding, followed by oxidative and biosynthetic metabolism and release of dissolved inorganic nitrogen, largely as ammonium. The possession of a non-nitrogen-fixing autotrophic symbiont superimposes three additional potential routes of nitrogen flux: (1) uptake of inorganic nitrogen, (2) translocation of nitrogen-bearing compounds from algae to animal host, and (3) retention of heterotroph products that might otherwise be excreted. The end result is an association that can potentially take up, retain, and recycle certain nutrient elements. These processes may be construed as tending to conserve these elements, thus contributing to a tolerance of nutrient deficiency in the environment.

There have been no detailed studies on the acquisition of nitrogen by uptake of dissolved organic nitrogen, nor has the availability of nitrogen fixed by algae in coral skeletons been investigated (see Crossland and Barnes, 1976). Most studies, and the remainder of this paper, address the uptake of dissolved inorganic nitrogen (DIN) such as nitrate and ammonia.

AVAILABILITY OF DIN

The cycle of nitrogen in the euphotic zone predicts that DIN should be available in surface waters in two principal forms, "new" nitrogen (i.e., nitrate-N from deep water; nitrogen newly fixed in the water column) and "regenerated" nitrogen, (i.e., ammonium-N regenerated from catabolism of organic nitrogen) (Dugdale and Goering, 1967; Dugdale, 1967). In tropical oceanic water, where the existence of a permanent thermocline blocks vertical transport from deep water, and where there is little N_2 fixation in the water column, new nitrogen sources are minimal. Thus, regenerated ammonium-N would appear to be the most important source of DIN (MacIsaac and Dugdale, 1972). In surface water bathing and traversing coral atoll and fringing reef communities, the sources and concentrations of DIN vary somewhat (table 1). In some cases there is evidence that nitrogen is fixed by reef-dwelling blue-green algae and subsequently disseminated as nitrate-N (Wiebe et al., 1975; Webb et al., 1975) along with ammonium-N regenerated from other substrates.

UPTAKE OF DIN

Unlike phytoplankton nutrient studies, which frequently employ ^{15}N to measure nutrient flux, virtually all investigations of coral nutrient flux have employed wet chemical methods to detect appearance or disappearance of a given nutrient in incubation media. Experiments have been designed as static or flow-through chamber incubations. Details of chamber incuba-

TABLE 1

SOME AMBIENT AMMONIUM AND NITRATE CONCENTRATIONS AT
VARIOUS CORAL REEF RESEARCH SITES

Location	NH$_4^+$–N (μM)	NO$_3^-$–N (μM)
Great Barrier Reef		
Lizard Island		
Transect	—	0.2–0.4
Oceanic	—	0.85
Marshall Islands		
Enewetak		
Transect	0.20–0.29	0.080–0.170
Oceanic	0.30	0.02
Lagoon	0.28	0.07
Eastern Tropical Pacific		
Revillagigedo Island	0.55–0.85	—
Hawaii		
Oceanic	0.4–0.8	0.07–0.12
Entrance Kaneohe Bay	—	0.16
Inner Kaneohe Bay	—	0.25–1.52
Jamaica		
Inner Discovery Bay	0.5–1.4	—

SOURCE: Data from Limer 1975 Expedition (1976), Muscatine and D'Elia (1978), Webb et al. (1975), and Muscatine et al. (1979)

tion methods are discussed by D'Elia (1978). The extent to which nutrient flux in symbiotic corals is influenced by zooxanthellae is deduced from various categories of controls, including, in order of preference, aposymbiotic corals, nonsymbiotic corals (e.g., ahermatypes), and symbiotic corals in darkness.

Uptake and accumulation of nitrate and nitrite by reef corals was first reported by Franzisket (1973, 1974), who observed that the concentration of nitrate in sea water decreased in a flow-through chamber apparatus containing individuals of *Fungia scutaria* and *Montipora verrucosa*. Since 1000× the sea water concentration of nitrate and nitrite was detected in the tissues of another coral species *Pocillopora damicornis*, and since the ratio of nitrate to nitrite in tissues shifted from 3:1 at dawn to 1:3 in the afternoon, Franzisket concluded that nitrate was taken up and reduced at least to nitrite.

Nitrate uptake by reef corals has recently been confirmed by D'Elia and Webb (1977) and Webb and Wiebe (1978). *Pocillopora elegans*, *Fungia* sp., *Acropora* spp. and the foraminiferan *Marinopora vertebralis*, incubated in the light for 2–3 hr in sea water enriched with 1.0–1.5 μmoles per liter nitrate-nitrogen rapidly removed this nutrient from the medium. Nitrate concentration approached 0.2 μM as a minimum depletion value, possibly resulting from either efflux of nitrate from the coral at a rate equal to that for uptake at low concentrations, or nitrification by bacteria, or inhibition of uptake at low ambient concentrations. Evidence from these studies indicated that nitrite was neither taken up nor released. Incubations with cleaned skeleton controls established only that uptake of nitrate was due to the presence of animal tissue plus zooxanthellae. To determine the extent to which the zooxanthellae alone influenced uptake, Webb and Wiebe (1978) compared the rate of nitrate uptake by *Acropora* branches with the uptake rate by excised *Acropora* terminal polyps. The latter contained few zooxanthellae and upon extraction yielded negligible chlorophyll *a*, although they continued to respire and calcify after excision. The rate of nitrate uptake by these tips was about 40% of that for normal branches, suggesting that either the few zooxanthellae in the tips were highly efficient at scavenging nitrate or that the animal tissue directly influenced nitrate uptake.

Kawaguti (1953) first demonstrated that in the light a variety of reef corals took up ammonium from enriched sea water, although he used lengthy incubations and unnaturally high concentrations. Nevertheless, his observations have been confirmed recently by Muscatine and D'Elia (1978). Of several genera of Pacific reef corals tested, only those with zooxanthellae took up and retained ammonium nitrogen. In most instances ammonium depletion reached the limits of detection. Incubations with ^{15}N-labeled ammonium and subsequent analyses of the coral tissues revealed that concomitant with the disappearance of ammonium from the medium ^{15}N-ammonium was detected in the coral tissues. Since nonsymbiotic controls (*Tubastrea aurea*) only released ammonium, uptake was provisionally attributed to the presence of zooxanthellae. Recently, Muscatine and colleagues (1979) compared uptake of ammonium by symbiotic and aposymbiotic corals of the same species using *Madracis mirabilis*. Since only the symbiotic corals took up ammonium, uptake was attributed directly to the activities of the zooxanthellae. However, isolated *M. mirabilis* zooxanthellae exhibited much lower uptake rates than those in the intact coral, suggesting that the structural integrity of the symbiosis was a factor in maintaining high uptake rates. Another possibility is that zooxanthellae were damaged by the isolation procedure.

FACTORS INFLUENCING UPTAKE OF DIN

The uptake of DIN by phytoplankton is stimulated by environmental factors such as light and nutrient concentration (see reviews of Healey and Stewart, 1973; Dugdale, 1976). Any of three mechanisms may be involved in light stimulation: (1) direct photoreduction, (2) supply of energy for uptake via photophosphorylation, and (3) supply of carbon skeletons from photosynthesis as acceptors for nitrogen. With appropriate experimental design, ammonium uptake by corals can be shown to be influenced by light. Thus, Muscatine and D'Elia (1978) observed that in 2-hr incubations in light or dark, *Pocillopora damicornis* showed no net release of ammonium nitrogen. When the medium was enriched with 5 μM ammonium nitrogen, corals took up ammonium in light and dark within 2 hours. However, during longer incubations in dark, ammonium was released (Kawaguti, 1953). If energy is required for ammonium uptake and retention, the difference between short and long incubations may be due to energy reserves that can fuel short incubations but not long ones. Muscatine and D'Elia (1978) tested this hypothesis by preincubating corals in darkness for up to 18 hours. In an ensuing 4-hr dark incubation, ammonium was released. If the dark incubation was preceded by a short light period, ammonium release was depressed concomitant with the length of light exposure. Muscatine and D'Elia (1978) also demonstrated that the light period of a natural light:dark diel cycle is sufficient to fuel ammonium retention during the nighttime period. These observations indicate that irradiance history dramatically affects the outcome of dark uptake experiments with reef corals, and may explain why short-term nitrate uptake experiments reported by Franzisket (1974), D'Elia and Webb (1977), and Webb and Wiebe (1978) might not be expected to show light:dark differences.

EFFECT OF CONCENTRATION

Following the suggestion of Dugdale (1967), the Michaelis-Menten equation has been widely used to describe nutrient uptake kinetics by marine phytoplankton (see review of Dugdale, 1976). The expression is:

$$V = \frac{V_{max}S}{K_s + S} \tag{1}$$

where V = net uptake rate of limiting nutrient, V_{max} = maximum uptake rate of limiting nutrient; S = concentration of limiting nutrient; and K_s = concentration of limiting nutrient at which $V = V_{max}$ (fig. 1a). It is assumed that the expression treats the movement of nutrient from outside to inside the cell, i.e., the transport phase. Since a specific "carrier" molecule is thought to effect entry of the nutrient into the cell, K_s is therefore thought to measure the affinity of the "carrier" for the nutrient. The magnitude of K_s can be thus taken as a measure of the ability of a phytoplankter to compete for a limiting nutrient. Thus, values of K_s and V_{max} for oligotrophic phytoplankton are about an order of magnitude less than those for eutrophic phytoplankton, with K_s for nitrate and ammonium ranging from less than 1.0 μM for the former, and greater than 1.0 μM for the latter (Dugdale, 1976).

The Michaelis-Menten expression has been applied to nitrate, ammonium, and phosphate uptake by reef corals (D'Elia and Webb, 1977; Webb and Wiebe, 1978; Muscatine and D'Elia, 1978; D'Elia, 1977). In some cases the curves for V vs. S do not strictly follow the Michaelis-Menten expression. Thus, a plot of V vs. S for nitrate uptake gives a hyperbola that intercepts the abscissa at some positive value for S rather than at zero (fig. 1b). Caperon and Meyer (1972) also observed a nonzero intercept in studies of nitrate uptake by phytoplankton. They interpreted the nonzero intercept to indicate a concentration below which uptake did not occur. They treated such curves by the expression:

$$V = \frac{V_{max}\,(S\text{-}S_o)}{K_s + (S\text{-}S_o)} \tag{2}$$

where, in their experiments, $S_o =$ a threshold concentration ranging from 50 to 70 nM. D'Elia (1977), in studies of phosphorus flux in reef corals, presented strong evidence that a positive abscissa intercept was the nutrient concentration at which the concentration-dependent uptake rate equaled the concentration-independent efflux rate. Thus,

$$V = \frac{V_{max}\,S}{K_s + S} - V' \tag{3}$$

where V' is a term describing phosphate efflux (fig. 1b). A slightly different departure from Michaelis-Menten kinetics was observed by Muscatine and D'Elia for ammonium uptake by reef corals (1978). The curve describing V vs. S had a strong diffusive component which could satisfactorily be expressed by:

$$V = \frac{V_{max}S}{K_s + S} + K_D \cdot S \tag{4}$$

where K_D is a diffusion constant (cf. Neame and Richards, 1972) (fig. 1c), and corrected by the expression (fig. 1d),

$$v = V - K_D \cdot S \tag{5}$$

where v is the rate of carrier mediated uptake.

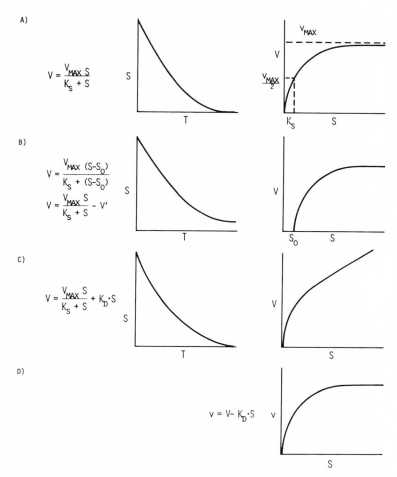

Fig. 1. Conventional and modified Michaelis-Menten kinetic equations and curves applied to nutrient uptake by symbiotic reef corals. (see text for explanations.)

Values for the various parameters of nitrate and ammonium uptake by reef corals are shown in table 2. It may be seen that virtually all the values for K_s are less than $1.0 \mu M$, suggesting that zooxanthellae behave as oligotrophic phytoplankton, exhibiting a relatively high affinity for these forms of dissolved inorganic nitrogen. D'Elia and Webb (1977, p. 329) have compared idealized uptake curves for ammonium, nitrate, and phosphate for *Pocillopora* spp. At ambient concentrations, ammonium uptake occurs at more than double the rate of nitrate uptake, consistent with the suggestion that ammonium is very likely the more significant of the several sources of dissolved inorganic nutrients.

TABLE 2

K_s AND V_{max} VALUES FOR NITRATE AND AMMONIUM UPTAKE
BY SYMBIOTIC REEF CORALS

	K_s (μM)	V_{max} (μg atom N/mg Chl a/hr)
Nitrate		
Pocillopora elegans	0.65	1.79
Fungia sp.	0.26	0.20
Acropora sp.	0.48	0.11
Ammonium		
P. damicornis	0.80	3.62
	1.05	5.26
	0.29	3.57
P. meandrina	0.39	1.64
	0.68	2.70
Porites lutea	0.55	—
	0.33	—

SOURCE: Data from Muscatine and D'Elia (1978) and Webb and Wiebe (1978).

MECHANISM OF UPTAKE OF DIN

The mechanism of uptake of nitrate by reef corals has not yet been studied in detail. By analogy with phytoplankton studies, it is presumed to involve transport, reduction to ammonium, and incorporation of ammonium into amino acids. Transport of inorganic nutrients is thought to be governed by cell surface "carrier" molecules. These putative catalysts control the rate of entry of specific nutrients and increase the intracellular concentration relative to that outside the cells so that enzyme systems can then operate more efficiently (see reviews of Dugdale, 1976). The mechanism of transport of nitrate by the diatom *Skeletonema costatum* may depend on the presence of a membrane-bound (nitrate, chloride)-activated ATPase (Falkowski, 1975a). In *S. costatum* Photosystem I generates most of the ATP required to supply energy to the carrier (Falkowski and Stone, 1975). Curiously, no (nitrate-chloride)-activated ATPase could be detected in the only dinoflagellate tested thus far (Falkowski, 1975b), so that extrapolation from diatoms to zooxanthellae may not be justified. Reduction of nitrate and nitrite to ammonium is mediated by nitrate and nitrite reductases, respectively. These enzymes have been relatively widely studied in algae and higher plants (Syrett and Leftley, 1976). Crossland and Barnes (1977) detected assimilatory nitrate reductase (NiR) activity in zooxanthellae from *Acropora acuminata* and *Goniastrea australensis*. The extraction medium required a source of reduced sulfur and NADH as an electron donor. Generally, but not always, NiR activity in phytoplankton and other plants is depressed during growth on reduced nitrogen such as ammonium, induced by nitrate in the absence of ammonium, and lost after nitrate is depleted (reviewed by Healey and Stewart, 1973). Thus, in extracts prepared from zooxanthellae, NiR activity could be induced by preincubation with nitrate, and by a suitable light period, and depressed by 10 μM ammonium. No NiR activity could be detected in algae-free coral tissue.

As with nitrate uptake, little is known of ammonium uptake mechanisms. Based on observations of uptake kinetics for phosphorus and ammonium, D'Elia (1977) and Muscatine and D'Elia (1978) suggested that, in some corals, nutrient uptake involves both carrier-mediated transport and diffusion. They hypothesized that the zooxanthellae are loci for carrier-mediated nutrient uptake and net removal of these nutrients, and that they deplete or reduce the concentration of nutrients in coral animal tissues in their vicinity. Ambient or endogenous nutrients diffuse down the resulting concentration gradient through these tissues to the sites of active transport. Since coral animal tissues produce ammonium nitrogen as an excretory product, it might be supposed that the tissues would have a high ammonium concentration and, in turn, the algae would exhibit a much higher value for K_s. In fact, levels of ammonium in coral tissue have been assayed at 5–50 μM, but such high levels are thought to result from deamination during the extraction procedure (Crossland and Barnes, 1977). Actual levels ought to be very low as a result of scavenging by the zooxanthellae. This assumption is consistent with the observed low K_s system for ammonium uptake. There are several alternative pathways for the incorporation of ammonium into amino acids in marine phytoplankton: into glutamate via NAD(P)H-dependent glutamate dehydrogenase (GDH) and into glutamine via glutamine synthetase (GS) or glutamate synthase (GOGAT) (Syrett and Leftley, 1976). Falkowski and Rivkin (1976), on the basis of enzyme activity and substrate affinities (K_mGS = 29 μM; GDH = 28 mM) in *Skeletonema costatum*, suggest that, in view of the low concentrations of environmental ammonium encountered by the organisms, GS is the major assimilatory enzyme (fig. 2). Perhaps consistent with this interpretation are the preliminary observations of Muscatine and Benson (unpublished). They incubated isolated coral zooxanthellae concurrently with $^{14}CO_2$ and increasing amounts of ammonium-N. Radiochromatograms of ethanolic extracts of the algae subsequently revealed that as the concentration of ammonium-N was increased, more ^{14}C was incorporated into ^{14}C-glutamine (fig. 3).

TRANSLOCATION AND RECYCLING OF NITROGEN

There is evidence from a wide range of studies that a substantial fraction of the carbon fixed by zooxanthellae *in situ* is translocated to host animal

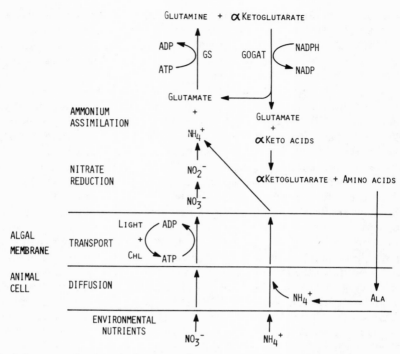

Fig. 2. Some steps in a hypothetical scheme for nutrient uptake, assimilation, and recycling in symbiotic reef corals.

tissue (Muscatine and Cernichiari, 1969; Lewis and Smith, 1971; Trench, 1971a,b,c; Muscatine et al., 1972; Trench, 1974). Such evidence comes from experiments of two types: *in vitro* and *in vivo*. In the former, zooxanthellae are isolated and incubated with $^{14}CO_2$ and the incubation medium analyzed for labeled products released selectively by the algae. In the latter, the whole organism is incubated in $^{14}CO_2$-enriched sea water and subsequently separated into algae and animal tissue constituents. Levels of ^{14}C in the animal tissue are taken as evidence for the extent of translocation. Invariably, labeled glycerol figures prominently in the released products along with lesser amounts of labeled glucose and alanine. Exactly how much total alanine is released under a given set of conditions has not yet been established, but the fact of alanine release suggests that alanine translocation is a means by which nitrogen moves from algae to host tissues.

In a series of *in vivo* "inhibition" experiments with reef corals, Lewis and Smith (1971) demonstrated that the amount of fixed ^{14}C incorporated by animal tissue as alanine was significantly increased if the corals were

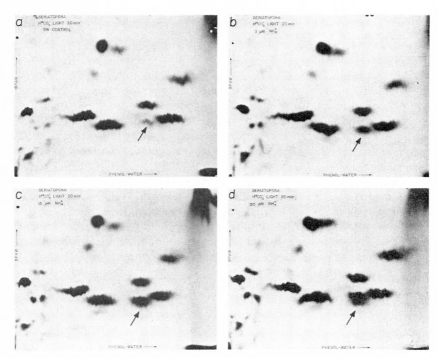

Fig. 3. Radiochromatogram of 80% ethanol extracts of zooxanthellae isolated from *Seriatopora* sp. and incubated for 20 min in light simultaneously with $^{14}CO_2$ and increasing amounts of ammonium (as NH_4Cl). (a) Control; (b) 3 μM ammonium; (c) 15 μM ammonium; and (d) 30 μM ammonium. Arrow denotes ^{14}C-glutamine in extracts.

preincubated with sea water enriched with 5 mM ammonium. These data suggest that ammonium taken up by the algae is converted to alanine, which then becomes available to the animal through translocation. Whereas the foregoing experiments offer indirect evidence for translocation of nitrogen, direct evidence for nitrogen translocation comes from experiments of Marian (1979). He demonstrated that $^{15}NO_3$-N is taken up by algae in the reef coral *Pocillopora damicornis* and accumulates in organic form in the animal tissue fraction.

With the demonstration of dissolved inorganic nitrogen uptake by algae in reef corals, and translocation of organic nitrogen to host tissues, one need only invoke conventional protein metabolism (i.e., incorporation of amino acids into protein, subsequent catabolism and release of ammonium, and uptake of ammonium by the algae) to complete a recycling circuit. Direct evidence for the latter may be forthcoming from back-translocation studies on the sea anemone *Anthopleura elegantissima*

(Marian, 1979) in which ^{15}N fed to the host anemone as labeled protein is later detected in the algal fraction.

NUTRITIONAL SIGNIFICANCE OF DIN UPTAKE

Nutritional significance of DIN uptake must be viewed in part within the context of the total nitrogen (TN) demand for a given symbiotic association and in terms of both the quantity and quality of the nitrogen required by the associants. It is thus necessary to have knowledge of total nitrogen turnover and release rates, and essential amino acid requirements (see Marian, 1979). Unfortunately few studies have addressed these specific parameters, and a detailed nitrogen budget for a reef coral remains to be established. Toward this end D'Elia and Webb (1977) measured TN and DIN flux in *Pocillopora elegans* incubated in the light for several hours. A comparison of TN and DIN uptake rates suggested some efflux of organic nitrogen (see also Johannes et al. 1969; Johannes and Webb, 1970), perhaps enough to offset DIN gains. These investigators view DIN as only a supplement to the nitrogen required by *P. elegans*, the remainder coming from zooplankton or perhaps from net uptake of dissolved organic nitrogen when the latter is available.

RELEASE OF DIN

In contrast to the foregoing observations that tropical symbiotic corals release relatively little DIN, temperate corals and both tropical and temperate sea anemones exhibit substantial efflux of DIN, mostly as ammonium nitrogen. One explanation for this difference is that the ratio of algae to animal tissue on a dry weight or protein basis in temperate sea anemones is approximately 0.003 (Taylor, 1969). This ratio is much lower than the ratio of .03–.08 observed in some tropical corals (Muscatine, McCloskey, and Marian, in preparation). Excretion by the relatively large proportion of animal tissue in anemones may simply exceed the uptake capacity of the relatively few algae and result in net release rather than net uptake. Nevertheless, the release rates are profoundly influenced by the presence of symbiotic algae, and by feeding and irradiance histories.

Szmant-Froelich and Pilson (1977) compared rates of total nitrogen excretion in fed and starved symbiotic and aposymbiotic colonies of the coral *Astrangia danae*, collected in the temperate Atlantic Ocean. Total nitrogen excretion rates of starved colonies were relatively constant, but the rates were higher for aposymbiotic than for symbiotic specimens, especially at high light intensities. Relatively more nitrogen was excreted by fed colonies than by starved ones, but rates were still higher for aposymbiotic corals, the difference being attributed to some uptake and retention by algae in the symbiotic corals. Only in certain colonies with extremely high zooxanthellae content was net uptake observed. In general, excretion rates were influenced more by the quantity of food ingested and less by the zooxanthellae content. However, the function of zooxanthellae in influencing nitrogen uptake and retention was dependent on light, and particularly on the length of a dark preincubation period.

Similarly, Cates and McLaughlin (1976) describe patterns of ammonium excretion by symbiotic and aposymbiotic anemones (*Condylactis* sp.) and jellyfish (*Cassiopeia* sp.). In both cases, in the light, much less ammonium was excreted by the symbiotic organisms than by aposymbiotic ones.

FLUX OF OTHER NUTRIENTS

Fluxes of dissolved phosphorus in reef corals have been investigated by Yamazato (1970) and D'Elia (1977). In the latter study, uptake of reactive phosphorus was demonstrated, but the rates were insufficent to offset simultaneous losses of organic phosphorus that resulted in a net loss of total phosphorus from the corals studied. Reactive phosphorus uptake was influenced by light and could be described by Michaelis-Menten kinetics modified to account for some efflux of reactive phosphorus. Presumably, uptake of reactive phosphorus serves to minimize losses from the system, a fact that may be of selective advantage in the long run in a nutrient-poor environment.

SUMMARY

In this paper I have reviewed the literature on flux of nutrients in marine algae-invertebrate associations, with emphasis on symbiotic reef corals. The factors, including the presence of symbiotic algae, that influence uptake and release of nutrients, and the presumed significance of nutrient fluxes are discussed. Although much remains to be learned about nutrient flux in symbiotic systems, it seems that ensuing frontiers of interest are likely to be in the general areas of transport mechanisms, nutrient metabolism, nutrient recycling, and the nature and sources of essential amino acids.

ACKNOWLEDGMENTS

Original research described in this paper was supported by a grant from the National Science Foundation (GB 41458) and a Biomedical Research Support Grant to the University of California at Los Angeles. I thank Dr. C. F. D'Elia and Mr. R. Marian for reviewing the manuscript.

LITERATURE CITED

Caperon, J., and J. Meyer. 1972. Nitrogen-limited growth of marine phytoplankton. II. Uptake kinetics and their role in nutrient limited growth of phytoplankton. Deep-Sea Research 19:619–32.

Cates, N., and J. J. A. McLaughlin. 1976. Differences of ammonia metabolism in symbiotic and aposymbiotic *Condylactus* and *Cassiopea* spp. J. Exp. Mar. Biol. Ecol. 21:1–5.

Crossland, C. J., and D. J. Barnes. 1976. Acetylene reduction by coral skeletons Limnol. Oceanogr. 21:153–55.

Crossland, C. J., and D. J. Barnes. 1977. Nitrate assimilation enzymes from two hard corals, *Acropora acuminata* and *Goniastrea australensis*. Comp. Biochem. Physiol. 57B:151–57.

D'Elia, C. F. 1977. The uptake and release of dissolved phosphorus by reef corals. Limnol. Oceanogr. 22:301–15.

D'Elia, C. F. 1978. Dissolved nitrogen, phosphorus and organic carbon. *In* Coral reefs: research methods, D. R. Stoddard and R. E. Johannes (eds), pp. 485–97. UNESCO Monographs on Oceanographic Methodology, No. 5.

D'Elia, C. F., and K. L. Webb. 1977. The dissolved nitrogen flux of reef corals. Proc. 3d Int. Coral Reef Symp., 1:325–30.

Dugdale, R. C. 1967. Nutrient limitation in the sea; dynamics, identification, and significance. Limnol. Oceanogr. 12:685–95.

Dugdale, R. D. 1976. Nutrient cycles. *In* D. H. Cushing and J. J. Walsh (eds.), The ecology of the seas, pp. 141–72. Blackwell, Oxford.

Dugdale, R. C., and J. J. Goering. 1967. Uptake of new and regenerated forms of nitrogen in primary productivity. Limnol. Oceanogr. 12:196–206.

Falkowski, P. G. 1975a. Nitrate uptake in marine phytoplankton: (nitrate, chloride)-activated adenosine triphosphatase from *Skeletonema costatum* (Bacillariophyceae). J. Phycol. 11:323–26.

Falkowski, P. G. 1975b. Nitrate uptake in marine phytoplankton: comparison of half-saturation constants from seven species. Limnol. Oceanogr. 20:412–17.

Falkowski, P. G., and R. B. Rivkin. 1976. The role of glutamine synthetase (EC-6.3.1.2.) in the incorporation of ammonium in *Skeletonema costatum* (Bacillariophyceae.) J. Phycol. 12:448–50.

Falkowski, P. G., and D. P. Stone. 1975. Nitrate uptake in marine phytoplankton: energy sources and the interaction with carbon fixation. Mar. Biol. 32:77–84.

Franzisket, L. 1973. Uptake and accumulation of nitrate and nitrite by reef corals. Naturwiss. 12:552.

Franzisket, L. 1974. Nitrate uptake by reef corals. Int. Revue ges. Hydrobiol. 59:1–7.

Healey, F. P., and W. P. D. Stewart. 1973. Inorganic nutrient uptake and deficiency in algae. CRC Critical reviews in Microbiology 3:69–113.

Johannes, R. E. 1974. Sources of nutritional energy for reef corals. Proc. 2d Int. Coral Reef Symp. 1:133–37.

Johannes, R. E., S. L. Coles, and N. T. Kuenzel. 190. The role of zooplankton in the nutrition of some scleractinian corals. Limnol. Oceanogr. 15:579–86.

Johannes, R. E., S. J. Coward, and K. L. Webb. 1969. Are dissolved amino acids an energy source for marine invertebrates? Comp. Biochem. Physiol. 29:283–88.

Johannes, R. E., and K. L. Webb. 1970. Release of dissolved organic compounds by marine and freshwater invertebrates. *In* D. W. Hood (ed.), Organic matter in natural waters, pp. 257–73. University of Alaska Press, College, Alaska.

Kawaguti, S. 1953. Ammonium metabolism of the reef corals. Biol. Jour. Okayama Univ. 1:171–76.

Lewis, D. H., and D. C. Smith. 1971. The autotrophic nutrition of symbiotic marine coelenterates with special reference to hermatypic corals. I. Movement of photosynthetic products between the symbionts. Proc. R. Soc. Lond. B. 178:111–29.

LIMER 1975 Expedition Team. 1976. Metabolic processes of coral reef communities at Lizard Island, Queensland. Search 7:463–68.

MacIsaac, J. J., and R. C. Dugdale. 1972. Interactions of light and inorganic nitrogen in controlling nitrogen uptake in the sea. Deep Sea Res. 19:209–32.

Marian, R. 1979. Aspects of nitrogen metabolism in marine algal-invertebrate symbiotic associations. Ph.D. diss. University of California, Los Angeles (in preparation).

Muscatine, L., and E. Cernichiari. 1969. Assimilation of photosynthetic products of zooxanthellae by a reef coral. Biol. Bull. 137:506–23.

Muscatine, L., and C. F. D'Elia 1978. The uptake, retention, and release of ammonium by reef corals. Limnol. Oceanogr. 23:725–34.

Muscatine, L., R. R. Pool, and E. Cernichiari. 1972. Some factors influencing selective release of soluble organic material by zooxanthellae from reef corals. Mar. Biol. 13:298–308.

Muscatine, L., and J. W. Porter. 1977. Reef corals.: mutualistic symbioses adapted to nutrient-poor environments. Bioscience 27:454–59.

Muscatine, L., H. Masuda, and R. Burnap. 1979. Ammonium uptake by symbiotic and aposymbiotic reef corals. Bull. Mar. Sci. 29 (in press).

Neame, K. D., and T. G. Richards. 1972. Elementary kinetics of membrane carrier transport. Blackwell, Oxford. 120 pp.

Porter, J. W. 1974. Zooplankton feeding by the Caribbean reef-building coral *Montastrea cavernosa*. Proc. 2d Int. Coral Reef Symp. 1:111–25.

Syrett, P. J., and J. W. Leftley. 1976. Nitrate and urea assimilation by algae. *In* Perspectives in experimental biology, vol. 2, N. Sunderland (ed.), Botany, pp. 221–34. Pergamon Press, Oxford.

Szmant-Froelich, A., and M. E. Q. Pilson. 1977. Nitrogen excretion by colonies of the temperate coral *Astrangia danae* with and without zooxanthellae. Proc. 3d Int. Coral Reef Symp. 1:417–23.

Taylor, D. L. 1969. On the regulation and maintenance of algal numbers in zooxanthellae-coelenterate symbiosis, with a note on the nutrional relationship in *Anemonia sulcata*. J. Mar. Biol. Ass., U.K. 49:1057–65.

Thomas, W. H. 1970. On nitrogen deficiency in tropical Pacific Ocean phytoplankton: photosynthetic parameters in rich and poor water. Limnol. Oceanogr. 15:380–85.

Trench, R. K. 1971a. The physiology and biochemistry of zooxanthellae symbiotic with marine coelenterates. I. The assimilation of photosynthetic products of zooxanthellae by two marine coelenterates. Proc. R. Soc. Lond. B. 177:225–35.

Trench, R. K. 1971b. The physiology and biochemistry of zooxanthellae symbiotic with marine coelenterates. II. Liberation of fixed ^{14}C by zooxanthellae *in vitro*. Proc. R. Soc. Lond. B. 177:237–50.

Trench, R. K. 1971c. The physiology and biochemistry of zooxanthellae symbiotic with marine coelenterates. III. The effect of homogenates of host tissues of the excretion of photosynthetic products *in vitro* by zooxanthellae from two marine coelenterates. Proc. R. Soc. Lond. B. 177:251–64.

Trench, R. K. 1974. Nutritional potentials in *Zoanthus sociatus* (Coelenterata, Anthozoa). Helgolander wiss. Meeresunters. 26:174–216.

Webb, K. L., and W. J. Wiebe. 1978. The kinetics and possible significance of nitrate uptake by several algal-invertebrate symbioses. Mar. Biol. 47:21–27.

Webb, K. L., W. D. DuPaul, W. Wiebe, W. Sottile, and R. E. Johannes. (1975) Enewetak (Eniwetok) Atoll: aspects of the nitrogen cycle on a coral reef. Limnol. Oceanogr. 20:198–210.

KWANG W. JEON

Symbiosis of Bacteria with *Amoeba*

11

INTRODUCTION

Like many other free-living organisms, amoebae are known to harbor various parasitic or symbiotic microorganisms within their cytoplasm (Leiner et al., 1954; Kudo, 1957; Roth and Daniels, 1961; Chapman-Andresen, 1962, 1971; Rabinovitch and Plaut, 1962a, b; Drozanski, 1963; Daniels et al., 1966). In most instances neither the symbiotic history of the infective organisms nor the physiological and metabolic relationship between the hosts and symbionts is known. Meanwhile, it is believed that symbionts play an important role in the hosts' metabolism (Cummins and Plaut, 1964) and hereditary processes (Wolstenholme, 1966). In particular, the giant amoebae, *Pelomyxa palustris*, have bacterial symbionts as constant cell components (Kudo, 1957; Daniels et al., 1966), while lacking mitochondria (Andresen et al., 1965; Daniels and Breyer, 1965). A putative role for these symbionts as filling the normal functions of mitochondria has been assigned (Daniels et al., 1966).

Recently, there has been a series of studies on a unique instance of symbiosis between an unidentified strain of rod-shaped bacteria and a strain of *Amoeba proteus*. The symbiotic relationship was established under the laboratory conditions (i.e., with known history), and the symbionts were integrated as required cell components within 5 years (fewer than 1,000 host cell generations). In this article the results of these studies will be reviewed.

SYMBIOSIS BETWEEN X-BACTERIA AND AMOEBA

The D strain of *A. proteus* was first noted to be infected with a large number of rod-shaped bacteria (60,000–150,000 bacteria per amoeba) in 1966 (Jeon and Lorch, 1967). These amoebae had been maintained in mass culture with *Tetrahymena* as food organisms at the time of initial observation. At first the bacterial infection had an adverse effect on the host amoebae; the amoebae exhibited various unhealthy symptoms such as reduced growth, poor clone formation (when grown singly), extreme sensitivity to starvation, reduced size, and increased fragility. The infective bacteria were called "X" bacteria, since their origin and indentity were unknown, and the infected amoebae were designated as the xD strain, in contrast to the original D strain (cf. Jeon and Lorch, 1973). The original D amoebae kept in stock cultures according to the method of Lorch and Danielli (1953) remained uninfected. Apparently, the xD amoebae arose as a result of spontaneous infection during the laboratory culture.

The X-bacterial infection is different from other reported cases of infection in two ways. First, the time of its initial establishment is known. Second, the number of infective bacteria per amoeba is unusually large. For example, Rabinovitch and Plaut (1962b) found the average number of DNA-containing bodies per amoeba to be between 5,400 and 11,000. Chapman-Andresen and Hayward (1963) counted 3,000–4,000 bacteria per amoeba from the strain most heavily infected among several strains examined.

At the time of its discovery, the X-bacteria's dependence on the host amoebae appeared to be well established, since it was not possible to culture isolated X-bacteria *in vitro*. Interestingly, the isolated X-bacteria remain viable for several hours outside the host cytoplasm, as evidenced by their ability to infect D amoebae when introduced either by intracellular injection or by exposure in suspension (Jeon and Lorch, 1967; Ahn and Jeon, 1979).

Preliminary attempts to "cure" the infected xD amoebae with antibiotics failed. Due to their fragility and sensitivity to nutritional conditions, extreme care was needed initially to maintain cultures of infected amoebae. However, with time, the adverse effects of infection gradually diminished and the infected amoebae grew better. The growth rates of infected amoebae continued to improve, such that after another few years of culture infected amoebae grew at rates comparable to those of uninfected cells under normal conditions. In addition, the infected amoebae became dependent on their endosymbionts for survival, as shown by the results of micrurgical studies (Jeon, 1972; Jeon and Jeon, 1976). Thus, the bacterial endosymbionts that started out as harmful parasites became required cytoplasmic components within a few years.

The symbiotic X-bacteria are Gram-negative, about 0.5×2 μm in size, and are enclosed in vesicles of varying sizes (fig. 1). At present their average number per amoeba is about 42,000 (Ahn and Jeon, 1979). Small vesicles normally contain only the symbionts. When many bacteria are within the same vesicle, the interbacterial space is filled with membrane-bounded vesicles containing fibrous material of unknown nature and function (fig. 2).

The ultrastructure of individual symbionts (fig. 3) is similar to that of symbiotic bacteria present in other cells (e.g., Chang, 1974) or that of *Escherichia coli* (e.g., Morgan et al., 1967). Symbionts are sensitive to antibacterial agents such as chloramphenicol or trimethoprim. For example, when xD amoebae are grown in the presence of chloramphenicol at concentrations of 500 μg/ml or higher, the average number of X-bacteria per amoeba decreases to about 30% within two weeks, and to below 10% of normal by the fourth week (Jeon and Hah, 1977). After two weeks in chloramphenicol, individual bacteria become deformed and exhibit an expansion of the nuclear zones (fig. 4). Trimethoprim also reduces the number of symbionts per amoeba to the same extent at a lower concentration (80 μg/ml; Jeon, 1977), although its effect on the ultrastructure of symbionts has not been studied. The effects of trimethoprim, an inhibitor of dihydrofolate reductase, can be related to the fact that the X-bacteria do not utilize exogenous thymine during DNA synthesis, apparently synthesizing this DNA precursor *de novo* (J. Freesh and Jeon, unpublished data).

EXPERIMENTAL INFECTION

Noninfected D amoebae are easily infected with X-bacteria following intracellular injection or after exposure to bacteria in suspension. In a series of experiments (Ahn and Jeon, 1979), D amoebae were forced to phagocytose in different concentrations of X-bacteria, using hydra extract as a phagocytosis inducer (see Jeon and Bell, 1965). Following 1.5 hrs of phagocytosis in a suspension of X-bacteria at concentrations of $3 \times 10^8/$ ml or higher, 100% of the amoebae were found to be infected with X-bacteria (fig. 5). During a 1.5 hr-period of induced phagocytosis in a bacterial suspension of $10^9/$ml (three times the minimal level needed to achieve 100% infection), each amoeba ingested an average of 300 X-bacteria, as determined by direct counts using a Petroff-Hausser bacteria counter. Assuming that amoebae take in the same volume of solution by

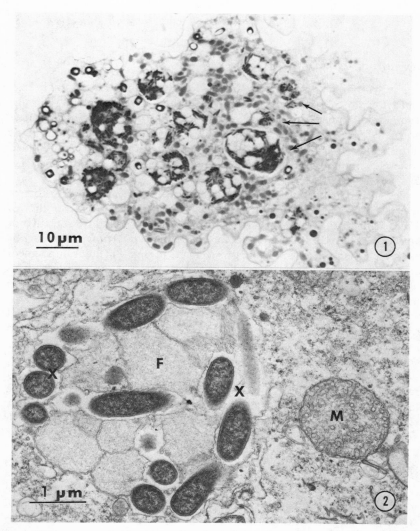

Fig. 1. Photomicrograph of a thin section of an xD amoeba to show symbiont-containing vesicles of various sizes (arrows). The total area occupied by the vesicles is about 15% of the total area of the amoeba section. Reproduced from Jeon and Lorch (1967) by permission.

Fig. 2. Electron micrograph of a vesicle containing many X-bacteria (X). Note the fibrous matter (F) contained in irregular membrane-bound vesicles filling the interbacterial space.

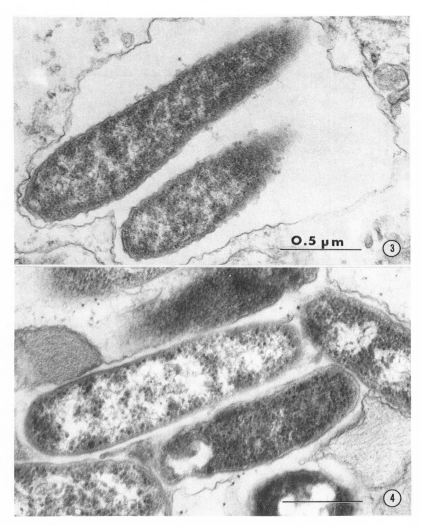

Fig. 3. Electron micrograph of a thin section of X-bacteria in a normal xD amoeba. Note the plasma membrane, electron-dense cytoplasm, and electron-lucent nuclear zones. Reproduced from Jeon and Hah (1977) by permission.

Fig. 4. Electron micrograph of a thin section of X-bacteria in an xD amoeba grown for 2 weeks in the presence of 0.5 mg/ml chloramphenicol. Note the expanded nuclear zones. Reproduced from Jeon and Hah (1977) by permission.

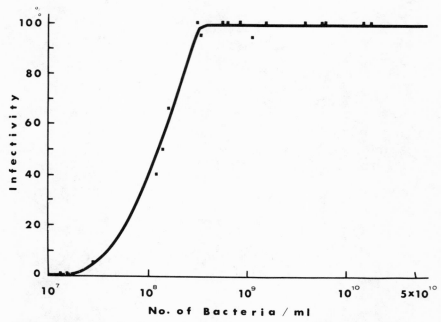

Fig. 5. Graph to show the effect of the concentration of X-bacteria on their infectivity when introduced into D amoebae by induced phagocytosis. All exposed amoebae become infected when exposed to bacterial suspensions of 3×10^8 or higher for 1.5 hrs. Reproduced from Ahn and Jeon (1979) by permission.

phagocytosis, at a bacterial concentration of 10^8/ml, each amoeba should ingest an average of 30 X-bacteria. Yet, only 40% of exposed D amoebae actually became infected (fig. 5), indicating that not all the ingested X-bacteria are able to establish symbiosis in amoebae. Consequently, the 10^9/ml X-bacteria suspension has been employed routinely in later experiments to ensure maximum infectivity.

Isolated X-bacteria retain their infectivity for over 72 hrs at 20°, over 7 days at 4°, and for many months at minus 20° C (Ahn and Jeon, 1979). However, since the X-bacteria suspension normally becomes contaminated with other bacteria at 4° or 20°C, it is best to use freshly isolated bacteria for infection experiments.

The development of bacteria-containing vesicles has been studied after induced infection by light and electron microscopy. If amoebae are fixed immediately after infection, X-bacteria are found inside phagocytic vacuoles together with other ingested material (fig. 6A). After two days, the X-bacteria are localized in separate vesicles, which can be detected either under a phase-contrast microscope or in the electron microscope (fig. 6B). The interbacterial space within the vesicle becomes filled with vesicles containing fibrous material as the number of X-bacteria increases (fig. 6C).

The symbiotic bacteria are unique in that they are not digested by amoebae during initial infection, whereas other ingested bacteria or symbiotic algae from green hydra are invariably digested by amoebae (fig. 7; see also Savanat and Pavillard, 1964; Casley-Smith and Savanat, 1966; Jeon and Lorch, 1970; Jeon and Jeon, 1976). In addition, isolated X-bacteria are not lysed by enzymes commonly used for lysing bacteria, such as lysozyme, various proteases, lipases, and amylases (Han and Jeon, 1979). Detergents such as Triton X-100 and Brij are also ineffective. Furthermore, the cell walls of X-bacteria appear to be particularly tough, and they can be broken only by high-pressure treatments or by high-intensity ultrasonication.

Endosymbionts or intracellular parasites are believed to elude digestion by their host cells by preventing in some way the fusion of lysosomes with vesicles containing the symbionts (Armstrong and Hart, 1971; Jones and Hirsch, 1972). Indeed, vesicles containing established X-bacteria have not been found to fuse with the amoeba's lysosomes (Jeon and Jeon, 1976; T. I. Ahn and Jeon, unpublished data).

However, during experimental infection by induced phagocytosis, lysosomes do fuse freely with phagocytic vacuoles containing X-bacteria (fig. 6A). Some of these X-bacteria are later found in separate vesicles where they multiply. Therefore, prevention of fusion between lysosomes and phagocytic vacuoles cannot be the primary protective mechanism for X-bacteria during initial infection. It is not possible to determine if X-bacteria are not digested due to their inherent resistance to the digestive action of the amoeba's enzymes or due to the amoeba's inability to recognize them as foreign organisms.

In the meantime, it is of interest to note that X-bacteria contain two plasmid DNAs that can be detected after phenol extraction, CsCl banding and gel electrophoresis. One plasmid band migrates more slowly than those representing chromosomal DNAs in 0.7% agarose gels (Han and Jeon, 1979). It is possible that these plasmids are related to the bacteria's resistance to enzymatic digestion and the rigidity of their cell walls.

Once established within a host, the infecting bacteria multiply exponentially until the maximum carrying capacity is reached in about three weeks. In several series of experiments, the number of symbionts has been observed to increase in two phases; an initial rapid phase lasting 15 days, followed by a slower rate of increase until the maximum number is reached (fig. 8). During these first 15 days, each newly infected amoeba divides an average of 5.88 times to produce 59 cells. Since each amoeba

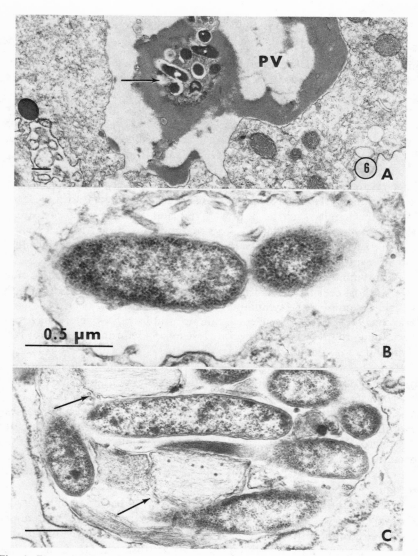

Fig. 6. Electron micrographs of thin sections of newly infected amoebae to show the subsequent development of symbiont-containing vesicles. The bars represent 0.5 μm in all figures. *A*. A phagocytic vacuole containing ingested X-bacteria (arrow) and condensed hydra material, both surrounded by lysosomal content. The amoeba was fixed immediately after induced infection. *B*. A small vesicle containing segregated X-bacteria. The amoeba was fixed two days after infection. *C*. A larger vesicle containing many X-bacteria and fibrous material (arrow) that fills the interbacterial space. The amoeba was fixed 9 days after infection. Reproduced from Ahn and Jeon (1979) by permission.

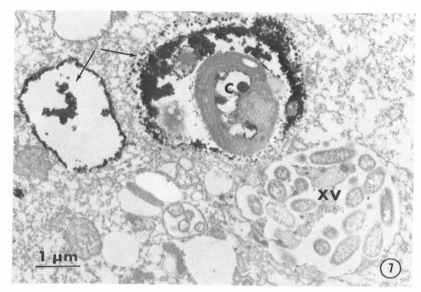

Fig. 7. Electron micrograph of a thin section of an xD amoeba fixed 16 hours after it had been induced to phagocytose symbiotic algae isolated from *Hydra viridis*, and treated for acid phosphatase activity before embedding in resin. Note the enzyme reaction product on and around the ingested chlorella (C). The nearby vesicle (XV) containing X-bacteria shows no reaction products. Reproduced from Jeon and Jeon (1976) by permission.

contains an average of 26,000 X-bacteria on the fifteenth day, the number of bacteria produced is 1.53×10^6. Assuming that each amoeba had ingested 300 X-bacteria during infection and that all the ingested bacteria had survived, then each bacterium would have divided an average of 12.3 times during this 15-day period; i.e., a rate twice that of the host cells. If only a portion of the ingested X-bacteria had survived, the surviving bacteria would have had to divide at an even greater frequency. After the maximum carrying number of bacteria (about 42,000 per amoeba) is reached three weeks after the initial infection, the growth rate of X-bacteria slows down and keeps pace with that of the host amoebae.

X-BACTERIA AS REQUIRED CYTOPLASMIC COMPONENTS IN xD AMOEBAE

Several years after the initial bacterial infection was noted, nuclei were transferred between xD and D amoebae to see if, as a result of bacterial infection, an incompatibility had developed between the two strains of amoebae (Jeon, 1972). The nuclei of D amoebae were viable in the cytoplasm of xD amoebae, whereas xD amoeba nuclei were unable to form

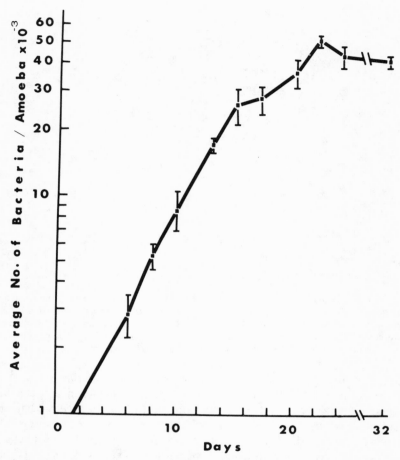

Fig. 8. Graph to show the increase in the average number of X-bacteria per newly infected D amoeba. The vertical bars represent standard deviation. The mean doubling time of the average number is 2.85 days for the first 15 days, and 3.94 days for the first 22 days. Reproduced from Ahn and Jeon (1979) by permission.

clones when transplanted into D amoeba cytoplasm, unless some of the xD cytoplasm was also transferred simultaneously or within a few days after nuclear transfer. Further micrurgical studies involving nuclear tansplantation and transfer of whole or fractionated xD cytoplasm revealed that live X-bacteria were needed for the nuclei of xD amoebae to function normally in the D amoeba cytoplasm (Jeon and Jeon, 1976). Thus, the X-bacteria had become a required cytoplasmic component within a few years after establishing symbiosis with amoebae.

The dependence of host amoebae on the symbionts for survival has also been shown by the death of xD amoebae following selective removal of X-bacteria with antibacterial drugs such as chloramphenicol (Jeon and Hah, 1977) or trimethoprim (Jeon, 1977). Recently, another way has been found to demonstrate that the survival of xD amoebae requires X-bacteria (Jeon and Ahn, 1978). Due to the psychrophile-like charcteristics of X-bacteria (they cannot grow at elevated temperatures; Morita, 1975), X-bacteria disappear from xD amoebae when the latter are cultured at or above 26°C (fig. 9). Following the disappearance of X-bacteria, xD amoebae stop growing and die within a few days (fig. 10). Thus, the dependence of the xD-amoebae on their symbionts has led to the acquisition of a new cell character: temperature sensitivity.

Using this trait as a criterion for dependence, an attempt was made to determine how long it takes for newly infected amoebae to become dependent on their symbionts. For this purpose, 5 groups of xD amoebae with different lengths of symbiotic history (one week to ten years) were grown at 26°C (table 1). Over 90% of amoebae that had been in symbiosis for longer than 18 months were unable to form clones when grown singly. Amoebae that had been in symbiosis for shorter periods of time were able to form clones to varying degrees. Thus, it appears that amoebae may develop dependence on their newly acquired endosymbionts in fewer than 200 cell generations (within 18 months) at 20°C.

DISCUSSION

The symbiotic relationship between X-bacteria and amoebae is unique in that its establishment took but a few years, thereby permitting a continual observation of how the symbiosis became established. The host's dependence on the symbionts is also unique, in contrast to other instances of cellular symbiosis studied where endosymbionts are not required for the host's survival (Karakashian, 1963; Muscatine and Lenhoff, 1965a; Chang, 1975). Unfortunately, the reasons why xD amoebae are dependent on their endosymbionts are unknown. Despite the host's inability to survive in the absence of X-bacteria, there is no clear evidence that the symbionts benefit the host. Surprisingly, the symbiotic association bestows obvious disadvantages on the hosts, such as loss in capacity to grow at temperatures above 26°C and a decreased resistance to starvation. This is not to say that no advantages exist, but they are as yet undetected. There are examples where endosymbionts confer specific advantages to the host under certain conditions, such as deficient food supply (Karakashian, 1963; Muscatine and Lenhoff, 1965b).

The precise mechanism protecting X-bacteria against digestion by

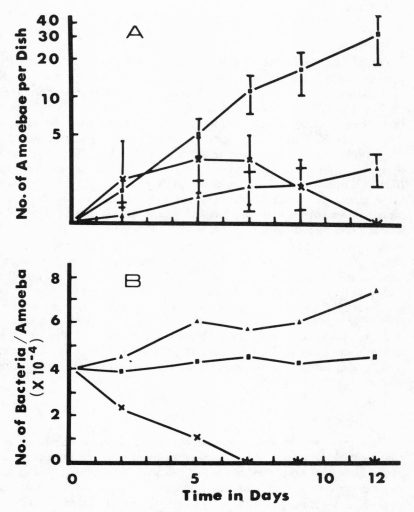

Fig. 9. Graphs to show the effects of different temperatures on growth of xD amoebae (A) and the average number of X-bacteria per amoeba (B). *A*. Growth rate of established xD amoebae at 13° (— ▲ —), 20° (— ■ —), and 27°C (— × —). During the 12-day culture period, the mean generation times of cells grown at 13° and 20° C were 8.1 and 2.3 days, respectively. Cells grown at 27°C died out completely by the 12th day. The vertical bars represent standard deviation. *B*. Changes in the average number of X-bacteria in amoebae grown at different temperatures (the symbols are the same as in A.) The average number of X-bacteria per amoeba kept at 20°C remained 4,250 ± 2,920 (S.D.) during the culture period. Reproduced from Jeon and Ahn (1978) by permission. Copyright 1978 by the American Association for the Advancement of Science.

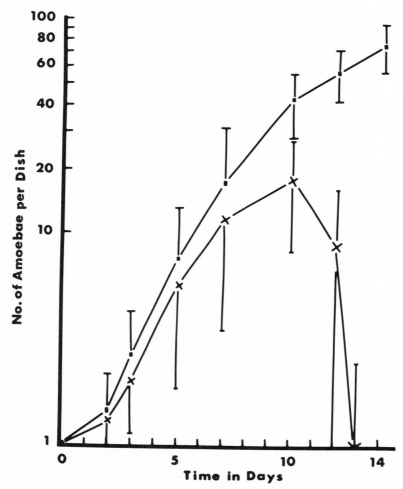

Fig. 10. Growth curves of D (— ■ —) and xD (— X —) amoebae at 26°C. All xD amoebae cytolyzed by the 13th day. The vertical bars represent standard deviation. Reproduced from Jeon and Ahn (1978) by permission. Copyright© 1978 by the American Association for the Advancement of Science.

amoebae is unknown. The presence of two plasmid groups in the endosymbionts holds promises as a useful key for the elucidation of such a mechanism. Work is in progress on the isolation and characterization of the plasmids (Han and Jeon, 1979), and to determine if they are involved in the stabilization of the host-symbiont relationship.

TABLE 1

GROWTH PATTERN OF AMOEBAE AT 26°C THAT HAVE BEEN IN SYMBIOSIS FOR
VARYING PERIODS OF TIME

Amoebae	Length of time in symbiosis	MGT at 26°C* in days	Percent of cells failing to form clones (No.)†
D	Nonsymbiotic	2.08	5.3% (4/76)
xD_1	One week	2.03	4.2 (1/24)
xD_2	6 months	1.82	18.2 (6/33)
xD_3	12 months	1.86	34.2 (13/38)
xD_4	18 months	2.22	90.4 (47/52)
xD_5	Over 10 years	2.47	100 (116/116)

SOURCE: From Jeon and Ahn, 1978.
* Mean generation time during the first 10-day culture period.
† The numbers represent cells cultured in 2 to 5 separate experiments at different times.

What makes the amoeba-bacteria symbiosis especially useful in the study of host-symbiont relationships is the presence of another kind of endosymbionts in the hosts, the so-called DNA-containing bodies (fig. 11; cf. Rabonivitch and Plaut, 1962a,b; Wolstenholme and Plaut, 1964; Wolstenholme, 1966). Although the DNA-bodies are bacterial in origin, they have not retained many of the properties characterizing free-living bacteria. Thus, they differ from X-bacteria with regard to several traits (table 2). For example, the DNA-bodies utilize exogenous thymine and disintegrate in 2N NaOH. Also, DNA-bodies are specifically dependent on the host's nuclear genome, and degenerate when placed under the influence of heterologous nuclei (fig. 12). In this respect, they are like mitochondria. In contrast, X-bacteria do not have specific relationship to the host's nuclear genome, evidenced by their ability to infect and to establish symbiosis with different strains of amoebae (Jeon and Jeon, 1976).

The symbiotic history of the DNA-bodies is unknown, but it is conceivable that their introduction into amoebae had a similar history to that of X-bacteria. Thus, DNA-bodies may be considered as a forerunner for the future fate of the newly established X-bacterial symbionts. Further studies on the symbiotic relations of these two endosymbionts might give useful insight into the mechanism of partner integration (in symbiosis) and to the possible origin of eukaryotic cell organelles (cf. Preer et al., 1974; Taylor, 1974; Margulis, 1976).

SUMMARY

Bacterial endosymbionts that spontaneously infected a strain of *Amoeba proteus* as harmful parasites 10 years ago have been integrated as

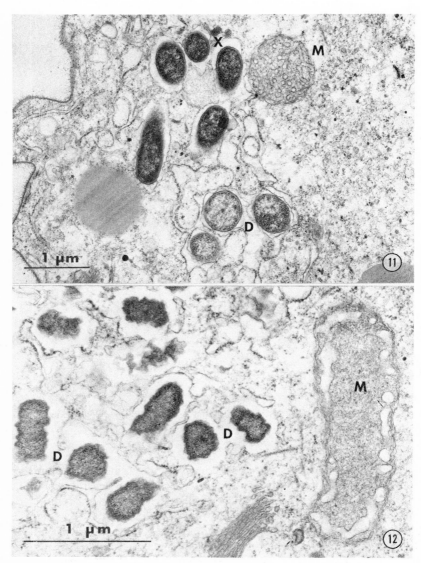

Fig. 11. Electron micrograph of a thin section of an xD amoeba to show the two kinds of endosymbionts; DNA-containing bodies (D) and X-bacteria (X). The DNA-bodies are round and individually enclosed in vesicles, and clusters of DNA-bodies are surrounded by another membrane that often has ribosome-like particles attached to it. The X-bacteria are rod-shaped, and have no membrane surrounding them individually.

Fig. 12. Electron micrograph of a thin section of an xD amoeba, the nucleus of which had been replaced with that from an amoeba of heterologous strain 10 days prior to fixation. Note the abnormal mitochondrion (M; compare with that in fig. 11) and the disfigured DNA-bodies (D; compare with those in fig. 11).

TABLE 2

COMPARISON OF KNOWN CHARACTERISTICS OF
THREE CYTOPLASMIC INCLUSIONS IN THE xD STRAIN OF A. PROTEUS

CHARACTERISTICS	COMPONENT		
	Mitochondria	DNA-bodies	xD-Bacteria
Shape	Round or rod-shaped[1]	Round[2]	Rod-shaped[3]
Size	~1 μm in diameter[1]	~0.5 μm in diameter[2]	~0.3 × 2 μm[3]
Limiting membranes	Double unit membrane[1]	One unit membrane plus surrounding vesicular membrane[4]	One-unit membrane plus surrounding vesicular membrane shared with others[3]
Inner cristae	Present[1]	Absent[4]	Absent[3]
Solubility of membranes in 2N NaOH	Yes[5]	Yes[5]	No[5]
Incorporation of exogenous thymine	Yes	Yes[1]	No[5]
Effect of chloramphenicol or trimethoprim	Reversible damage[3]	Reversible damage[3]	Irreversible damage[3]
Specific dependence on nuclear genome	Yes[7]	Yes[7]	No[7]
Required for the cell	Yes.	Unknown	Yes[8]
Remarks	True organelle	Quasi-organelle	Newly acquired cytoplasmic component

REFERENCES: 1. Flickinger, 1974; 2. Rabinovitch and Plaut, 1962a; 3. Jeon and Hah, 1977; 4. Wolstenholme and Plaut, 1964; 5. Freesh and Jeon, unpublished data; 6. Jeon, 1977; 7. Jeon, 1975; 8. Jeon and Jeon, 1976.

required cell components. As a consequence, the symbiotic amoebae lose viability when their symbionts are removed in one of several ways. The symbiotic bacteria can infect and establish symbiosis with noninfected amoebae of the originally same strain or amoebae of a different strain. Such newly infected amoebae may become dependent on the symbionts in fewer than 200 cell generations. The amoeba-bacteria symbiosis may serve as a model system in the study of partner integration at the subcellular level and the possible origin of eukaryotic cell organelles.

ACKNOWLEDGMENTS

I thank Dr. A. M. Jungreis for his critical reading of the manuscript. The original research work described in this article was supported by grants from the National Institute of General Medical Sciences, U. S. Public Health Service, and the National Science Foundation.

LITERATURE CITED

Ahn, T. I., and K. W. Jeon. 1979. Growth and electron microscopic studies on an experimentally established bacterial endosymbiosis in amoebae. J. Cell. Physiol. 98:49–58.

Andresen, N., C. Chapman-Andresen, and J. R. Nilsson. 1965. The fine structure of *Pelomyxa palustris*. Progr. Protozool. Excerpta. Med. Fndn., Intern. Congr. Ser. 91:258.

Armstrong, J. A., and P. D. Hart. 1971. Response of cultured macrophages to *Mycobacterium tuberculosis*, with observations on fusion of lysosomes with phagosomes. J. Exp. Med. 134:713–40.

Casley-Smith, J. R., and T. Savanat. 1966. The formation of membranes around microorganisms and particles injected into amoebae: support for the reticulosome concept. Aust. J. Exp. Biol. Med. Sci. 44:111–22.

Chang, K.-P. 1974. Ultrastructure of symbiotic bacteria in normal and antibiotic-treated *Blastocrithidia culicis* and *Crithidia onocopelti*. J. Protozool. 21: 699–707.

Chang, K.-P. 1975. Reduced growth of *Blastocrithidia culis* and *Crithidia oncopelti* freed of intracellular symbiotes by chloramphenicol. J. Protozool. 22:271–76.

Chapman-Andresen, C. 1962. Studies on pinocytosis in amoebae. Compt. rend. Lab. Carlsberg. 33:73–264.

Chapman-Andresen, C. 1971. Biology of the large amoebae. Ann. Rev. Microbiol. 25:27–48.

Chapman-Andresen, C., and A. F. Hayward. 1963. Bacterial complexes in *Amoeba proteus*. Presented at Soc. Exp. Biol. Mtg., Oxford.

Cummins, J. E., and W. Plaut. 1964. The distribution of ^{32}P in ribonucleic acid from nucleate and anucleate half cells of *Amoeba proteus*. Biochim. Biophys. Acta. 80:19–30.

Daniels, E. W., and E. P. Breyer. 1965. Fine structure of the giant, algae-eating amoeba, *Pelomyxa palustris*. ANL Annu. Rpt. 7136:216–20.

Daniels, E. W., E. P. Breyer, and R. R. Kudo. 1966. *Pelomyxa palustris*, Greeff. II. Its ultrastructure. Zeit. Zellforsch. 73:367–83.

Drozanski, W. 1963. Observations on intracellular infection of amoebae by bacteria. Acta Mirobiol. Pol. 12:9–24.

Flickinger, C. J. 1974. The fine structure of four species of *Amoeba*. J. Protozool. 21:59–68.

Han, J. H., and K. W. Jeon. 1979. Plasmids of bacterial endosymbionts of *Amoeba proteus*. (In preparation.)

Jeon, K. W. 1972. Development of cellular dependence on infective organisms: Micrurgical studies in amoebas. Science 176:1122–23.

Jeon, K. W. 1975. Selective effects of enucleation and transfer of heterologous nuclei on cytoplasmic organelles in *Amoeba proteus*. J. Protozool. 22:402–5.

Jeon, K. W. 1977. Further evidence for the host cell's dependence on newly acquired cytoplasmic components in amoebae. 5th Intern. Congr. Protozool., p. 445.

Jeon, K. W., and T. I. Ahn. 1978. Temperature sensitivity: a cell character determined by obligate endosymbionts in amoebae. Science 202:635–37.

Jeon, K. W., and L. G. E. Bell. 1965. Chemotaxis in large, free-living amoebae. Exp. Cell Res. 38:536–55.

Jeon, K. W., and J. C. Hah. 1977. Effect of chloramphenicol on bacterial endosmybiotes in a strain of *Amoeba proteus*. J. Protozool. 24:289–93.

Jeon, K. W., and M. S. Jeon, 1976. Endosymbiosis in amoebae: recently established endosymbionts have become required cytoplasmic components. J. Cell. Physiol. 89:337–44.

Jeon, K. W., and I. J. Lorch. 1967. Unusual intracellular bacterial infection in large, free-living amoebae. Exp. Cell Res. 48:236–40.

Jeon, K. W., and I. J. Lorch. 1970. Strain-specific mitotic inhibition in large mononucleate amoebae. J. Cell. Physiol. 75:193–98.

Jeon, K. W. and I. J. Lorch. 1973. Strain specificity in *Amoeba proteus. In* K. W. Jeon (ed.), The biology of Amoeba, pp. 549–67. Academic Press, New York.

Jones, T. C., and J. G. Hirsch. 1972. The interaction between *Toxoplasma gondii* and mammalian cells. II. The absence of lysosomal fusion with phagocytic vacuoles containing living parasites. J. Exp. Med. 136:1173–94.

Kudo, R. R. 1957. *Pelomyxa palustris* Greeff. I. Cultivation and general observations. J. Protozool. 4:154–64.

Karakashian, S. J. 1963. Growth of *Paramecium bursaria* as influenced by the presence of algal symbionts. Physiol. Zool. 36:52–68.

Leiner, M., M. Wohlfeil, and D. Schmidt. 1954. *Pelomyxa palustris* Greeff. Ann. Sci. Nat. Zoologie. Ser. II, 16:537–94.

Lorch, I. J., and J. F. Danielli. 1953. Nuclear transplantation in amoebae. I. Some species characters of *Amoeba proteus* and *Amoeba discoides.* Quart. J. Micros. Sci. 94:445–60.

Margulis, L. 1976. Genetic and evolutionary consequences of symbiosis. Exp. Parasitol. 39:277–349.

Morgan, C., H. S. Rosenkranz, and H. S. Carr. 1967. Electron microscopy of chloramphenicol-treated *Eschericichia coli.* J. Bacterol. 93:1987–2002.

Morita, R. Y. 1975. Psychrophilic bacteria. Bacteriol. Rev. 39:144–67.

Muscatine, L., and H. M. Lenhoff. 1965a. Symbiosis of hydra and algae. I. Effects of some environmental cations on growth of symbiotic and aposymbiotic hydra. Biol. Bull. 128:415–24.

Muscatine, L., and H. M. Lenhoff. 1965b. Smybiosis of hydra and algae. II. Effects of limited food and starvation on growth of symbiotic and aposymbiotic hydra. Biol. Bull. 129:316–28.

Preer, J. R., L. B. Preer, and A. Jurand 1974. Kappa and other endosymbionts in *Paramecium aurelia.* Bacteriol. Rev. 38:113–63.

Rabinovitch, M., and W. Plaut. 1962a. Cytoplasmic DNA synthesis in *Amoeba proteus.* I. On the particulate nature of the DNA-containing elements. J. Cell Biol. 15:525–34.

Rabinovitch, M., and W. Plaut. 1962b. Cytoplasmic DNA synthesis in *Amoeba proteus.* II. On the behavior and possible nature of the DNA-containing elements. J. Cell Biol. 15:535–40.

Roth, L. E., and E. W. Daniels. 1961. Infective organisms in the cytoplasm of *Amoeba proteus.* J. Biophys. Biochem. 9:317–23.

Savanat, T., and E. R. J. Pavillard. 1964. The ability of *A. proteus* to kill *Salmonella enteritidis* introduced by microinjection and the influence of opsonins on intracellular killing. Aust. J. Exp. Biol. Med. Sci. 42:615–24.

Taylor, F. J. R. 1974. Implications and extensions of the serial endosymbiosis theory of the origin of eukaryotes. Taxon 23:229–58.

Wolstenholme, D. R. 1966. Cytoplasmic deoxyribonucleic acid-containing bodies in amoebae. Nature 211:652–53.

Wolstenholme, D. R., and W. Plaut. 1964. Cytoplasmic DNA sythesis in *Amoeba proteus.* III. Further studies on the nature of the DNA-containing elements. J. Cell Biol. 22:505–13.

Symbiosis as Parasexuality

12

The genetic and evolutionary consequences of many kinds of symbioses have been reviewed recently (Margulis, 1976). The purpose of this more restricted essay is to examine the single theme that symbiosis may be viewed as a parasexual phenomenon that has had evolutionary consequences comparable to those of the biparental systems of eukaryotes. If all eukaryotic cells are products of symbioses between prokaryotes, then the evolutionary consequences of symbioses have been immensely important in the origin of biological innovation. A corollary of this theme is that a thorough understanding of symbioses requires genetic and molecular analyses. Such analyses must consider both the number of genomes present and the extent to which each genome acts as an independent unit. The work on *Cyanophora* by Trench, which is outlined elsewhere in this volume (Trench, 1980), demonstrates the potential of this approach.

ANALOGY BETWEEN PARASEXUALITY AND
ESTABLISHMENT OF SYMBIOSES

Symbiosis can be defined as a consistent association between individuals of different species that exists for significant fractions of the life cycle of each. This is essentially what De Bary (1879) meant when he defined symbiosis as "the living together of dissimilarly named organisms." On the other hand, sexuality and parasexuality appear to have little in common with this definition. Sexuality may be defined as the process by which individuals containing genes from more than one parent are produced; in the case of sexual eukaryotes, the new individual contains roughly half of the genes of each parent. The haploid condition (the result of meiosis) alternates with the diploid; the diploid is produced by the fusion of two haploids in fertilization or syngamy. In organisms in which reproduction involves sexuality, fertilization and meiosis must occur at some regular time in the life cycle. Parasexuality is an occasional process that also results in the combination of separate genomes in a single individual, but this process does not involve regular haploid/meiosis diploid/fertilization cycles, and the genetic contributions of the parents need not be equal. The term may be used to describe bacterial conjugation, transduction, and transformation (Rieger et al., 1976), and has been applied to mitotic recombination in heterokaryotic fungi that results in viable recombinant monokaryons (Pontocorvo and Kafer, 1958).

The thesis of this paper is that the dissolution and reestablishment of many symbiotic partnerships differ from the alternation of meiosis and fertilization of sexual cycles primarily in terms of time and timing. Both sexual and parasexual processes produce new individuals from parents that have a close genetic relationship, whereas the establishment of symbioses involves formation of new individuals from parents that are only remotely related. In sexuality merging on the molecular level is as intimate as is conceivable; for example, if crossing-over occurs, each newly combined DNA molecule contains nucleotide sequences from both of the parents. The transcription of this hybrid DNA signals the completion of the merger. In some cases fertilization involves complete cytoplasmic and nuclear fusion (e.g., in *Saccharomyces*), whereas in other cases one parent donates far more cytoplasm than the other (e.g., mammals and the "double fertilization" of angiosperms). In yet other cases only nuclear exchange precedes karyogamy; cytoplasmic fusion does not occur (e.g., ciliates). Yet in all cases of regularly sexual eukaryotes, a sequence of elegantly coordinated processes has evolved so that the intrinsic problems of doubleness that result from cell fusion are reduced. In fungal sexuality the details of how genomic and organellar doubleness or multiplicity are reduced are not understood. These processes involve a low probability of recombinant recovery, and are not as elegantly organized as meiotic processes.

In what sense are symbioses parasexual? They bring together into one individual genes that were previously separated and divergent. Unlike eukaryote sexuality, there is no contact and merger of two partners to form precise genomic doubleness, and regular mechanisms that restore genome singleness are lacking. Symbioses are generally subject to far fewer controls than are the meiotic events that are the *sine qua non* of eukaryote sexuality. Regular symbioses retained by natural selection share features with parasexuality: e.g., no essential DNA may be lost; DNA of neither partner may accumulate to excess; and, at least in intracellular symbioses, protein-synthesizing systems of the two partners must be compatible. During the early stages of the evolution of a symbiosis, indirect methods of controlling separate genetic functions must develop. In intracellular symbioses behavioral, reproductive, and nutritional modes of regulation are the results of selection, so that the transmission of symbionts to succeeding generations is ensured. As symbioses evolve, more direct interactions between the genetic systems ensue. Redundancies become lost; barriers such as cell walls and cell membranes diminish and may disappear. It is only required that the genetic functions needed for the propagation of the genomes of both partners be retained.

In surveying the literature I find many variations on the theme of solving the intrinsic problems of association and dissociation. The formal equivalents of fertilization and meiosis may not exist; there often is no fusion of individuals to form the associated state. Usually symbioses do not represent the mere doubling of genomic content; symbiont-host ratios may vary from 1:1 to nearly any other value. For example, a *Cyanophora paradoxa* cell may contain from one to eight cyanelles (Trench, 1980), whereas a ml of the hindgut fluid of a dry wood termite may contain up to 10^{11} bacteria and 10^5 protists (To, 1978). Analogous to the simultaneous fusion of several different mating types, several different species may merge to form symbioses. On an individual basis humans are outnumbered by skin, mouth, and gut symbionts by 10^{14} to 1; symbionts may contribute a significant percentage of our body weight (Savage, 1977).

By De Bary's definition symbionts are different species and often belong to different kingdoms. This usually precludes the complete and harmonious nuclear fusion that is typical of fertilization. Yet, symbioses are held together by analogous and delightfully varied processes. Interrelationships on several different levels—genetic, metabolic, and behavioral—occur between the partners. Partnerships may or may not cyclically dissolve.

The formal analogy is presented in figure 1. In an idealized case, in which association and dissociation regularly occur, the existence of the free-living partners is analogous to the haplophase of the sexual cycle. Examples of this stage include leguminous plants and soil rhizobia prior to nodulation. Examples of algae and invertebrates may be cited as well. Sea anemones regularly expel their dinoflagellates (Taylor, 1969; Steele, 1976), although the fate of these expelled symbionts is not known. Newly hatched *Convoluta roscoffensis* flatworms lack algal symbionts; they subsequently reassociate with free-living algae of the genus *Platymonas*. The formation

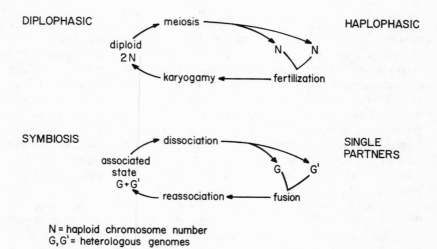

N = haploid chromosome number
G, G' = heterologous genomes

Fig. 1. Diagram showing the formal analogy between meiotic sex and associating and dissociating symbioses.

of such associations is comparable to fertilization in that it involves organismic invasion and often cell penetration. For example, in the green hydra (*Hydra viridis*) the penetration of host cells by chlorellae involves endocytosis, recognition, and intracellular transport (see Cook, 1980). Following establishment, the associated partners are analogous to the diplophase of the sexual cycle. Similarly, the dissociation of the partners is analogous to meotic events that produce haploidy. *H. viridis* provides an example of natural dissociation. Asexual reproduction occurs by budding; gastrodermal cells containing chlorellae are passed on to the bud. However, during sexual maturation the hydra forms eggs that lack algae; normally these eggs produce symbiotic hydra, but occasionally they develop into aposymbiotic hydra. We have found that during the development of sex organs on these hydra, membrane-bound vesicles are formed that contain both chlorellae and associated bacteria (Margulis et al., 1979). These vesicles are ejected into the medium (fig. 2), and often surround or stick to the algae-free eggs. Presumably this makes the chlorellae and bacteria available for ingestion by the newly hatched hydrids (Thorington, 1979). Aposymbiotic hydra become reinfected with chlorellae and bacteria easily. The algae are transported to the basal ends of gastrodermal cells by a process that involves hydra microtubule polymerization (Fracek and Margulis, 1979). During many life cycles of symbiotic partners, associated stages alternate with dissociated stages. Just as the relative length of the haplophase and diplophase varies between different species, in symbiotic partnerships the fraction of time spent associated varies with partner species.

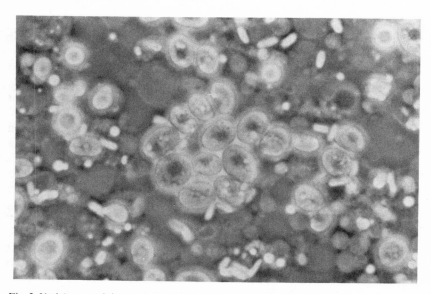

Fig. 2. Vesicles containing bacteria and algae that are expelled into the medium during sexual maturation of *Hydra viridis* (Ohio strain).

Just as meiotic sexuality does not occur in some protists that may yet fuse (Cleveland, 1947), some associations never seem to dissociate; the partners may have been associated continuously throughout the history of the complex. For example, dry wood and subterranean termites are continuously associated with their cellulose-digesting, anaerobic bacteria for tens of millions of years (To et al., 1979). Just as sexuality has been lost in some formerly sexual groups (e.g., *fungi imperfecti*), the tendency to dissociate and reassociate has probably been lost in some symbiotic partnerships (e.g., lichens that do not develop from ascospores: the fungi do not produce ascospores, and the lichen is dispersed through soredia that contain bits of associated algae). Another example of failure to dissociate is the ovarian transmission of mycetomes in leaf hoppers; these bodies are interpreted to contain degenerate microbes (Schwemmler, 1973; Körner, 1978). The former symbionts may now be degenerate microbes from an evolutionary point of view, but since they are required for leaf hopper survival and morphogenesis, they are also at present insect organelles (Schwemmler, 1975).

IMPLICATIONS OF THE ANALOGY

The parasexuality point of view has important implications for traditional classification schemes of symbiotic relationships. Certain of these classifications are based on the selective advantages of the association to the partners, and include terms such as phoresy, mutualism, parasitism, pathogenicity, and commensalism (Henry, 1966). By the argument introduced here such distinctions are comparable to organizing and naming life cycles by their relative success. Labeling a lichen association "mutualistic" is analogous to deciding whether some fungal conidia and primary hyphae (asexual, diploid) are more successful stages than are ascogenous hyphae or ascospores (sexual, dikaryotic, haploid). Calling a flatworm a seagull parasite is comparable to classifying certain fish by whether small males are nutritionally dependent on large females. Associations have been categorized by nutritional mode (e.g., necrotrophy versus biotrophy; Lewis, 1975). This is analogous to categorizing life history phenomena according to the degree of nutritional dependence of one stage on the other (e.g., some fern and lycopod gametophytes, embryo sacs and ovoviviparous larvae). A more recent classification scheme takes into account the length of time two partners are associated (Starr, 1975). Although all of these classifications provide insights into the "normative" interactions between the symbionts, they refer to the ecological rather than the genetic natures of associations. A mutualistic association may become a parasitic association when stressed. For example, this occurs when lichen algae are used as food by the fungus after a lichen is excessively watered and breaks down (V. Ahmadjian, oral communication). Hence, none of these classifications approach the essential nature of symbiosis itself.

In analyzing the semiautonomous organelles of eukaryotic cells, Taylor (1974) suggested that the word *cell* is misleading. Eukaryotes are composed of multiple genome-directed, protein-synthesizing units, whereas prokaryotes are usually single. Taylor suggested that the word *cell* be replaced by the series: monad (single genome-based protein-synthetic unit), dyad (double genome-based protein-synthetic unit), triad, and tetrad. By this terminology polymonads and polydyads describe "multicellular" states.

This terminology might be applied as follows. *Bacillus subtilis* is a unicellular bacterium, and would be considered a monad, whereas the filamentous prokaryote *Actinomyces* is a polymonad. *Trypanosoma*, a flagellate protist, is a dyad (nucleoplasm plus mitochondria), whereas fruitflies are polydyads. The unicellular chlorophyte *Chlamydomonas* is a triad (nucleoplasm, mitochondria, chloroplasts), whereas oak trees are polytriads. This terminology has been applied to microbial symbioses (Margulis, 1976), and can be applied to the analysis of other symbioses. Such analyses require that the number of nonhomologous protein-synthetic units be ascertained, and can be used to organize information about particular associations.

One advantage of this point of view is that it is relatively insensitive to ecological setting. As an example, bacteria endosymbiotic with *Amoeba* initially are pathogenic but become benign symbionts within five years, possibly in as short a time as eighteen months (Jeon, 1980). Based on the number of genomes and associated protein-synthetic systems, these amoebae are at least triads (nucleoplasm, mitochondria, bacteria) and possibly are tetrads, since the "DNA-containing bodies" found in the cytoplasm of these amoebae may represent still a fourth heterologous genome. The changing interactions in this system emphasize the futility of classifying symbioses on what are essentially ecological, and therefore labile, distinctions.

"Genome counting" also points out what is a new field in the study of cellular interactions in symbiosis, the interactions between heterologous symbiont genomes. This field is analogous to the genetic interactions between nucleoplasm and organelle DNA (e.g., the "biogenesis of organelles": Birky et al., 1975; Bucher et al., 1976). Nuclear-organellar interactions include the contribution of nuclear genes to mitochondrial and chloroplast proteins, the source of tRNA, the organization of the organelle genome, and other questions about the structure and function of organelle DNA. For example, both the F_1 mitochondrial ATPase and the ribulose biphosphocarboxylase of chloroplasts have subunits that are coded by nuclear DNA as well as by organelle DNA (Gillham, 1978). In some cases elements of the organelle protein-synthetic systems are products of nuclear genes (e.g., the mitochondrial RNA polymerase of *Neurospora*: Barath and Kuntzel, 1972).

These relationships have parallels in symbioses. Symbioses begin with free-living partners that each contain complete protein-synthetic units. It may then be asked how these nonhomologous genomes have come to interact during the evolutionary history of the associations. Do the metabolic products of one symbiont serve as the substrates for the other? For example, the pyruvate that is produced through cytoplasmic glycolysis and becomes the substrate for mitochondrial oxidation is analogous to the carbohydrate that is translocated from endosymbiotic algae to invertebrate hosts. Does the symbiosis produce "hybrid" gene products, either proteins or RNA? This kind of integration is seen in legume nodules, in which the leghemoglobin is a "hybrid" molecule: the heme is a bacterial product, but the globin is of plant origin (Cutting and Shulman, 1971; see Berringer, 1979, for review). The phycobiliproteins of the cyanellas of *Cyanophora paradoxa* may represent a similar situation (Trench, 1980). The participation of two separate partner genomes is analogous to cells in which subunits of a protein are coded by both nuclear and organelle genes. Still more intimate levels may exist, mainly nuclear and recombinational DNA. In the nuclear case genetic material originating from one partner comes to lie within the nuclear membrane of the second, and the nuclear transcript of one partner is translated on the ribosomes of the second. The relationship between infective microsporidians and salivary gland cells of the dipteran *Rhyncosciara* appears to be an example of this kind of interaction. The only polysomes seen in electron micrographs of these cells belong to the microsporidians; it has been found that the synthesis of polytene chromosomes in single-cell tumors may require the use of these ribosomes (Pavan, 1971, and pers. comm.). Actual genetic recombination between symbionts requires the integration of the DNA of one partner into the linear structure of the DNA of the other. Of course, such DNA recombination normally occurs in sexual processes, but is integration in a symbiosis ever that tight?

Perhaps the best-analyzed case of gene interaction between partners is the plant-bacterial association responsible for the crown gall tumor. The bacteria (*Agrobacterium*) bear DNA plasmids that are transmitted to both partners. The plasmids pass through wounded stems or leaves into plant cells. Inside the plant they replicate and spread to other cells, sometimes far from the site of entry. They carry with them genetic information for the production of distinctive amino acids such as nopaline and octapine, as well as for their own replication. The plasmid DNA is transcribed into plant messenger RNA, and proteins are subsequently transcribed on the plant ribosomes. These proteins include several enzymes that catalyze reactions synthesizing octapine and other nonprotein amino acids. The amino acids, which are essentially the products of bacterial piracy of plant metabolism, leak out of the roots into the soil. Because they may be the sole source of carbon and nitrogen, the octapines support large numbers of free-living agrobacteria in the rhizosphere (Schell, et al., 1979). The exact site of plasmid replication (chloroplast, mitochondria, cytoplasm, or nucleus) is not known.

These are but a few of the many examples of partnership levels; certainly one does not preclude another. As the complexity of the metabolism and, especially, the behavior of the partners increase, the possibility for multiple levels of interaction increases as well, as does the task of understanding the interaction between the partners. As one approaches the typical vertebrate, with its extensive and diverse oral, digestive tract and skin microbiota, the task becomes formidable. It must be recognized at the outset that what was once thought to be an individual is in reality a community of interacting symbionts containing different genomes.

IMPLICATIONS

When symbioses are considered in the context suggested above, it becomes obvious that the understanding of partner genetic interrelations in symbiosis is critical for cell biology; it is central to the understanding of both the physiology of the eukaryotic cell and the origins of innovation in evolution. Symbioses derive from independently evolved genomes that have been casually or intimately fused for relatively brief interludes or for long periods. They provide models for the origin, evolution, and function of multigenomic systems, including all eukaryotes. For example, mycorrhizal associations with plant roots were probably essential for the evolution of land plants (Pirozynski and Malloch, 1975). Perhaps 18,000 of the estimated 33,000 ascomycetes are found lichenized in nature (D. C. Smith, pers. comm.). In view of their importance to the biosphere and its past evolution, symbioses should be the objects of more research attention. Certainly no gnotobiotic polydyad or species of higher-level organization has been documented in nature—animal, plant, or fungus—to exist without its associates. Whether in casual roles or in intimate endocellular symbiosis, divergent species interact, coexist, and coevolve. Reproduction of so-called individuals actually depends on the contributions of gene and metabolic products from heterologous genomes. Nature abhors a pure culture; symbioses, whether ecto- or endocellular, are the rule rather than the exception.

ACKNOWLEDGMENTS

I am grateful to Prof. D. C. Smith, Brian Berger, John Stolz, Susan Francis, Glynne Thorington, and Stephen Fracek for aid with the research and the manuscript. This work was supported by NASA grant NGR-004-025, the NSF Cell Biology Program, and the Boston University Graduate School.

REFERENCES

Barath, Z., and H. Küntzel. 1972. Cooperation of mitochondrial and nuclear genes specifying the mitochondrial genetic apparatus in *Neurospora crassa*. Proc. Nat. Acad. Sci. USA 69:1371–74.

Beringer, J. E., N. J. Brewin, A. W. B. Johnston, H. M. Shulman, and D. A. Hopwood. 1979. The *Rhizobium*-legume symbioses. Proc. R. Soc. Lond. Ser. B. 204:219–33.

Bucher, T., W. Neupert, W. Sebald, and S. Werner. 1976. Genetics and biogenesis of chloroplasts and mitochondria. North-Holland, Amsterdam.

Cleveland, L. R. 1947. The origin and evolution of meiosis. Science 105:14–17.

Cook, C. B. 1980. The infection of invertebrates with algae. In this volume.

Cutting, J. A., and H. M. Shulman. 1971. Biogenesis of leghemoglobin. The determinant in the *Rhizobium*-legume symbiosis for leghemoglobin specificity. Biochim. Biophys. Acta 229:58–62.

Fracek, S., and L. Margulis. 1979. Colchicine, nocodazole, and trifluarin: different effects of microtubule polymerization inhibitors on the uptake and migration of endosymbiotic algae in *hydra viridis*. Cytobios. (In press.)

Gillham, N. 1978. Organelle heredity. Raven Press, New York.

Henry, S. M. (ed.). 1966. Symbiosis, 2 vols. Academic Press, New York.

Jeon, K. W. 1972. Development of cellular dependence in infective organisms: microsurgical studies in amoebas. Science 176:1122–23.

Jeon, K. W. 1980. Symbiosis of bacteria with *Amoeba*. In this volume.

Korner, H. H. 1978. Intraovarially transmitted symboints of leaf hoppers. In K. Herter (ed.), Zoologische Beitrage, pp. 59–68. Dunker and Humblot, Berlin.

Lewis, D. H. 1975. The relevance of symbiosis to taxonomy and ecology, with particular reference to mutualistic symbiosis and the exploitation of marginal habitats. In V. H. Heywood (ed.), Taxonomy and ecology, pp. 151–72. Academic Press, New York.

Margulis, L. 1976. Genetic and evolutionary consequences of symbiosis. Exp. Parasitol. 39:277–349.

Margulis, L., G. Thorington, B. Berger, and J. Stolz. 1979. Endosymbiotic bacteria associated with intracellular algae in *Hydra viridis*. Curr. Microbiol. 1:227–232.

Pavan, C., J. Biesele, R. W. Reiss, and A. V. Wertz. 1971. Changes in the ultrastructure of *Ryhncosciara* cells by microsporidia. In Studies in genetics, 6:241–71. University of Texas Publication 7103.

Pontecorvo, G., and E. Kafer. 1958. Genetic analysis based on mitotic recombination. Adv. in Genetics 9:71–104.

Pyrozinski, K. A., and D. W. Mulloch. 1975. The origin and the evolution of land plants: a question of mycotrophy. Biosystems 6:153–73.

Rieger, R., a. Michaelis, and M. M. Green. 1976. Glossary of genetics and cytogenetics. Springer-Verlag, New York.

Savage, D. C. 1977. Microbial community of the gastrointestinal tract. Ann. Rev. Microbiol. 31:107–33.

Schell, J., M. van Montagu, M. de Beuckeleer, M. de Block, A. Depicker, M. de Wilde. G. Engler, C. Genetello, J. P. Hernalsteens, M. Holsters, J. Seurinck, B. Silva, F. van Vliet and R. Villaroel. 1979. Interactions and DNA transfer between *Agrobacterium tumefaciens*, the Ti-plasmid and the plant host. Proc. R. Soc. Lond. Ser. B. 204:251–266.

Schwemmler, W. 1973. Ecological significance of endosymbiosis: an overall concept. Acta Biotheoretica 22:113–19.

Schwemmler, W. 1975. Control mechanisms of leafhopper symbiosis. In E. L. Cooper (ed.), Contemporary topics in immunology, 4:179–87. Plenum Press, New York.

Starr, M. P. 1975. A generalized scheme for classifying organismic associations Symp. Soc. Exp. Biol. 29:1–20.

Steele, R. D. 1976. Stages in the life history of a symbiotic zooxanthella in pellets extruded by its host, *Aiptasia tagetes* (Duch. and Mich.) (Coelenterata:Anthozoa). Biol. Bull. 149:590–600.

Taylor, D. L. 1969. On the maintenance and regulation of algal numbers in zooxanthellae-coelenterate symbiosis, with a note on the nutritional relationship in *Anemonia sulcata*. J. Mar. Biol. Ass. U.K. 51:227–34.

Taylor, F. J. R. 1974. Implications and extensions of the serial endosymbiosis theory of the origin of eukaryotes. Taxon 23:229–58.

Thorington, G. 1979. Metabolic relationships between *Hydra viridis* and its endosymbiotic algae and ultrastructural studies of transfer of endosymbionts through the sexual cycle of the host. Ph.D. diss., Boston University.

To, L. P. 1978. Spirochaetes and other hindgut microbiota of dry wood termites. Ph.D. diss., Boston University.

To, L., L. Margulis, W. Nutting, and D. Chase. 1979. *Pterotermes occidentalis* and its hindgut microbial community. (In preparation.)

Trench, R. K. 1980. Integrative mechanisms in mutualistic symbioses. *In* this volume.

ROBERT K. TRENCH

Integrative Mechanisms in Mutualistic Symbioses

13

INTRODUCTION

Biological integration encompasses all those processes whereby two organisms of divergent evolutionary origins come together and coexist as a stable, self-regulating unit. Within the context of intracellular symbioses between algae and invertebrates, the process of integration begins with intercellular contact between the two potential partners and may be regarded as complete when the initially independent components of the association have become irreversibly dependent on each other. Intracellular mutualistic symbioses could therefore progress to a stage wherein it becomes difficult to distinguish between an intracellular symbiont and a *de facto* organelle (Margulis, 1970, 1976; Schnepf and Brown, 1971; Trench et al., 1978).

In the establishment of intracellular symbioses, one cell (by convention, "the symbiont") makes contact with, and somehow gains entry into, another cell (by convention, "the host"). When two such genetically distinct cells come into contact, they are usually able to distinguish between self and nonself. This capability appears to be a property of all living cells (Davis et al., 1973). Therefore, recognition as being foreign during initial contact between two cells of divergent evolutionary histories represents the first hurdle that the "invading entity" must overcome. If the invading cell is recognized as foreign, it may be destroyed intracellularly. Clearly, the recognition system and the defense mechanisms of the host must be circumvented in order to enhance the probability for survival of the symbiont and hence the perpetuation of the association.

There are probably no examples available, with the possible exception of the bacteria-*Amoeba* association described by Jeon (1972, 1980), wherein a given investigator could trace, in the same association, the progression from initial inception to complete integration and the formation of a mutually obligate association. However, within the spectrum of known symbioses, it is possible to recognize through experimental analysis of extant associations a range of levels of interdependences between the interacting associants.

If it is assumed that an unstable association, such as that resulting in the destruction of the potential symbiont or host, represents a "primitive" condition in symbiosis (as in many parasitic, oncogenic, or any pathologic condition), then a stable, mutually obligate association would imply the most integrated state. Between these two extremes would be a range of "levels of integration" between the partners in association (see fig. 1). It is not assumed that a given association must progress through each step or level to achieve complete integration. This would presuppose directionality in the evolution of symbiosis. The origins of most extant symbioses are obscure, and the future direction an association may follow with respect to the degree of integration between the interacting partners is unpredictable (Dubos and Kessler, 1963; Trench, 1979). Unpredictability could be regarded as the "Heisenberg Uncertainty Principle" of symbiosis, and suggests that associations are discrete and not necessarily part of a continuum. Nonetheless, it is a useful exercise, in attempting to decipher uniformity in the plethora of symbiotic associations, to construct hypothetical "barriers" that would have to be overcome before two genetically different cells could become obligatorily and mutually dependent on one another. Models of intercellular interactions in the establishment of intracellular symbioses could help define more clearly the many different aspects of partner integration. I shall restrict myself here to two different associations that lend themselves to the experimental analysis of several different aspects of partner integration: (a) the association between the "gymnodinioid" dinoflagellate *Symbiodinium* (=*Gymnodinium*) *microadriaticum* and marine invertebrates (see fig. 2) and (b) the relation between the cyanobacterium *Cyanocyta korschikoffiana* and *Cyanophora paradoxa* (see fig. 3). Several aspects of partner integration in symbioses involving animals and plants are also discussed elsewhere in this volume (see Cook, 1980; Margulis, 1980; Muscatine, 1980), and, where relevant, reference will be made to analogous symbiotic systems not discussed in detail in this paper (cf. Trench, 1979).

HOST ACQUISITION OF SYMBIONTS

Host Location by Symbionts

In the initiation of any intracellular symbiosis, there must have been a

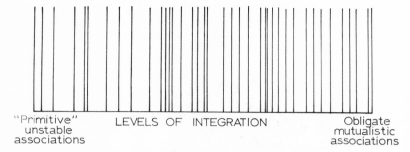

Fig. 1. "Line spectrum" depicting the range of different levels of biologic integration that exists between associated organisms. Each vertical line represents an individual association at one point in evolutionary time. No inferences or implications on the origins of the associations or predictions of their final destiny are intended.

process whereby the two potential partners, initially separated by space and evolutionary time, would have had to have overcome at least the distance barrier. If one cell is to enter the other, the two cells must first come into physical contact. How this may have been accomplished in the first establishment of an intracellular symbiosis is beyond the scope of experimental analysis and may never be established. However, it is reasonable to conceive that in those extant symbiotic associations wherein the symbionts are not passed on from one host generation to another through maternal inheritance that, at least with respect to host offspring, an association is being established for the first time and is therefore analogous to, but not homologous with, the initiation of a symbiosis between the two organisms. Some symbiotic associations lend themselves naturally to the study of events involving host location by symbiotic algae. In this respect, the association between *S. (=G.) microadriaticum* and marine invertebrates is particularly useful, since several animal hosts can be brought into culture, rendered aposymbiotic, and cloned. In addition, motile clones of *S. (=G.) microadriaticum* can be produced in axenic culture (see Schoenberg and Trench, 1979a, b, c). Cloned hosts and cloned symbionts provide excellent systems for experimental manipulation.

In several instances in the association between the symbiotic dinoflagellate *S. (=G.) microadriaticum* and marine invertebrates, convincing evidence exists that illustrates that the algae are passed on from one generation to the other through maternal inheritance (Trench, 1979). But in several other instances equally incontrovertible evidence exists that such is not the case, and that juvenile hosts acquire the symbionts from the ambient environment (Schoenberg, 1976). Since, in most cases, the hosts are sessile and benthic, the algae would have to be mobile in order to

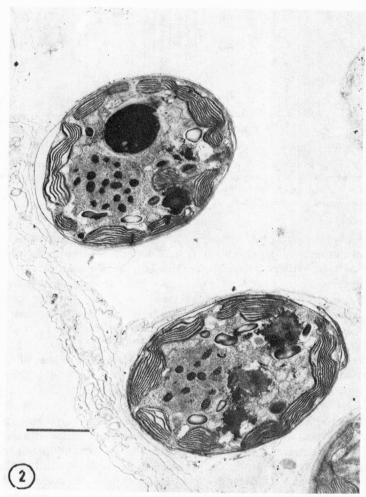

Fig. 2. Transmission electron micrograph of *S.* (=*G.*) *microadriaticum in situ* in the marine hydroid *M. amboinense*. Scale bar approximately 1.5 μm.

"locate" an appropriate host. It is known that in the life cycle of *S.* (=*G.*) *microadriaticum*, there is an alternation between a nonmotile coccoid stage and a swimming "gymnodinioid" stage. It is assumed that the latter stage serves as a vehicle of dispersal and as the "infectious" stage. It should be borne in mind, however, that nonmotile stages of this alga do infect appropriate hosts when experimentally placed in the host (Trench, 1971c; Schoenberg, 1976; Schoenberg and Trench, 1979c).

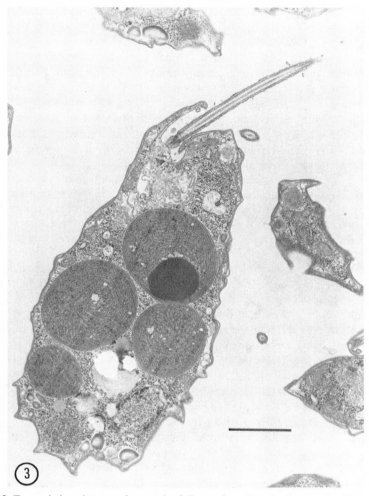

Fig. 3. Transmission electron micrograph of *C. paradoxa*. Longitudinal section illustrating four cyanelles in the host cytoplasm. Scale bar approximately 1.5 μm.

An example of host location by free swimming *S.* (=*G.*) *microadriaticum* comes from studies on the scyphistomae of the rhizostome jellyfish *Cassiopeia xamachana*. The planulae produced as a result of sexual reproduction by the medusae are devoid of algae and metamorphose into the polyp stage after settlement. It is therefore possible to use these aposymbiotic scyphistomae as test hosts through which the mechanism of host location may be analyzed.

Algal taxes do not seem to play a significant role in location of these scyphistomae by *S.* (=*G.*) *microadriaticum.* When motile symbionts isolated from *C. xamachana* are placed in finger bowls containing the polyps, the contact between host and algae appears to occur by random chance rather than by any directed movement (W. Fitt, unpublished). The chemosensory attraction of *Platymonas convolutae* to the egg cases of *Convoluta roscoffensis* has been postulated (Holligan and Gooday, 1975), but we have found no evidence for such a taxis in our system. Kinzie (1974) described swimming *S.* (=*G.*) *microadriaticum* "accumulating around the oral disc" of aposymbiotic polyps of *Pseudopterogorgia bipinnata,* but he made no attempt to investigate any such taxis.

Having located the potential host, the algae must gain access to the coelenteric cavity (at least in the case of coelenterates) before they can be taken up by the endoderm cells. Feeding behavior in coelenterates appears to be stimulated by reduced glutathione, a few selected amino acids, or by a combination of these (see Lenhoff, 1974). These substances are usually released to the environment after puncture of the prey animal by nematocysts. Hence, feeding behavior in coelenterates is usually associated with capture and engulfment of animal prey, and coelenterates are generally regarded as specialized carnivores (see Yonge, 1963).

In light of the specializations for the capture of animal prey, it was surprising to observe that the scyphistomae of *C. xamachana* respond to the motile *S.* (=*G.*) *microadriaticum* in a manner closely resembling the capture of food particles (W. Fitt, unpublished), i.e., the algae became adhered to tentacles, which transported them to the mouth. Ultimately, the algae could be observed passing through the opened mouth and into the stomodaeum. These algae always became incorporated into the tissues of the animal. It is therefore possible that the motile algae, like their coccoid counterparts in animal tissue, release organic compounds that elicit a feeding response on the part of the potential host. It is not known what these substances are, as motile "zooxanthellae" have never been assayed for released organic carbon.

Uptake of Algae by Host Endoderm Cells

The processes of recognition, endocytosis, and sequestration in symbioses between *S.* (=*G.*) *microadriaticum* and marine invertebrates have not yet been studied in as much detail as have similar phenomena in the *Hydra-Chlorella* system (see Pardy and Muscatine, 1973; Muscatine, Cook, Pardy, and Pool, 1975; Muscatine, Pool, and Trench 1975; Pool, 1976; Cook et al., 1978; and Cook, 1980), wherein Muscatine et al. (1975a) recognized five phases in the process of reinfection of *Hydra* by symbiotic

Chlorella; (1) the contact phase; (2) the engulfment phase; (3) a recognition phase; (4) sequestration; and (5) repopulation.

The cellular events occurring during intercellular contact and endocytosis have not yet been elucidated in the association between *S.* (=*G.*) *microadriaticum* and marine invertebrates. However, the evidence for a system of intercellular recognition and discrimination during infection is becoming increasingly strong (see Schoenberg and Trench, 1979a, b, c). Other systems demonstrating selective uptake of algae by invertebrate hosts have also been described (see Provasoli et al., 1968; Taylor, 1971).

S. (=*G.*) *microadriaticum* has been isolated from a wide range of invertebrate hosts, rendered axenic, and grown through several asexual generations. The nonmotile coccoid stages of these algae have been introduced into aposymbiotic clones of the sea anemone *Aiptasia tagetes*, and their infectivity and repopulation of the animals measured by monitoring the increase in the number of algae per unit of animal tissue. Algae originally isolated from the same host species were used as controls.

A summary of the results (table 1) indicates that (1) algae originally isolated from *Aiptasia* reinfect and repopulate *Aiptasia* most rapidly, and that (2) algae from different hosts either infected and repopulated *Aiptasia* polyps at a slower rate or did not infect at all, suggesting maximum compatibility between *Aiptasia* and algae originally isolated from *Aiptasia*. These results are consistent with the definition of biological specificity as defined by Weiss (1953) and Dubos and Kessler (1963).

Analogous experiments have been conducted using clones of the scyphistomae of *Cassiopeia xamachana* as the test host organism. W. Fitt and R. Trench (unpublished) found (table 2) that motile and nonmotile algae from *Cassiopeia* (strain C) infected the scyphistomae and stimulated strobilation and the production of ephyrae more rapidly than algae isolated from other hosts, many of which did not infect at all when introduced either as the motile or the nonmotile stages. It should be pointed out that it has been shown repeatedly that the presence of the algae is indispensable for the completion of the life history of symbiotic rhizostomes (Bigelow, 1900; Sugiura, 1964; Ludwig, 1969). This phenomenon has been confirmed in our laboratory, and it now appears that infection by a specific type or strain of algae is equally important.

In order to study the process of recognition in symbiotic systems, it is necessary first to establish that there are significant differences in the infecting organisms. The existence of stable recognizable differences between strains of *S.* (=*G.*) *microadriaticum* has only recently been demonstrated (Schoenberg, 1976; Schoenberg and Trench, 1976; 1979a, b, c). When axenic cultures of these algae (the same cultures used in the infection experiments described above) were analyzed using isoenzyme

TABLE 1

Infectivity and Repopulation of Aposymbiotic *Aiptasia tagetes* by Different "Strains" of S. (=G) *microadriaticum* Isolated From Different Hosts and Grown Axenically under Identical Conditions

Algal "Strain"	Original Host	Mean No. of Algae/mm Tentacle, 0 + 5 days	N	Mean No. of Algae/mm Tentacle, 0 + 50 Days	N
		($\pm$ range)		($\pm$ range)	
A	*Aiptasia tagetes*	3×10^2 ($\pm$100)	12	2×10^4 ($\pm$1600)	12
N	*Mussa angulosa*	1×10^1 ($\pm$5)	5	7×10^3 ($\pm$400)	5
B	*Bartholomea annulata*	1×10^1 ($\pm$2)	5	2.5×10^3 ($\pm$370)	5
H	*Heteractis lucida*	7×10^1 ($\pm$31)	5	2.3×10^2 ($\pm$80)	5
O	*Oculina diffusa*	1.5×10^2 ($\pm$10)	5	2×10^2 ($\pm$90)	5
Z	*Zoanthus sociatus*	0	5	0	5

Source: Data from Schoenberg, 1976. See also Schoenberg and Trench, 1979c.

TABLE 2

Preliminary Observations on the Infectivity of Different "Strains" of S. (=G) *microadriaticum* in Aposymbiotic Scyphistomae of C. *xamachana* and Their Specificity in stimulating Strobilation

Algal "Strain"	% Infection	Days to Initial Strobilation	% Individuals Strobilated	N
B	33	56	6.0	22
C	100	7–8	99.8	34
O	0	∞	0.0	20
T	0	∞	0.0	18
Z	0	∞	0.0	20

N = number of scyphistomae tested.

electrophoretic patterns and patterns of migration of general soluble proteins in polyacrylamide gels, differences in the algae at the macromolecular level were apparent (see fig. 4). Within the limits of the sensitivity of the techniques employed, there was a very strong indication that the algae present in a given host were equivalent to clones. Finally, algae that were biochemically closely related were similar in their infectivity and capacity to repopulate a test host organism. Morphological studies also demonstrated characteristic differences among the algae, particularly with respect to their size and details of the ultrastructure of the cell covering or amphiesma (see Schoenberg and Trench, 1979b). However, there was no direct correlation between morphology and infectivity (see table 3), suggesting that perhaps surface properties of the algae such as the distribution of antigenic determinants or other molecular binding sites specific to the different algae are involved in surface recognition. This phenomenon has been shown to be important in the recognition system in *Hydra* and *Chlorella* (Pool, 1976).

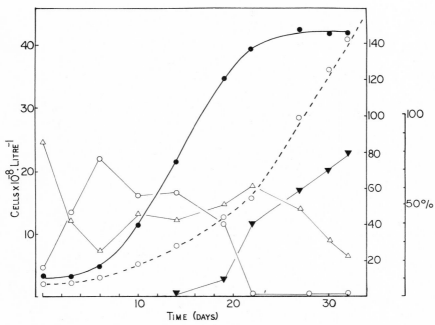

Fig. 4. A dendrogram illustrating the relatedness of the different "strains" of *S.* (=*G.*) *microadriaticum*. Relatedness is based on relative similarity of isoenzyme patterns obtained following electrophoresis of extracts of cultured algae (see Schoenberg and Trench, 1976, 1979a).

TABLE 3

COMPARISON OF BIOCHEMICAL AND MORPHOLOGICAL CHARACTERISTICS OF "STRAINS"
OF S. (=G) *microadriaticum* WITH THEIR ABILITY TO
INFECT AND REPOPULATE A. *tagetes*

Electrophoretic Group	Size Group	Strain	Infectivity and Repopulation (with Respect to A. tagetes)
1	Small ($< 10\mu m \times 10\mu m$)	A	High
		N	Intermediate
		M, O, T	Low
		D	None
2	Large ($> 10\mu m \times 10\mu m$)	B, C, H, Q	Low
		P	None
3	Large ($> 10\mu m \times 10\mu m$)	Z	None

SOURCE: Data from Schoenberg, 1976, and Schoenberg and Trench, 1976.

Since the algae in *C. xamachana* and in *A. tagetes* are intracellular, it must be assumed that the data presented on infection and repopulation indicate selection or discrimination on the part of the host's endoderm cells, the algae, or both. Usually the assumption is made that the host is the discriminating partner in such associations

The "zooxanthellae" that eventually come to reside in individual vacuoles in the animals' cells are undoubtedly phagocytosed, as are the "zoochlorellae" in *Hydra* (see Cook et al., 1978). The location of the algae in host endoderm cells appears to be very markedly polarized, since, for example in *Myrionema* and *Briarium*, functional, morphologically intact algae are displaced in individual vacuoles toward the base or proximal end of the cells, as determined by tissue maceration and light and electron microscopic methods (see fig. 5, and cf. Muscatine, Cook, Pardy, and Pool, 1975). Within this context of recognition phenomena, it is also of interest to note that pycnotic zooxanthellae observed in the digestive cells of many coelenterates occur at the distal end of the cells and may be exocytosed, suggesting that the animal cells can distinguish differences in the physiological state of their endosymbionts (Trench, 1974; Cook, 1980).

The number of algae per host cell varies dramatically from one association to another and from one morphological location to another in the same host. For example, W. Fitt and R. Trench (unpublished) observed that in the marine hydroid *Myrionema amboinense*, the number of algae per host endoderm cell ranged from one to fifty-six, the highest densities occurring in the tentacles with intermediate to low densities in the cells of the hydranth and the stalk. The endoderm cells of *Briarium asbestinum* contained about eight algae per cell, whereas the endoderm cells of *Condylactis gigantea* contained only one or two.

Once the algae are within the animal cells and begin to proliferate, their dispersal through the host's tissues becomes important. Very little is clearly understood about how additional animal cells become occupied by algae as the algal population increases. There are two possibilities: (1) that the animal cells containing the algae divide synchronously with the algae, as may be the case in *Hydra* (Pardy, 1974a,b); or (2) that algae are exocytosed by overpopulated cells, transported in the coelenteron, only to be endocytosed by another digestive cell, wherein the algae may continue to divide (cf. Trench, 1974). Although not proved, this latter possibility is suggested by the observation of large numbers of nonmotile *S.*(=*G.*) *microadriaticum* circulating within the coelenteric cavity of the alcyonarian *Xenia*, the coral *Plerogyra*, and the anemone *Aiptasia* (Trench, unpublished).

Circumventing Host Digestion

The symbiotic algae entering the cell of an animal host can be viewed in two different ways. First, the algae can be regarded as potential food

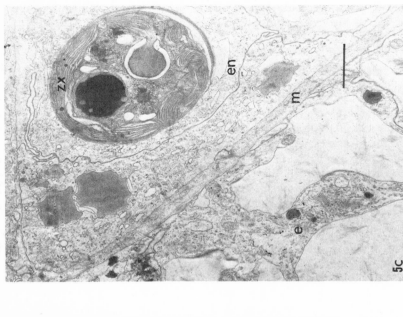

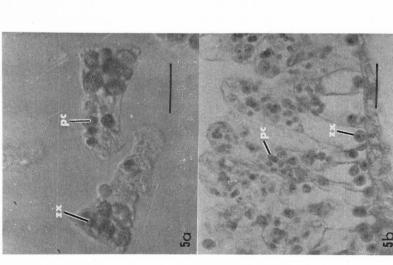

Fig. 5. (a). Light micrograph of two digestive cells of *M. amboinense* isolated by tissue maceration and photographed using Nomarski interference optics. The morphologically intact algae (zx) may be seen at the base of the cells; pycnotic algae (pc) are at the apical end of one cell. Scale bar approximately 25 μm. (b) Light micrograph of a paraffin-embedded section of *M. amboinense* showing the endoderm cells with morphologically intact algae at the base (zx) and pycnotic algae (pc) at the apical ends. Scale bar approximately 20 μm. (c) Transmission electron micrograph of *M. amboinense* showing an intact alga at the base of a digestive cell. e = ectoderm; m = mesoglea; en = endoderm; zx = algal endosymbiont. Scale bar approximately 2 μm.

particles being phagocytosed by a phagotrophic cell. It is significant that most of the invertebrates with intracellular algal symbionts demonstrate phagotrophy and intracellular digestion of food particles (Yonge, 1944). Second, the algae may be viewed as "foreign" infectious agents that, if recognized as such, should be eliminated by the animal's defense system (see, for example, Allen, 1969).

In the past, digestion of algal symbionts has usually been discussed within the context of the nutrition of the host animal (see Muscatine, 1973, 1974). Despite several suggestions to the contrary (Fankboner, 1971; Steele and Goreau, 1977), there is no really good evidence that demonstrates either *in situ* intracellular or extracellular digestion of symbiotic dinoflagellates. Nonetheless, several investigators have reported the presence of pycnotic algae in the cells of animal hosts, but whether these algae represent the result of host digestion or some autolytic degradative process related to senescence as suggested by Trench (1974), remains to be resolved. Thus, the establishment of intracellular symbioses can be viewed as failure of digestive processes within host cells. Yet in many instances it is known that the hosts are competent at intracellular digestion.

The possibility that algae are able to avoid host digestive attack by modifying the molecular basis of recognition as seen in some parasitic associations (Terry and Smithers, 1975) has not been ascertained. However, there is the possibility that algae may be able to counteract fusion of lysosomes with vacuoles containing them, as has been suggested by Karakashian (1975) and Muscatine, Pool, and Trench (1975). These phenomena are not unlike those described in the process of infection of mammalian cells (Jones, 1974) and indeed may suggest that establishment of intracellular symbioses may be the result of the failure of host defense systems.

Finally, the algae may possess cell coverings or walls that are resistant to hydrolytic degradation by host enzymes. The possible existence of the polycarotenoid sporopollenin in the cell wall of symbiotic *Chlorella* has been cited as an explanation for the resistance of these algae to digestion (Muscatine, Pool, and Trench, 1975). Apparently sporopollenin is also a major component of the encysted stage of many dinoflagellates (F. J. R. Taylor, pers. comm.). However, the majority of intracellular symbiotic dinoflagellates studied appear to possess a reduced or nonexistent amphiesma, and should not be regarded as being encysted when in the host cell (cf. figs. 6a and b, and see Schoenberg and Trench, 1979b). The reduced amphiesma is a feature often viewed as an adaptation enhancing the intercellular transport of metabolites (Smith et al., 1969; Smith, 1974). Other morphological modifications that symbiotic dinoflagellates may undergo coincident with establishing symbioses are discussed in greater detail elsewhere (Trench, 1979).

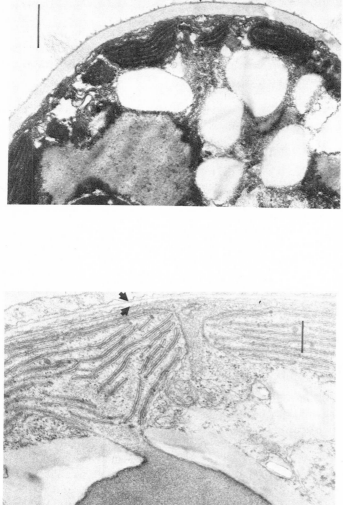

Fig. 6. (a) Transmission electron micrograph of S. (= G.) microadriaticum in situ in Protopalythoa sp. illustrating the highly reduced amphiesma (arrows). Scale bar approximately 0.2 μm. (b) Transmission electron micrograph of S. (= G.) microadriaticum isolated from Zoanthus sociatus and grown in axenic culture. Note the prominent amphiesma (arrows). Scale bar approximately 0.6 μm.

INTERCELLULAR TRANSPORT OF METABOLITES

Intermediary Metabolites

Once established, an association between an alga and a host organism must be perpetuated through time to satisfy the criteria of a symbiosis. It is likely that such perpetuation would be favored if there were selective advantages for the consortium that are greater than those experienced by either partner existing separately.

In the case of the association of *S.*(=*G.*) *microadriaticum* with marine invertebrates, there is abundant evidence that the algae play an important role in the nutrition of the animals with which they are associated by photosynthesizing and transporting large quantities of photosynthetic products in the form of sugars, organic acids, and amino acids to the host (Muscatine, 1973, 1974, 1980; and Trench, 1971a, b, 1980). The movement of organic metabolites in the opposite direction, from the animal to the algae, has yet to be firmly established, but more and more evidence is being accumulated which suggests that such might be the case (Muscatine, 1980). In addition, unpublished studies conducted in our laboratory on the sea anemone *Anthopleura elegantissima* showed that when organic-^{35}S was fed to the animal, up to 40% of the ^{35}S became associated with soluble and insoluble pools within the algae over a period of six days. There was also strong indication that some of these metabolites, identified chromatographically as cystine, methionine, and cystathione, were shunted back and forth between algae and animal tissues. Cook (1971) also found evidence that up to 40% of the ^{35}S ingested by *Aiptasia* as organic-^{35}S became associated with its algae, but no evidence was presented that organic-S was translocated from the animals to the algae.

Several investigators (Lewis and Smith, 1971; Trench, 1974; Muscatine and Porter, 1977) have drawn attention to the importance of the "short-circuiting" of normal biogeochemical cycles by symbiotic associations as being of selective advantage and a mechanism for enhancing survival in nutrient-depauperate environments. Nonetheless, it is apparent that some isolates of *S.*(=*G.*) *microadriaticum* can survive and grow completely autotrophically (Taylor, 1973a,b), but apparently retain the capability for the heterotrophic utilization of organic nitrogen, sulphur, and possibly phosphorus. The selective advantage to the consortium may well be based on the conservation and recycling of potentially limiting nutrients.

In the obligatorily mutualistic association between the cyanobacterium *Cyanocyta korschikoffiana* and the cryptomonad *Cyanophora paradoxa*, it has been demonstrated (Trench et al., 1978) that the intact association possesses very little heterotrophic capability and indeed may be regarded as an obligate photoautotroph. Under these circumstances, the sole source of reduced carbon for the association comes from photosynthesis by the symbiotic cyanobacteria.

Experimental analysis of carbon translocation *in situ* and *in vitro* have illustrated that about 20% of the carbon fixed photosynthetically by the cyanobacteria is made available to the host mostly as glucose. It was pointed out by Trench and colleagues (1978), and it is reiterated now, that this could not possibly represent the total qualitative or quantitative contribution of the symbionts to the association.

Macromolecular Transport and Control of Gene Expression

The regulation of gene expression in mutualistic symbioses is an area that has received very little attention; yet, at the cellular level, the expression of the genome of both partners in a consortium must somehow be controlled in a coordinated manner so that both organisms may coexist in a state of equilibrium. Hence, the regulation of intermediary metabolism, cell division, and the reproductive cycles of both partners must ultimately be a function of the cooperative coordination of the expression of the genomes of both partners (see Margulis, 1979).

One of the few reports of possible genetic interaction between symbiotic algae and their hosts comes from the study of von Holt (1968) on the association between *S.(=G.) microadriaticum* and *Zoanthus flos-marinus* (=*sociatus*), wherein the observation of movement of nucleoside polyphosphate from the algae to the animal host was interpreted as illustrating the transport of genetic information from one partner to the other. Unfortunately, this study has never been repeated, and the conclusions drawn have yet to be corroborated.

The release of soluble photosynthetic products by symbiotic *S.(=G.) microadriaticum* has been shown to be enhanced *in vitro* by a factor present in homogenates of the tissues of their hosts (Muscatine, 1967; Muscatine et al., 1972; Trench, 1971c). It has been suggested that this factor is a protein-like substance.

In his studies with *A. elegantissima*, Trench (1971c) found that the factor was present in animal tissues when those animals possessed algal symbionts. However, homogenates of aposymbiotic animals did not demonstrate the capability to enhance transport from the algae until such animals had become infected with algae, suggesting that the presence of the algae, in some unknown manner, had "induced" the production of a substance that in turn enhanced the flow of intermediary metabolites from the algae to the animal tissues.

Evidence that the presence of symbiotic algae may influence host gene expression comes from data on the role of *S.*(= *G.*) *microadriaticum* in the morphogenesis of rhizostome medusae such as *Mastigias* (Sugiura, 1964) and *Cassiopeia* (Bigelow, 1900; Ludwig, 1969). In several scyphozoans, iodine and thyroxin stimulate strobilation of the scyphistomae and the production of ephyrae (Spangenberg, 1971; Olman and Webb, 1974). In the case of the symbiotic rhizostomes, these substances have been found not to stimulate this morphogenetic sequence, and only the presence of an appropriate strain of *S.*(=*G.*) *microadriaticum* in the scyphistomae appears to be effective (Fitt and Trench, unpublished). It is unlikely that the release of small molecular weight substances such as glucose, glycerol, or alanine would be important in triggering such a fundamental morphogenetic process, particularly since all of the symbiotic dino-flagellates assayed release these same metabolites *in vitro*, yet only one stimulates the morphogenetic sequence of events.

Further evidence for intergenomic interaction between host and symbionts comes from studies of *C. paradoxa* and *C. korschikoffiana*. Trench and Ronzio (1978) demonstrated that all the photosynthetic pigments in *C. paradoxa* were located in the symbiotic cyanobacteria. Among these pigments are two phycobiliproteins, *c*-phycocyanin and allo-phycocyanin. The latter pigment was shown to be a homodimer, and the former was found to be heterodimeric.

Siebens and Trench (1978) and Trench and Siebens (1978) studied the ribosomal RNAs from the cyanobacteria and from the host and found the cyanobacterial r-RNAs to be typically prokaryotic and those of the host to be typically eukaryotic. Using a series of metabolic inhibitors that have selective efficacy against either the prokaryotic or the eukaryotic partner, Trench and Siebens (1978) found that the synthesis of the prokaryotic r-RNAs was under the control of the cyanobacterium, whereas synthesis of eukaryotic r-RNAs was under the control of the host.

However, when studies were conducted on the effect of cycloheximide (an inhibitor of protein synthesis in eukaryotic systems shown also to affect eukaryotic r-RNA synthesis) on the synthesis of chlorophyll *a*, the data illustrated that some process involving macromolecular synthesis in the eukaryotic host was closely linked to chlorophyll synthesis in the symbionts because chlorophyll synthesis was markedly reduced in the presence of the inhibitor. These observations are not unlike those reported from studies of some plant cell-chloroplast systems (Kirk and Tilney-Bassett, 1967; Bogorad and Woodcock, 1971).

In further studies on the control of macromolecular synthesis in *C. paradoxa*, the influence of the inhibitors rifampicin and cycloheximide on the synthesis of the phycobiliproteins was determined. Rifampicin, an inhibitor shown to block r-RNA synthesis *in vivo* in *C. korschikoffiana*, also blocked the synthesis of *c*- and allo-phycocyanin. Cycloheximide, which was found to have no detectable effect on r-RNA synthesis in *C. korschikoffiana*, did not influence the synthesis of allo-phycocyanin but reduced the incorporation of ^{14}C into c-phycocyanin. When the purified protein was further analyzed by SDS polyacrylamide gel disc electrophoresis, it was found that cycloheximide inhibited most dramatically the synthesis of the small subunit of the heterodimer of c-phycocyanin. These observations indicate either that the small subunit of c-phycocyanin in *C. korschikoffiana* is synthesized on the host's ribosomes and then transferred to the symbiotic cyanobacteria, or that some protein essential to the synthesis of the biliprotein by the cyanobacteria is snythesized by the host. If the former possibility is correct, then the synthesis of *c*-phycocyanin would be closely analogous to the synthesis of ribulose diphosphate carboxylase in plant cells, wherein the small subunit of the protein is synthesized on cytoplasmic ribosomes (Criddle et al., 1970; Kawashima, 1970; Boulter et al., 1972).

REGULATION OF ALGAL POPULATION DENSITIES

Very little is known about how cell numbers are regulated in multualistic symbioses. It is quite apparent that one of the unique features of these associations, in contrast to many pathologic conditions, is that the two components of the consortium proliferate in harmony; the algae do not overgrow the hosts, and the hosts' growth does not outstrip that of the symbionts. It may, therefore, be assumed that either the host provides constraints on the population increase of the algal symbionts, or the algae themselves, possibly through some unknown feedback mechanism of regulation, keep their population in check.

Studies on the growth of *C. paradoxa* and *C. korschikoffiana* illustrate the phenomenon (Trench et al., 1978). When cultures of *C. paradoxa* have achieved stationary phase of growth, the number of cyanobacteria per host cell may be as high as eight. When these cultures are inoculated into fresh media, there is a marked readjustment in the distribution of cyanobacteria in the host cells (see fig. 7), such that most of the cells contain one or two cyanobacteria. Such redistribution of symbiont numbers continues throughout the lag phase of growth of the host population. Exponential growth of the host population is not synchronized with that of the symbionts, which do not increase exponentially until the host population approaches its asymptote. At this stage, there is a marked increase in the number of host cells containing two or more cyanobacteria.

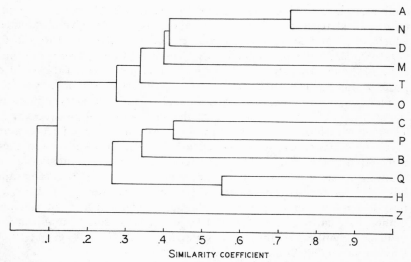

Fig. 7. Growth characteristics of *C. paradoxa* and *C. korschikoffiana*. Closed circles = *C. paradoxa* cells (lefthand scale); open circles, broken line = *C. korschikoffiana* (inner right-hand scale, also 10^{-8} cells · litre $^{-1}$); open circles, solid line = percentage *C. paradoxa* cells containing one cyanelle; open triangles = percentage of *C. paradoxa* cells containing two cyanelles; closed triangles = percentage of *C. paradoxa* cells containing four cyanelles. Data from Trench et al. (1978), by permission of the Royal Society of London.

Although almost no experimental work has been done on the regulation of algal numbers in associations between *S.*(=*G*) *microadriaticum* and marine invertebrates, it has been suggested (Smith et al., 1969) that the transport of metabolites from the algae to the hosts may be a factor that regulates growth and division in the algae. The possibilities that the animals may somehow regulate the flow of critical nutrients to the algae, thereby regulating their proliferation, or that the animals might produce some substance that regulates cell division by the algae have not been investigated.

CONCLUSIONS

From the data presented in this paper, it is clear that intracellular symbioses between unicellular algae and animal cells provide excellent model systems through which the experimental analysis of several phenomena important to cell biology can be conducted. These phenomena include intercellular recognition and specificity, endocytosis, intracellular transport, metabolite exchange, intergenomic interactions, and regulation.

Over-all, these phenomena describe the processes involved in the integration of genetically distinct biologic entities into a stable functional unit.

Although over the past decade much new information has become available, the most significant advance has been the new insights into problems of symbiosis that have been forthcoming. These new insights provide added impetus and clearer definition of the phenomena that need investigation. One of the major outstanding problems that has yet to receive attention is the genetic basis of infection and specificity.

Another aspect of the study of symbiosis that has improved dramatically is the selection of appropriate experimental systems through which to investigate particular phenomena. The introduction of concepts, methods and techniques from other disciplines such as cell and molecular biology will probably go a long way in increasing our knowlege of symbiosis in future.

ACKNOWLEDGMENTS

I would like to thank my many colleagues who, through many discussions, have unwittingly contributed to the formation and crystalization of much of what has been presented in this paper.

Financial assistance was provided by the Biomedical Sciences Support Fund (NIH, RR-07099), from the Department of Biological Sciences, and from the Marine Science Institute, University of California, Santa Barbara.

LITERATURE CITED

Allen, J. M. 1969. Lysosomes in bacterial infection. *In* S. T. Drogle and H. B. Fell (eds.), Lysosomes in biology and pathology, pp. 41–68. North Holland, Amsterdam.

Bigelow, R. P. 1900. The anatomy and development of *Cassiopeia xamachana*. Mem. Boston Soc. of Nat. Hist. 5:191–236.

Bogorad, L., and C. L. F. Woodcock. 1971. Rifamycins: the inhibition of plastid RNA synthesis *in vivo* and *in vitro* and variable effects on chlorophyll formation in maize leaves. *In* N. K. Boardman, A. W. Linnane, and R. M. Smillie (eds.), Autonomy and biogenesis of mitochondria and chloroplasts, pp. 92–97. North Holland, Amsterdam, London.

Boulter, D., R. J. Ellis, and A. Yarwood. 1972. Biochemistry of protein synthesis in plants. Biol. Rev. 47:113–75.

Cook, C. B. 1971. Transfer of ^{35}S-labelled material from food ingested by *Aiptasia* sp. to its endosymbiotic zooxanthellae. *In* H. M. Lenhoff, L. Muscatine, and L. V. Davis (eds.), Experimental coelenterate biology, pp. 218–24. U. of Hawaii Press, Honolulu.

Cook, C. B., 1980. The infection of invertebrates by algae. *In* this volume.

Cook, C. B., C. F. D'Elia, and L. Muscatine 1978. Endocytic mechanisms of the digestive cells of *Hydra viridis*. I. Morphological aspects. Cytobios 23:17–31.

Criddle, R. S., B. Dan, G. E. Kleinkopf, and R. C. Huffaker. 1970. Differential synthesis of ribulose diphosphate carboxylase subunits. Biochem., Biophys. Res. Commun. 41:621–27.

Davis, B. D., R. Dulbecco, H. N. Eisen, H. S. Ginsberg, and W. B. Wood. 1973. Microbiology (2d ed.). Harper and Row, New York. 1,562 pages.

Dubos, R., and A. W. Kessler. 1963. Integrative and disintegrative factors in symbiotic associations. Symp. Soc. Gen. Microbiol. 13: 1–11.

Fankboner, P. V. 1971. Intracellular digestion of symbiotic zooxanthellae by host amoebocytes in giant clams (*Bivalvia, Tridacnidae*) with a note on the nutritional role of the hypertrophied siphonal epidermis. Biol. Bull. 141:222–34.

Holligan, P. M., and G. W. Gooday. 1975. Symbiosis in *Convoluta roscoffensis*. Symp. Soc. Exp. Biol. 29:205–27.

Jeon, K. W. 1972. Development of cellular dependence on infective organisms: microsurgical studies in Amoebas. Science 176:1122–23.

Jeon, K. W. 1980. Symbiosis of bacteria with *Amoeba. In* this volume.

Jones, T. C. 1974. Macrophages and intracellular parasitism. J. Reticulothelial Soc. 15:439–50.

Karakashian, M. W. 1975. Symbiosis in *Paramecium bursaria*. Symp. Soc. Exp. Biol. 29: 145–73.

Kawashima, N. 1970. Non-synchronous incorporation of ^{14}C into amino acids of the two subunits of fraction I protein. Biochem. Biophys. Res. Commun. 38:119–24.

Kinzie, R. A., III. 1974. Experimental infection of aposymbiotic gorgonian polyps with zooxanthellae. J. Exp. Mar. Biol. Ecol. 15:335–45.

Kirk, J.T.D., and R.A.E. Tilney-Bassett. 1967. The plastids. W. H. Freeman, London and San Francisco. 608 pp.

Lenhoff, H. M. 1974. On the mechanism of action and evolution of receptors associated with feeding and digestion. *In* L. Muscatine and H. M. Lenhoff (eds.), Coelenterate biology: reviews and new perspectives. Academic Press, New York.

Lewis, D. H., and D. C. Smith. 1971. The autotrophic nutrition of symbiotic marine coelenterates with special reference to hermatypic corals. I. Movement of photosynthetic products between the symbionts. Proc. R. Soc. Lond. (B) 178:111–29.

Ludwig, L. D. 1969. Die Zooxanthellen bei *Cassiopeia andromeda* Eschsholz 1929 (Polyp-Stadium) und ihre Bedeutung fur die Strobilation. Zool. Jb. (Anat) 86:238–77.

Margulis, L. 1970. The origin of eukaryotic cells. Yale University Press, New Haven, Conn. 349 pp.

Margulis, L. 1976. Genetic and evolutionary consequences of symbiosis. Exp. Parasit. 39:277–349.

Margulis, L. 1980. Symbiosis as parasexuality. *In* this volume.

Muscatine, L. 1967. Glycerol excretion by symbiotic algae from corals and *Tridacna* and its control by the host. Science 156:516–19.

Muscatine, L. 1973. Nutrition of corals. *In* O. A. Jones and R. Endean (eds.), Biology and geology of coral reefs, 2:77–115. Academic Press, New York and London.

Muscatine, L. 1974. Endosymbiosis of cnidarians and algae. *In* L. Muscatine and H. M. Lenhoff (eds.), Coelenterate biology: reviews and new perspectives, pp. 359–95. Academic Press, New York.

Muscatine, L. 1980. Uptake, retention, and release of dissolved inorganic nutrients by marine algae-invertebrate associations. *In* this volume.

Muscatine, L., C. B. Cook, R. L. Pardy, and R. R. Pool. 1975a. Uptake, recognition, and maintenance of symbiotic *Chlorella* by *Hydra viridis*. Symp. Soc. Exp. Biol. 29:175–203.

Muscatine, L., R. R. Pool, and E. Cernichiari. 1972. Some factors influencing selective release of soluble organic material by zooxanthellae from reef corals. Mar. Biol. 13: 298–308.

Muscatine, L., R. R. Pool, and R. K. Trench. 1975b. Symbiosis of algae and invertebrates: aspects of the symbiont surface and the host-symbiont interface. Trans. Amer. Micros. Soc. 94:450–69.

Muscatine, L., and J. W. Porter. 1977. Reef corals: mutualistic symbioses adapted to nutrient-poor environments. BioScience 27: 454–60.

Olman, J. E., and K. L. Webb. 1974. Metabolism of ^{131}I in relation to strobilation in *Aurelia aurita* L. (Scyphozoa). J. Exp. Mar. Biol. Ecol. 16:113–22.

Pardy, R. L. 1974a. Some factors affecting the growth and distribution of the algal endosymbionts of *Hydra viridis*. Biol. Bull. 147:105–18.

Pardy, R. L. 1974b. Regulation of the endosymbiotic algae in *Hydra* by digestive cells and tissue growth. Amer. Zool. 14:583–88.

Pardy, R. L., and L. Muscatine. 1973. Recognition and uptake of symbiotic algae by *Hydra viridis*. A quantitative study of the uptake of living algae by aposymbiotic *Hydra viridis*. Biol. Bull. 145:565–79.

Pool, R. R. 1976. Symbiosis of *Chlorella* and *Chlorohydra viridissima*. Ph. D. diss., University of California, Los Angeles. 122 pp.

Provasoli, L., T. Yamasu, and I. Manton. 1968. Experiments on the resynthesis of symbiosis in *Convoluta roscoffensis* with different flagellate cultures. J. Mar. Biol. Ass. U.K. 48:456–79.

Schnepf, E., and R. M. Brown, Jr. 1971. On relationships between endosymbiosis and the origins of plastids and mitochondria. Pages 299–322 *In* J. Reinert and H. Ursprung (eds.), Origins and continuity of cell organelles, pp. 299–322. Springer-Verlag, New York.

Schoenberg, D. A. 1976. Genetic variation and host specificity in the zooxanthella. *Symbiodinium microadriaticum* Freudenthal, an algal symbiont of marine invertebrates. Ph. D. diss., Yale University. 183 pp.

Schoenberg, D. A., and R. K. Trench. 1976. Specificity of symbioses between marine cnidarians and zooxanthellae. *In* G. O. Mackie (ed.), Coelenterate ecology and behavior, pp. 423–32. Plenum Publ., New York.

Schoenberg, D. A., and R. K. Trench. 1979a. Genetic variation in *Symbiodinium* (=*Gymnodinium*) *microadriaticum* Freudenthal and specificity in its symbiosis with marine invertebrates. I. Isozyme and soluble protein patterns of axenic cultures of *S. microadriaticum*. Proc. R. Soc. Lond. Ser. B. (in press).

Schoenberg, D. A., and R. K. Trench. 1979b. Genetic variation in *Symbiodinium* (=*Gymnodinium*) *microadriaticum* Freudenthal and specificity in its symbiosis with marine invertebrates. II. Morphological variation in *S. microadriaticum*. Proc. R. Soc. Lond. Ser. B. (in press).

Schoenberg, D. A., and R. K. Trench. 1979c. Genetic variation in *Symbiodinium* (=*Gymnodinium*) *microadriaticum* Freudenthal and specificity in its symbiosis with marine invertebrates. III. Specificity of infectivity in *S. microadriaticum*. Proc. P. Soc. Lond. Ser. B. (in press).

Siebens, H. B., and R. K. Trench. 1978. Aspects of the relation between *Cyanophora paradoxa* (Korschikoff) and its endosymbiotic cyanelles *Cyanocyta korschikoffiana* (Hall and Claus). III. Characterization of ribosomal ribonucleic acids. Proc. R. Soc. Lond. (B) 202: 463–72.

Smith, D. C. 1974. Transport from symbiotic algae and symbiotic chloroplasts to host cells. Symp. Soc. Exp. Biol. 28:437–508.

Smith, D. C., L. Muscatine, and D. H. Lewis. 1969. Carbohydrate movement from autotrophs to heterotrophs in parasitic and mutualistic symbiosis. Biol. Rev. 44:17–90.

Spangenberg, D. B. 1971. Thyroxin induced metamorphosis in *Aurelia*. J. Exp. Zool. 178:183–94.

Steele, R. D., and N. I. Goreau. 1977. The breakdown of symbiotic zooxanthellae in the sea anemone *Phyllactis* (=*Oulactis*) *flosculifera*, (*Actiniaria*). J. Zool. Lond. 181:421–37.

Sugiura, Y. 1964. On the life history of rhizostome medusae. II. Indispensability of zooxanthellae for strobilation in *Mastigias papua*. Embriologia 8:223–33.

Taylor, D. L. 1971. On the symbiosis between *Amphidinium klebsii* (Dinophyceae) and *Amphiscolops langerhansi* (Turbellaria: Acoela). J. Mar. Biol. Ass. U.K. 51: 301–13.

Taylor, D. L. 1973a. Cellular interactions of algae-intertebrate symbiosis. Adv. Mar. Biol. 11:1–56.

Taylor, D. L. 1973b. Algal symbionts of invertebrates. Ann. Rev. Microbiol. 27: 171–87.

Terry, R. J., and S. R. Smithers. 1975. Evasion of the immune response by parasites. Symp. Soc. Exp. Biol. 29:453–65.

Trench, R. K. 1971a. The physiology and biochemistry of zooxanthellae symbiotic with marine coelenterates. I. Assimilation of photosynthetic products of zooxanthellae by two marine coelenterates. Proc. R. Soc. Lond. (B) 177:225–35.

Trench, R. K. 1971b. The physiology and biochemistry of zooxanthellae symbiotic with marine coelenterates. II. Liberation of fixed ^{14}C by zooxanthellae *in vitro*. Proc. R. Soc. Lond. (B) 177:237–50.

Trench, R. K. 1971c. The physiology and biochemistry of zooxanthellae symbiotic with marine coelenterates. III. The effects of homogenates of host tissues on the excretion of photosynthetic products *in vitro* by zooxanthellae from two marine coelenterates. Proc. R. Soc. Lond. (B) 177:251–64.

Trench, R. K. 1974. Nutritional potentials in *Zoanthus sociatus* (Coelenterata, Anthozoa). Helgolander. wiss. Meeresunters. 26:174–216.

Trench, R. K. 1979. The cell biology of plant-animal symbiosis. Ann. Revs. Pl. Physiol. 30:485–532.

Trench, R. K., R. R. Pool, M. Logan, and A. Engelland. 1978. Aspects of the relation between *Cyanophora paradoxa* (Korschikoff) and its endosymbiotic cyanelles *Cyanocyta korschikoffiana* (Hall and Claus). I. Growth, photosynthesis and the obligate nature of the association. Proc. R. Soc. Lond. (B) 202:423–44.

Trench, R. K., and G. S. Ronzio. 1978. Aspects of the relation between *Cyanophora paradoxa* (Korschikoff) and its endosymbiotic cyanelles *Cyanocyta korschikoffiana* (Hall and Claus). II. The photosynthetic pigments. Proc. R. Soc. Lond. (B) 202:445–62.

Trench, R. K., and H. B. Siebens. 1978. Aspects of the relation between *Cyanophora paradoxa* (Korschikoff) and its endosymbiotic cyanelles *Cyanocyta korschikoffiana* (Hall and Claus). IV. The effects of rifampicin, chloramphenicol and cycloheximide on the synthesis of ribosomal ribonucleic acids and chlorophyll. Proc. R. Soc. Lond. (B) 202:473–83.

von Holt, C. 1968. Uptake of glycine and release of nucloside polyphosphates by zooxanthellae. Comp. Biochem. Physiol. 26: 1071–79.

Weiss, P. 1953. Specificity in growth control. *In* E. G. Butler (ed.), Biological specificity and growth, pp. 195–206. Princeton University Press, Princeton, N.J.

Yonge, C. M. 1944. Experimental analysis of the association between invertebrates and unicellular algae. Biol. Rev. 19:68–80.

Yonge, C. M. 1963. The biology of coral reefs. Adv. Mar. Biol. 1:209–60.

Index

Acarospora, 8, 11, 91

Acaulospora, 174, 176

Acetate-polymalonate pathway, 8

Acid phosphatase, 62, 153, 221. *See also* Phospholydrolase

Acropora, 231, 232, 236

Acrosyphus, 9

Active site, 161, 163

Agrostis, 90

Aiptasia, 51, 55–57, 281–84

Alanine: release of, by zooxanthellae, 238, 290

Alectoria, 11

Alkaline phosphatase, 153, 183. *See also* Phosphohydrolase

Allium, 177

Allophycocyanin, 290

Amastigote, 31

Amino acids, 88, 219, 221; release of, by symbiotic algae, 288; uptake of, by helminths, 114, 115, 119, 155, 156

Aminopeptidase, 155–57

Ammonia, 90, 217, 219, 221

Ammonium: uptake of, by corals, 231–37

Amoeba proteus, 245–60

Amoeba-bacteria symbiosis, 245–60, 269, 275

Amphidinium klebsii, 49–53

Amphiesma: of dinoflagellates, 282, 286

Amphiscolops langerhansi, 49–53

Amylase, 159–62, 164, 165

Anaptychia, 9, 10

Antibody, 131–33, 135–37, 140, 141

Antigen, 131–41

Antigenic variation, 133, 135, 140

Anthopleura elegantissima, 51, 239,

Apical cap, 102

Aplanospore, 20

Appressorium, 5, 14, 16, 20, 174

Arabitol, 207

Arbuscule, 175, 187, 188

Arum-type infection, 176

Ascaris, 166

Ascogonium, 5

Ascospore, 6, 7, 20, 80

Aspicilia, 11

Astrangia danae, 240

ATPase (adenosine triphosphatase), 100, 119–22, 152, 210; mitochondrial, 269; nitrate-chloride activated, 236

Atranorin, 8, 10

Baeomyces, 8, 9

Baeomycesic acid, 8

Barbatic acid, 16

Basement lamina, 102, 108, 112, 119, 121

Bellidiflorin, 9

Biotin, 8

Biotrophy, 197, 198, 220, 223, 268

Blue-green algae, 6, 90, 199, 212, 213, 214, 217, 230 (reef-dwelling). *See also* Cyanelles

Bombliospora, 9

Briareum asbestinum, 284

Brij, 251

Brush border, 97, 100, 102, 105

Buellia, 11, 13

Calcium, 182, 186; role of, in microtubule polymerization, 66